Handbook of Robotic Manipulators

Handbook of Robotic Manipulators

Edited by **Dmitry Hofland**

NY RESEARCH PRESS

New York

Published by NY Research Press,
23 West, 55th Street, Suite 816,
New York, NY 10019, USA
www.nyresearchpress.com

Handbook of Robotic Manipulators
Edited by Dmitry Hofland

International Standard Book Number: 978-1-63238-278-8 (Hardback)

Printed in the United States of America.

Contents

Preface

This book provides extensive information regarding the rapidly growing field of robotics. Robotics form an essential part of modern engineering and is associated with various other fields namely mathematics, mechanism design, computer and electric & electronics. In the past few decades, interest in robotics has notably increased which has led to the advancement of theoretical research and products in this field. This book includes contributions of prominent researchers and engineers dealing with robotics and its related aspects. It extensively covers topics of serial manipulators such as kinematics & dynamics and control modeling.

All of the data presented henceforth, was collaborated in the wake of recent advancements in the field. The aim of this book is to present the diversified developments from across the globe in a comprehensible manner. The opinions expressed in each chapter belong solely to the contributing authors. Their interpretations of the topics are the integral part of this book, which I have carefully compiled for a better understanding of the readers.

At the end, I would like to thank all those who dedicated their time and efforts for the successful completion of this book. I also wish to convey my gratitude towards my friends and family who supported me at every step.

Editor

Part 1

Kinematics and Dynamics

1

Exploiting Higher Kinematic Performance – Using a 4-Legged Redundant PM Rather than Gough-Stewart Platforms

Mohammad H. Abedinnasab[1], Yong-Jin Yoon[2] and Hassan Zohoor[3]
[1]*Door To Door Company, Sharif University of Technology*
[2]*Nanyang Technological University*
[3]*Sharif University of Technology,*
The Academy of Sciences of IR Iran
[1,3]*Iran*
[2]*Singapore*

1. Introduction

It is believed that Gough and Whitehall (1962) first introduced parallel robots with an application in tire-testing equipments, followed by Stewart (1965) , who designed a parallel mechanism to be used in a flight simulator. With ever-increasing demand on the robot's rigidity, redundant mechanisms, which are stiffer than their non-redundant counterparts, are attracting more attention.

Actuation redundancy eliminates singularity, and greatly improves dexterity and manipulability. Redundant actuation increases the dynamical capability of a PM by increasing the load carrying capacity and acceleration of motion, optimizing the load distribution among the actuators and reducing the energy consumption of the drivers. Moreover, it enhances the transmission characteristics by increasing the homogeneity of the force transmission and the manipulator stiffness (Yi et al., 1989). From a kinematic viewpoint, redundant actuation eliminates singularities (Ropponen & Nakamura, 1990) which enlarge the usable workspace, as well. The kinematic analysis on general redundantly actuated parallel mechanisms was investigated by Müller (2005).

A number of redundantly full-actuated mechanisms have been proposed over the years and some of them which are more significant are listed in this section. The spatial octopod, which is a hexapod with 2 additional struts, is one of them (Tsai, 1999). A 5-DOF 3-legged mechanism was proposed by Lu et al. (2008), who studied its kinematics, statics, and workspace. Staicu (2009) introduced a new 3-DOF symmetric spherical 3-UPS/S parallel mechanism having three prismatic actuators. As another work of Lu et al. (2009), they introduced and used 2(SP+SPR+SPU) serial–parallel manipulators. Wang and Gosselin (2004) addressed an issue of singularity and designed three new types of kinematically redundant parallel mechanisms, including a new redundant 7-DOF Stewart platform. They concluded that such manipulators can be used to avoid singularities inside the workspace of non-redundant manipulators.

Choi et al. (2010) developed a new 4-DOF parallel mechanism with three translational and one rotational movements and found this mechanism to be ideal for high-speed machining.

Gao et al. (2010) proposed a novel 3DOF parallel manipulator and they increased the stiffness of their system, using an optimization technique. Lopes (2010) introduced a new 6-DOF moving base platform, which is capable of being used in micro robotic applications after processing serial combination with another industrial manipulator. It is in fact a non-redundant parallel mechanism with 6 linear actuators.

Deidda et al. (2010) presented a 3-DOF 3-leeged spherical robotic wrist. They analyzed mobility and singularity. Tale Masouleh et al. (2011) investigated the kinematic problem of a 5-DOF 5-RPUR mechanism with two different approaches, which differ by their concepts of eliminating passive variables. Zhao and Gao (2010) investigated the kinematic and dynamic properties of a 6-DOF 8-PSS redundant manipulator. They presented a series of new joint-capability indices, which are general and can be used for other types of parallel manipulators.

Li et al. (2007) worked on the singularity-free workspace analysis of the general Gough-Stewart platform. In a similar line of work, Jiang and Gosselin (Jiang & Gosselin, 2009a;b;c) determined the maximal singularity-free orientation workspace at a prescribed position of the Gough-Stewart platform. Alp and Ozkol (2008) described how to extend the workspace of the 6-3 and 6-4 Stewart platforms in a chosen direction by finding the optimal combination of leg lengths and joint angles. They showed that the workspace of the 6-3 Stewart platform is smaller than that of the 6-4 one.

Mayer and Gosselin (2000) developed a mathematical technique to analytically derive the singularity loci of the Gough-Stewart platform. Their method is based on deriving an explicit expression for the determinant of the jacobian matrix of the manipulator.

To demonstrate the redundancy effects, Wu et al. (2010) compared a planar 2-DOF redundant mechanism with its non-redundant counterpart. Arata et al. (2011) proposed a new 3-DOF redundant parallel mechanism entitled as Delta-R, based on its famous non-redundant counterpart, Delta, which was developed by Vischer & Clavel (1998).

Sadjadian and Taghirad (2006) compared a 3-DOF redundant mechanism, hydraulic shoulder, to its non-redundant counterpart. They concluded that the actuator redundancy in the mechanism is the major element to improve the Cartesian stiffness and hence the dexterity of the hydraulic shoulder. They also found that losing one limb reduces the stiffness of the manipulator significantly.

The rest of the chapter is organized as follows. In Section 2, in addition to introduction and comparison of non-redundant 3-legged and redundant 4-legged UPS PMs, four different architectures of the Gough-Stewart platforms are depicted. The kinematics of the abovementioned mechanisms are reviewed in Section 3. The jacobian matrices using the screw theory is derived in Section 4. In Sections 5 and 6, the performances of the redundant and non-redundant mechanisms are studied and compared with four well-known architectures of hexapods. Finally we conclude in Section 7.

2. Mechanisms description

The schematics of the 6-DOF non-redundant,3-legged and redundant 4-legged mechanisms are shown in Figs. 1 and 2, respectively.

The non-redundant 3-legged mechanism was first introduced by Beji & Pascal (1999). The redundant 4-legged mechanism has the similar structure, with a single leg added to the 3-legged system, which keeps symmetry. Each leg is composed of three joints; universal,

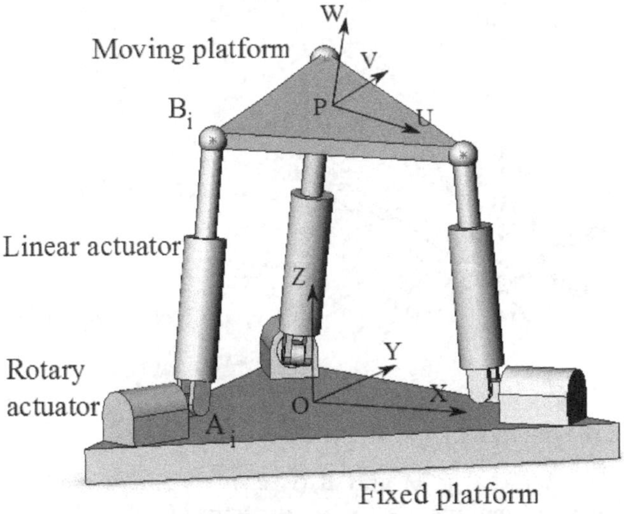

Fig. 1. Schematic of the non-redundant mechanism.

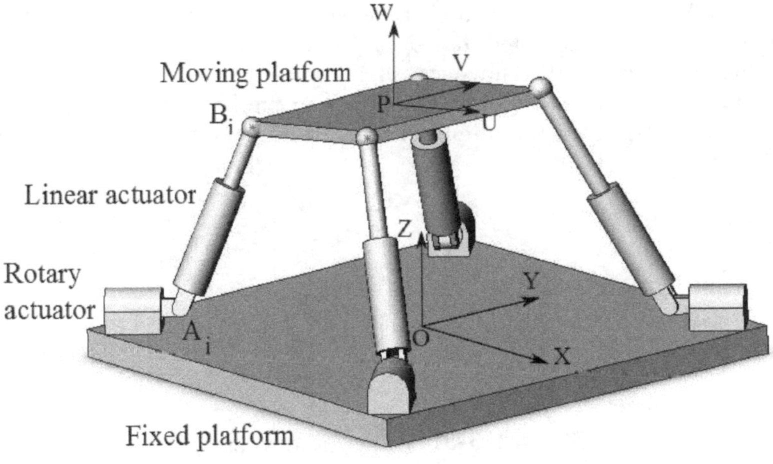

Fig. 2. Schematic of the redundant mechanism.

prismatic, and spherical (Fig. 3). A rotary actuator and a linear actuator are used to actuate each leg. The rotary actuators, whose shafts are attached to the lower parts of the linear actuators through the universal joints, are placed on the corners of the fixed platform (Abedinnasab & Vossoughi, 2009; Aghababai, 2005). The spherical joints connect the upper parts of the linear actuators to the moving platform.

Rotary actuators are situated on the corners A_i (for $i=1, 2, 3, 4$) of the base platform and each shaft is connected to the lower part of linear actuators through a universal joint (Figs. 1 and 2). The upper parts of linear actuators are connected to the corners of the moving platform, B_i points, through spherical joints (Fig. 3).

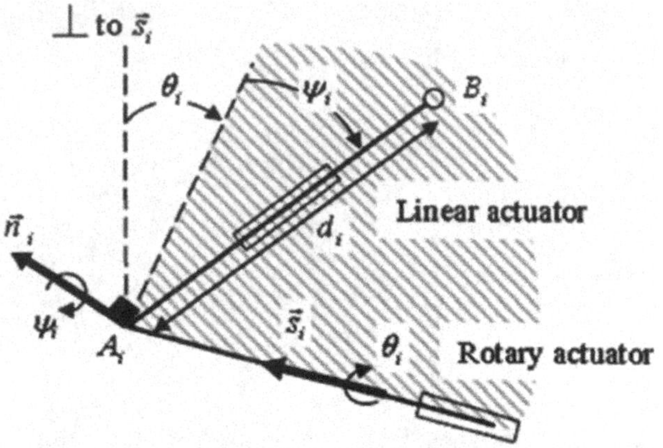

Fig. 3. Schematic of the universal joint, and the joints variables.

Cartesian coordinates A (O, x ,y, z) and B (P, u, v, w) represented by {A} and {B} are connected to the base and moving platforms, respectively. In Fig.3, $\vec{s}_i$ represents the unit vector along the axes of ith rotary actuator and $\vec{d_i}$ is the vector along $\overrightarrow{A_iB_i}$ with the length of d_i . Assuming that each limb is connected to the fixed base by a universal joint, the orientation of ith limb with respect to the fixed base can be described by two Euler angles, rotation θ_i around the axis $\vec{s}_i$, followed by rotation ψ_i around $\vec{n}_i$, which is perpendicular to $\vec{d_i}$ and $\vec{s}_i$ (Fig. 3). It is to be noted that θ_i and d_i are active joints actuated by the rotary and linear actuators, respectively. However, ψ_i is inactive.

By replacing the passive universal joints in the Stewart mechanism with active joints in the above mentioned mechanisms, the number of legs could be reduced from 6 to 3 or 4. This

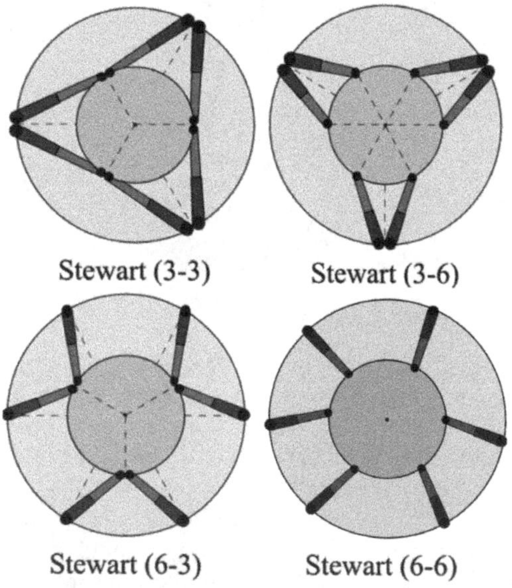

Fig. 4. Schematics of well-known Stewart platforms.

change makes the mechanism to be lighter, since the rotary actuators are resting on the fixed platform, which causes higher accelerations to be available due to smaller inertial effects.
The applications of this type of robots can be found in flight simulators, high precision surgical tools, positioning devices, motion generators, ultra-fast pick and place robots, haptic devices, entertainment, multi-axis machine tools, micro manipulators, rehabilitation devices, etc.
Advantages of high rigidity and low inertia make these PMs ideal for force feedback control, assembly, manufacturing, underground projects, space technologies, and biology projects.

3. Kinematic analysis

One of the most important issues in the study of parallel mechanisms is the kinematic analysis, where the generated results form the base for the application of the mechanism.
$\vec{a}_i$ represents the vector $\overline{OA_i}$ (Fig. 1). $\vec{a}_i = g[\cos\gamma_i \quad \sin\gamma_i \quad 0]^T$, in which,

$$\gamma_1^{Red.} = -45^\circ,\ \gamma_2^{Red.} = 45^\circ,\ \gamma_3^{Red.} = 135^\circ,\ and\ \gamma_4^{Red.} = -135^\circ,$$

and

$$\gamma_1^{Non-Red.} = 0^\circ,\ \gamma_2^{Non-Red.} = 120^\circ,\ and\ \gamma_3^{Non-Red.} = -120^\circ,$$

where g and r are the radius of the fixed and moving platforms, respectively. $^B\vec{b}_i$ represents the position of the ith joint on the platform in the moving frame {B}, $^B\vec{b}_i = \overrightarrow{PB_i}\big)_B$. $^B\vec{b}_i$ is constant and is equal to $^B\vec{b}_i = h[\cos\gamma_i \quad \sin\gamma_i \quad 0]^T$. We can represent $^A_BR = \left[r_{ij}\right]$, the rotation matrix from $\{B\}$ to $\{A\}$, using Euler angles as

$$^A_BR = \begin{bmatrix} c\alpha_2 c\alpha_3 & c\alpha_3 s\alpha_2 s\alpha_1 - s\alpha_3 c\alpha_1 & c\alpha_3 s\alpha_2 c\alpha_1 + s\alpha_3 s\alpha_1 \\ c\alpha_2 s\alpha_3 & s\alpha_3 s\alpha_2 s\alpha_1 + c\alpha_3 c\alpha_1 & s\alpha_3 s\alpha_2 c\alpha_1 - c\alpha_3 s\alpha_1 \\ -s\alpha_2 & c\alpha_2 s\alpha_1 & c\alpha_2 c\alpha_1 \end{bmatrix}, \tag{1}$$

where $s\alpha_1 = \sin\alpha_1$ and $c\alpha_1 = \cos\alpha_1$, and so on. α_1, α_2, and α_3 are three Euler angles defined according to the $z-y-x$ convention. Thus, the vector $^B\vec{b}_i$ would be expressed in the fixed frame {A} as

$$\vec{b}_i = {}^A_BR\,\overrightarrow{PB_i}\big)_B. \tag{2}$$

Let $\vec{p}$ and $\vec{r}_i$ denote the position vectors for P and B_i in the reference frame $\{A\}$, respectively. From the geometry, it is obvious that

$$\vec{r}_i = \vec{p} + \vec{b}_i. \tag{3}$$

Subtracting vector $\vec{a}_i$ from both sides of (3) one obtains

$$\vec{r}_i - \vec{a}_i = \vec{p} + \vec{b}_i - \vec{a}_i. \tag{4}$$

Left hand side of (4) is the definition of $\overrightarrow{d_i}$, therefore

$$d_i^2 = (\vec{p} + \vec{b}_i - \vec{a}_i) \cdot (\vec{p} + \vec{b}_i - \vec{a}_i) . \tag{5}$$

Using Euclidean norm, d_i can be expressed as:

$$d_i = \sqrt{\left((x - x_i)^2 + (y - y_i)^2 + (z - z_i)^2\right)} , \tag{6}$$

where,

$$\begin{cases} x_i = -h\left(\cos\gamma_i r_{11} + \sin\gamma_i r_{21}\right) + g\cos\gamma_i \\ y_i = -h\left(\cos\gamma_i r_{12} + \sin\gamma_i r_{22}\right) + g\sin\gamma_i \\ z_i = -h\left(\cos\gamma_i r_{13} + \sin\gamma_i r_{23}\right) \end{cases} . \tag{7}$$

Coordinates $C_i(A_i, x_i, y_i, z_i)$ are connected to the base platform with their x_i axes aligned with the rotary actuators in the $\vec{s}_i$ directions, with their z_i axes perpendicular to the fixed platform. Thus, one can express vector $\overrightarrow{d_i}$ in $\{C_i\}$ as

$$^{C_i}\overrightarrow{d_i} = d_i \begin{bmatrix} \sin\psi_i \\ -\sin\theta_i \cos\psi_i \\ \cos\theta_i \cos\psi_i \end{bmatrix} , \tag{8}$$

and from the geometry is clear that

$$\vec{r}_i = \vec{a}_i + {}^{A}_{C_i}R \, {}^{C_i}\overrightarrow{d_i} , \tag{9}$$

where ${}^{A}_{C_i}R$ is the rotation matrix from $\{C_i\}$ to $\{A\}$,

$${}^{A}_{C_i}R = \begin{bmatrix} \cos\gamma_i & -\sin\gamma_i & 0 \\ \sin\gamma_i & \cos\gamma_i & 0 \\ 0 & 0 & 1 \end{bmatrix} . \tag{10}$$

By equating the right sides of (3) and (4), and solving the obtained equation, ψ_i and θ_i can be calculated as follows:

$$\psi_i = \sin^{-1}\left(\frac{\cos\gamma_i (x - x_i) + \sin\gamma_i (y - y_i)}{d_i} \right) , \tag{11}$$

$$\theta_i = \sin^{-1}\left(\frac{\sin\gamma_i (x - x_i) - \cos\gamma_i (y - y_i)}{d_i \cos\psi_i} \right) . \tag{12}$$

4. Jacobian analysis using screw theory

Singularities of a PM pose substantially more complicated problems, compared to a serial manipulator. One of the first attempts to provide a general framework and classification may be traced back to Gosselin and Angeles (1990) ,who derived the input-output velocity map for a generic mechanism by differentiating the implicit equation relating the input and output configuration variables. In this way, distinct jacobian matrices are obtained for the inverse and the direct kinematics, and different roles played by the corresponding singularities are clearly shown.

For singularity analysis other methods rather than dealing with jacobian matrix are also available. Pendar et al. (2011) introduced a geometrical method, namely *constraint plane method,* where one can obtain the singular configurations in many parallel manipulators with their mathematical technique. Lu et al. (2010) proposed a novel analytic approach for determining the singularities of some 4-DOF parallel manipulators by using *translational/rotational jacobian matrices.* Piipponen (2009) studied kinematic singularities of planar mechanisms by means of *computational algebraic geometry* method. Zhao et al. (2005) have proposed *terminal constraint* method for analyzing the singularities based on the physical meaning of reciprocal screws.

Gosselin & Angeles (1990) have based their works on deriving the jacobian matrix. They performed this by defining three possible conditions. In these conditions the determinant of forward jacobian matrix or inverse jacobian matrix is investigated. They have shown that having dependent Plücker vectors in a parallel manipulator is equivalent to zero determinant of the forward jacobian matrix and then a platform singularity arises.

In this section the jacobian analysis of the proposed PMs are approached by using the theory of screws. Zhao et al. (2011) proposed a new approach using the screw theory for force analysis, and implemented it on a 3-DOF 3-RPS parallel mechanism. Gallardo-Alvarado et al. (2010) presented a new 5-DOF redundant parallel manipulator with two moving platforms, where the active limbs are attached to the fixed platform. They find the jacobian matrix by means of the screw theory.

A class of series-parallel manipulators known as 2(3-RPS) manipulators was studied by Gallardo-Alvarado et al. (2008) by means of the screw theory and the principle of virtual work. Gan et al. (2010) used the screw theory to obtain the kinematic solution of a new 6-DOF 3CCC parallel mechanism. Gallardo-Alvarado et al. (2006) analyzed singularity of a 4-DOF PM by means of the screw theory.

Hereafter we derive the jacobian matrices of the proposed mechanisms using screw theory. Joint velocity vector in the redundant mechanism, $\dot{\vec{q}}^{\mathrm{Red.}}$, is an 8×1 vector:

$$\dot{\vec{q}}^{\mathrm{Red.}} = [\,\dot{\theta}_1 \quad \dot{\theta}_2 \quad \dot{\theta}_3 \quad \dot{\theta}_4 \quad \dot{d}_1 \quad \dot{d}_2 \quad \dot{d}_3 \quad \dot{d}_4]^T \tag{13}$$

where $\dot{\theta}_i$ and $\dot{d}_i$ are the angular and linear velocities of the rotary and linear actuators, respectively. However, joint velocity vector in the non-redundant mechanism, $\dot{\vec{q}}^{Non-\mathrm{Red.}}$, is a 6×1 vector:

$$\dot{\vec{q}}^{Non-\mathrm{Red.}} = [\dot{\theta}_1 \quad \dot{\theta}_2 \quad \dot{\theta}_3 \quad \dot{d}_1 \quad \dot{d}_2 \quad \dot{d}_3]^T\,. \tag{14}$$

The linear and angular velocities of the moving platform are defined to be $\vec{v}$ and $\vec{\omega}$, respectively. Thus, $\dot{\vec{x}}$ can be written as a 6×1 velocity vector;

$$\dot{\vec{x}} = [\vec{v}^T \quad \vec{\omega}^T]. \tag{15}$$

Jacobian matrices relate $\dot{\vec{q}}$ and $\dot{\vec{x}}$ as follow;

$$J_x\dot{\vec{x}} = J_q\dot{\vec{q}}, \tag{16}$$

where J_x and J_q are forward and inverse jacobian matrices, respectively. If one defines J as follows;

$$J = J_q^{-1}J_x. \tag{17}$$

Thus $\dot{\vec{q}}$,and $\dot{\vec{x}}$ can be simply related as;

$$\dot{\vec{q}} = J\,\dot{\vec{x}}. \tag{18}$$

The concept of reciprocal screws is applied to derive J_x and J_q (Tsai, 1998; 1999). The reference frame of the screws is point P of the moving platform. Fig. 5 shows the kinematic chain of each leg, where universal joints are replaced by intersection of two unit screws, $\hat{\$}_1$ and $\hat{\$}_2$. $\hat{\$}_1 = \begin{bmatrix} \vec{s}_{1,i} \\ (\vec{b}_i - \vec{d}_i)\times\vec{s}_{1,i} \end{bmatrix}$ and $\hat{\$}_2 = \begin{bmatrix} \vec{s}_{2,i} \\ (\vec{b}_i - \vec{d}_i)_i\times\vec{s}_{2,i} \end{bmatrix}$ where $\vec{s}_{1,i}$ and $\vec{s}_{2,i}$ are unit vectors along the inactive and active joints of each universal joint, respectively. Spherical joints in each leg are replaced by intersection of three unit screws, $\hat{\$}_4$, $\hat{\$}_5$, and $\hat{\$}_6$. $\hat{\$}_4 = \begin{bmatrix} \vec{s}_{4,i} \\ \vec{b}_i\times\vec{s}_{4,i} \end{bmatrix}$, $\hat{\$}_5 = \begin{bmatrix} \vec{s}_{5,i} \\ \vec{b}_i\times\vec{s}_{5,i} \end{bmatrix}$, and $\hat{\$}_6 = \begin{bmatrix} \vec{s}_{6,i} \\ \vec{b}_i\times\vec{s}_{6,i} \end{bmatrix}$, where $\vec{s}_{4,i} = \vec{s}_{1,i}$, $\vec{s}_{6,i}$ is the unit vector along the linear actuator, and $\vec{s}_{5,i} = \vec{s}_{6,i}\times\vec{s}_{4,i}$. $\hat{\$}_3 = \begin{bmatrix} 0 \\ \vec{s}_{3,i} \end{bmatrix}$ explains the prismatic joint, as well. It is to be noted that $\vec{s}_{3,i} = \vec{s}_{6,i}$. Each leg can be assumed as an open-loop chain to express the instant twist of the moving platform by means of the joint screws:

$$\begin{aligned}\hat{\$}_P &= \dot{\psi}_i\hat{\$}_{1,i} + \dot{\theta}_i\hat{\$}_{2,i} + \dot{d}_i\hat{\$}_{3,i} + \dot{\phi}_{1,i}\hat{\$}_{4,i} \\ &+ \dot{\phi}_{2,i}\hat{\$}_{5,i} + \dot{\phi}_{3,i}\hat{\$}_{6,i}\end{aligned}. \tag{19}$$

By taking the orthogonal product of both sides of (19) with reciprocal screw $\hat{\$}_{r1,i} = \begin{bmatrix} \vec{s}_{3,i} \\ \vec{b}_i\times\vec{s}_{3,i} \end{bmatrix}$, one can eliminate the inactive joints and rotary actuator which yields to

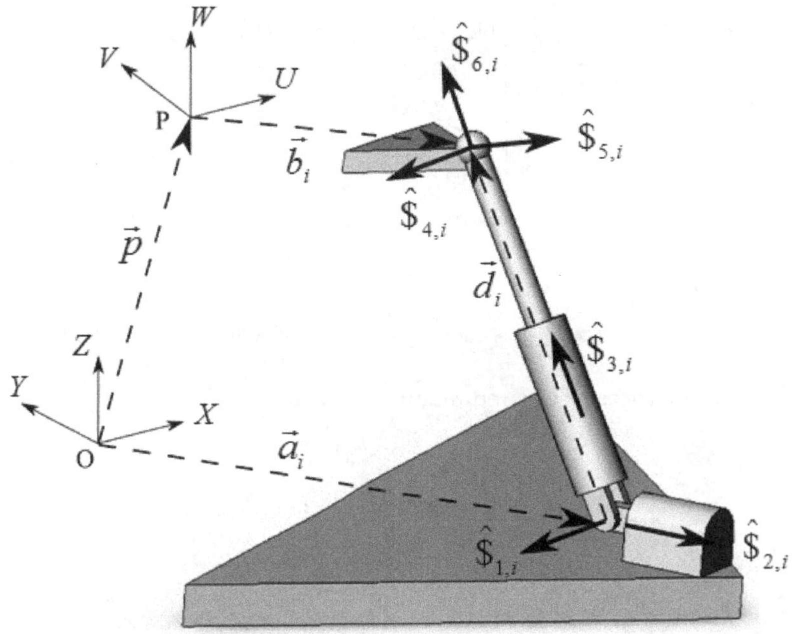

Fig. 5. Infinitesimal screws.

$$\begin{bmatrix} \dfrac{\vec{d}_i^{\,T}}{d_i} & \dfrac{(\vec{b}_i \times \vec{d}_i)^T}{d_i} \end{bmatrix} \dot{\vec{x}}_P = \dot{d}_i \,,\ i = 1,2,3,4 \,. \tag{20}$$

Similarly, if one takes the orthogonal product of both sides of (19) with reciprocal screw $\hat{\$}_{r6,i} = \begin{bmatrix} \dfrac{\vec{s}_i \times \vec{d}_i}{d_i \cos\psi_i} \\ \vec{b}_i \times \dfrac{\vec{s}_i \times \vec{d}_i}{d_i \cos\psi_i} \end{bmatrix}$ the resultant is as follows;

$$\begin{bmatrix} (\dfrac{\vec{s}_i \times \vec{d}_i}{d_i \cos\psi_i})^T & (\vec{b}_i \times \dfrac{\vec{s}_i \times \vec{d}_i}{d_i \cos\psi_i})^T \end{bmatrix} \dot{\vec{x}}_P = d_i \cos\psi_i \dot{\theta}_i \,. \tag{21}$$

Note that in (21), $\left|\vec{s}_i \times \vec{d}_i\right| = d_i \cos\psi_i$.

Finally, using (20) and (21), jacobian matrices $J_x^{Red.}$ and $J_q^{Red.}$ are expressed as;

$$J_x^{\mathrm{Red.}} = \begin{bmatrix} (\vec{s}_1 \times \overrightarrow{d_1})^T & (\vec{b}_1 \times (\vec{s}_1 \times \overrightarrow{d_1}))^T \\ (\vec{s}_2 \times \overrightarrow{d_2})^T & (\vec{b}_2 \times (\vec{s}_2 \times \overrightarrow{d_2}))^T \\ (\vec{s}_3 \times \overrightarrow{d_3})^T & (\vec{b}_3 \times (\vec{s}_3 \times \overrightarrow{d_3}))^T \\ (\vec{s}_4 \times \overrightarrow{d_4})^T & (\vec{b}_4 \times (\vec{s}_4 \times \overrightarrow{d_4}))^T \\ \overrightarrow{d_1}^T & (\vec{b}_1 \times \overrightarrow{d_1})^T \\ \overrightarrow{d_2}^T & (\vec{b}_2 \times \overrightarrow{d_2})^T \\ \overrightarrow{d_3}^T & (\vec{b}_3 \times \overrightarrow{d_3})^T \\ \overrightarrow{d_4}^T & (\vec{b}_4 \times \overrightarrow{d_4})^T \end{bmatrix}, \tag{22}$$

and

$$J_q^{\mathrm{Red.}} = diag(d_1^2 \cos^2 \psi_1, d_2^2 \cos^2 \psi_2, d_3^2 \cos^2 \psi_3, d_4^2 \cos^2 \psi_4, d_1, d_2, d_3, d_4). \tag{23}$$

Similarly, jacobian matrices of non-redundant mechanism ($J_x^{Non-\mathrm{Red.}}$ and $J_q^{Non-\mathrm{Red.}}$) can be expressed as;

$$J_x^{Non-\mathrm{Red.}} = \begin{bmatrix} (\vec{s}_1 \times \overrightarrow{d_1})^T & (\vec{b}_1 \times (\vec{s}_1 \times \overrightarrow{d_1}))^T \\ (\vec{s}_2 \times \overrightarrow{d_2})^T & (\vec{b}_2 \times (\vec{s}_2 \times \overrightarrow{d_2}))^T \\ (\vec{s}_3 \times \overrightarrow{d_3})^T & (\vec{b}_3 \times (\vec{s}_3 \times \overrightarrow{d_3}))^T \\ \overrightarrow{d_1}^T & (\vec{b}_1 \times \overrightarrow{d_1})^T \\ \overrightarrow{d_2}^T & (\vec{b}_2 \times \overrightarrow{d_2})^T \\ \overrightarrow{d_3}^T & (\vec{b}_3 \times \overrightarrow{d_3})^T \end{bmatrix}, \tag{24}$$

and

$$J_q^{Non-\mathrm{Red.}} = diag(d_1^2 \cos^2 \psi_1, d_2^2 \cos^2 \psi_2, d_3^2 \cos^2 \psi_3, d_1, d_2, d_3). \tag{25}$$

And for the Stewart platform one can have

$$J_x^{Stewart} = \begin{bmatrix} \overrightarrow{d_1}^T & (\vec{b}_1 \times \overrightarrow{d_1})^T \\ \overrightarrow{d_2}^T & (\vec{b}_2 \times \overrightarrow{d_2})^T \\ \overrightarrow{d_3}^T & (\vec{b}_3 \times \overrightarrow{d_3})^T \\ \overrightarrow{d_4}^T & (\vec{b}_4 \times \overrightarrow{d_4})^T \\ \overrightarrow{d_5}^T & (\vec{b}_5 \times \overrightarrow{d_5})^T \\ \overrightarrow{d_6}^T & (\vec{b}_6 \times \overrightarrow{d_6})^T \end{bmatrix}, \tag{26}$$

and

$$J_q^{Stewart} = diag(d_1, d_2, d_3, d_4, d_5, d_6). \tag{27}$$

Note that the joint velocity vector in stewart mechanism, $\dot{q}^{Stewart}$, is the following 6×1 vector:

$$\dot{q}^{Stewart} = [\dot{d}_1 \quad \dot{d}_2 \quad \dot{d}_3 \quad \dot{d}_4 \quad \dot{d}_5 \quad \dot{d}_6]^T . \tag{28}$$

Based on the existence of the two jacobian matrices above, the mechanism is at a singular configuration when either the determinant of J_x or J_q is either zero or infinity (Beji & Pascal, 1999).

Inverse kinematic singularity occurs when the determinant of J_q vanishes, namely;

$$\det(J_q) = 0 . \tag{29}$$

As it is clear from (25) and (27), the determinant of J_q in the workspace of this mechanism cannot be vanished. Therefore we do not have this type of singularity.

Forward kinematic singularity occurs when the rank of J_x is less than 6, namely;

$$rank(J_x) \leq 5 . \tag{30}$$

If (30) holds, the moving platform gains 1 or more degrees of freedom. In other words, at a forward kinematic singular configuration, the manipulator cannot resist forces or moments in some directions. As it can be seen, the redundancy can help us to avoid this kind of singularity, which is common in nearly all parallel mechanisms.

5. Performance indices

With the increasing demand for precise manipulators, search for a new manipulator with better performance has been intensive. Several indices have been proposed to evaluate the performance of a manipulator. Merlet reviewed the merits and weaknesses of these indices (Merlet, 2006). The dexterity indices have been commonly used in determining the dexterous regions of a manipulator workspace. The condition number of the jacobian matrix is known to be used as a measuring criterion of kinematic accuracy of manipulators. Salisbury & Craig (1982) used this criterion for the determination of the optimal dimensions for the fingers of an articulated hand. The condition number of the jacobian matrix has also been applied for designing a spatial manipulator (Angeles & Rojas, 1987).

The most performance indices are pose-dependant. For design, optimization and comparison purpose, Gosselin & Angeles (1991) proposed a global performance index, which is the integration of the local index over the workspace.

The performance indices are usually formed based on the evaluation of the determinant, norms, singular values and eigenvalues of the jacobian matrix. These indices have physical interpretation and useful application for control and optimization just when the elements of the jacobian matrix have the same units (Kucuk & Bingul, 2006). Otherwise, the stability of control systems, which are formed based on these jacobian matrices, will depend on the physical units of parameters chosen (Schwartz et al., 2002). Thus, different indices for rotations and translations should be defined (Cardou et al., 2010).

5.1 Manipulability

For evaluation of kinematic transmissibility of a manipulator, Yoshikawa (1984) defined the manipulator index,

$$\mu = \sqrt{\det(J^T J)} \,. \tag{31}$$

The manipulability can geometrically be interpreted as the volume of the ellipsoid obtained by mapping a unit n-dimensional sphere of joint space onto the Cartesian space (Cardou et al., 2010). It also can be interpreted as a measure of manipulator capability for transmitting a certain velocity to its end-effector (Mansouri & Ouali, 2011). To have a better performance for a manipulator, It is more suitable to have isotropic manipulability ellipsoid (Angeles & Lopez-Cajun, 1992). The isotropy index for manipulability can be defined as:

$$\mu_{iso} = \frac{\sigma_{Min}}{\sigma_{Max}} \,, \tag{32}$$

where σ_{Max} and σ_{Min} are maximum and minimum of singular values of jacobian matrix (J), respectively. The isotropy index is limited between 0 and 1. When the isotropy index is equal to 1, it indicates the ability of manipulator to transmit velocity uniformly from actuators to the end-effector along all directions. Inversely, when the isotropy index is equal to zero, the manipulator is at a singular configuration and cannot transmit velocity to the end-effector.

5.2 Dexterity

The accuracy of a mechanism is dependent on the condition number of the jacobian matrix, which is defined as follows:

$$k = \| J \| . \| J^{-1} \| \,, \tag{33}$$

where J is the jacobian matrix and $\|J\|$ denotes the norm of it and is defined as follows:

$$\| J \| = \sqrt{tr(\frac{1}{n} J J^T)} \,, \tag{34}$$

where n is the dimension of the square matrix J that is 3 for the manipulator under study. Gosselin (1992) defined the local dexterity (υ) as a criterion for measuring the kinematics accuracy of a manipulator,

$$\upsilon = \frac{1}{\| J \| . \| J^{-1} \|} \,. \tag{35}$$

The local dexterity can be changed between zero and one. The higher values indicate more accurate motion generated at given instance. When the local dexterity is equal to one, it denotes that the manipulator is isotropic at the given pose. Otherwise, it implies that the manipulator has reached a singular configuration pose.

To evaluate the dexterity of a manipulator over the entire workspace (W Gosselin & Angeles (1991) have introduced the global dexterity index (GDI) as:

$$GDI = \frac{\int_W \upsilon \; dW}{\int_W dW}, \tag{36}$$

which is the average value of local dexterity over the workspace. This global dexterity index can be used as design factor for the optimal design of manipulators (Bai, 2010; Li et al., 2010; Liu et al., 2010; Unal et al., 2008). Having a uniform dexterity is a desirable goal for almost all robotic systems. Uniformity of dexterity can be defined as another global performance index. It can be defined as the ratio of the minimum and maximum values of the dexterity index over the entire workspace,

$$\upsilon_{iso} = \frac{\upsilon_{Min}}{\upsilon_{Max}}. \tag{37}$$

5.3 Sensitivity

Evaluating of the kinematic sensitivity is another desirable concept in the performance analysis of a manipulator. The kinematic sensitivity of a manipulator can be interpreted as the effect of actuator displacements on the displacement of the end-effector. Cardou et al. (2010) defined two indices (τ_r, τ_p) for measuring the rotation and displacement sensitivity of a manipulator. They assumed the maximum-magnitude rotation and the displacement of the end-effector under a unit-norm actuator displacement as indices for calculating the sensitivity of a manipulator. The sensitivity indices can be defined as:

$$\tau_r = \| J_r \|, \tag{38}$$

and

$$\tau_p = \| J_p \|, \tag{39}$$

where J_r and J_p are rotation and translation jacobian martices (Cardou et al., 2010), respectively, where $\| \; \|$ stands for a *p*-norm of the matrix. Cardou et al. (2010) suggested to use 2-norm and ∞-norm for calculating the sensitivity.

6. Comparison between 4-legged mechanism and other mechanisms

In order to investigate the kinematic performance of 4-legged mechanism, the response of the mechanisms are compared in several different aspects; reachable points, performance indices, and singularity analysis.

6.1 Reachable points and workspace comparison

Consider the 3-legged and the 4-legged mechanisms with g=1 m and h=0.5 m, respectively; g and h are the radii of the fixed and moving platforms, respectively.

By assuming a cubic with 1m length, 1 m width and 1 m height located 0.25 m above the base platform, we are interested in determining the reachability percentage in which each mechanism can successfully reach to the locations within this cubic space.

Table 1 shows that adding one leg to the 3-legged mechanism reduces the *R.P.* (Reachable points Percentage) by 5.03%. However, it should be noted that, although the non-redundant mechanism has a wider workspace, it has much more singular points than the redundant mechanism, and actuator forces and torques are also less in the redundant mechanism. As it can be seen from Table 1, *RP*s in the Stewart platforms are smaller than 3-legged and 4-legged mechanisms. In the 6-legged Stewart-like UPS mechanisms (Stewart, 1965), the workspace is constructed by intersecting of 6 spheres. On the other hand, in the proposed 4-legged UPS mechanism, the workspace is constructed by intersecting of only 4 spheres.

Mechanism	% RP
Stewart (6-6)	76.35
Stewart (3-6)	74.21
Stewart (6-3)	73.70
Stewart (3-3)	63.46
3-legged mechanism	84.75
4-legged mechanism	79.72

Table 1. Reachable points

Assuming similar dimensions for the two mechanisms result in a larger workspace for the 4-legged mechanism.
To compare the workspace of the proposed 4-legged mechanism with the 3-legged mechanism and Steward platforms, the reachable points for them were calculated and obtained results are shown in Fig. 6 It is appear that the 3-legged mechanism has the largest workspace followed by 4-legged mechanism, 3-6 and 3-3 Steward Platforms.

6.2 Performance comparison

To compare the kinematic performance of the proposed 4-legged mechanism with the 3-legged mechanism and Steward platform, abovementioned performance indices were used. The results obtained are shown in Figs. 7 to 10. These figures show how performance indices vary on a plate (*Z=0.75*) within the workspace.
Fig. 7 depicts the Stewart 3-3 has the higher isotropy index for manipulability comparing with Stewart 3-6, 3-legged and 4-legged mechanisms. After Stewart 3-3, the 4-legged mechanism presents the better performance. This figure illustrates that adding a leg to a 3-legged mechanism can significantly improve the manipulability of the mechanism. Furthermore, a 4-legged mechanism can have a performance comparable with the Stewart platforms or even better.
Fig. 8 depicts the Stewart 3-3 compared with the other mechanisms under study has the higher dexterity index. After Stewart 3-3, again, the 4-legged mechanism presents the better performance. This figure also illustrates significant enhancement of the dexterity of the mechanism due to the additional leg. Also it shows that, in terms of dexterity, the 4-legged mechanism can have a comparable performance against the Stewart platforms or even better.
In the next step of performance comparison of manipulators, displacement and rotation sensitivities for mechanisms of our interest are compared. Fig. 9 shows the amount of displacement sensitivity indices of the abovementioned mechanisms on the selected plane

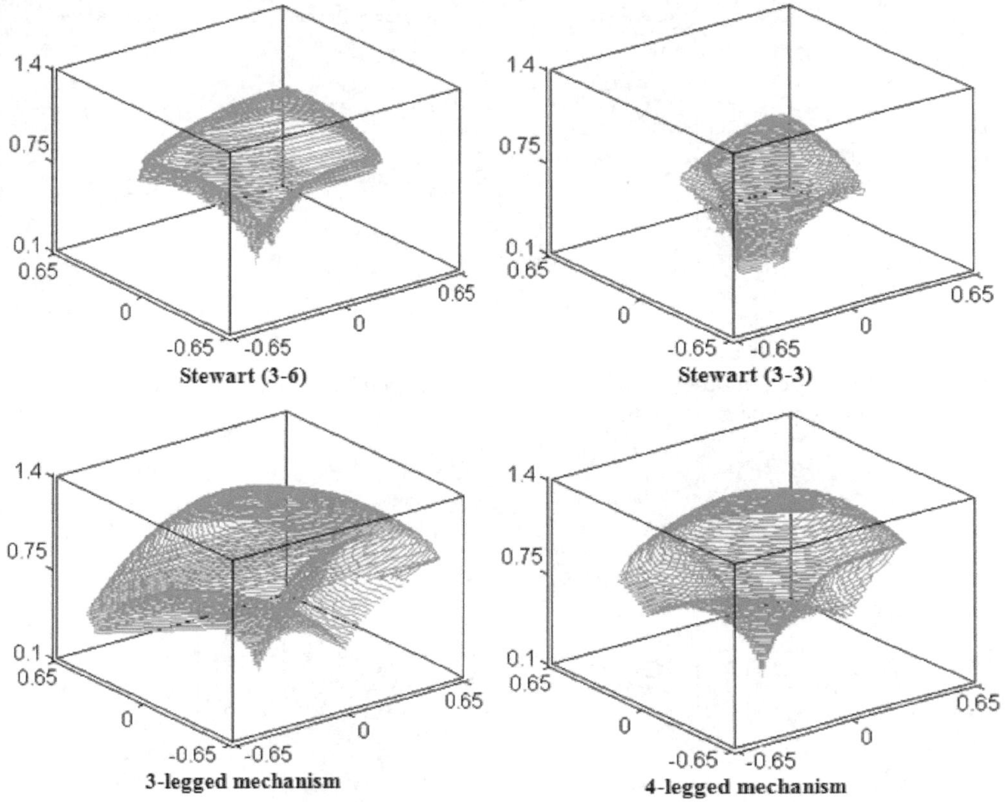

Fig. 6. Workspaces of Stewart (3-6), Stewart (3-3), 3-legged and 4-legged mechanisms.

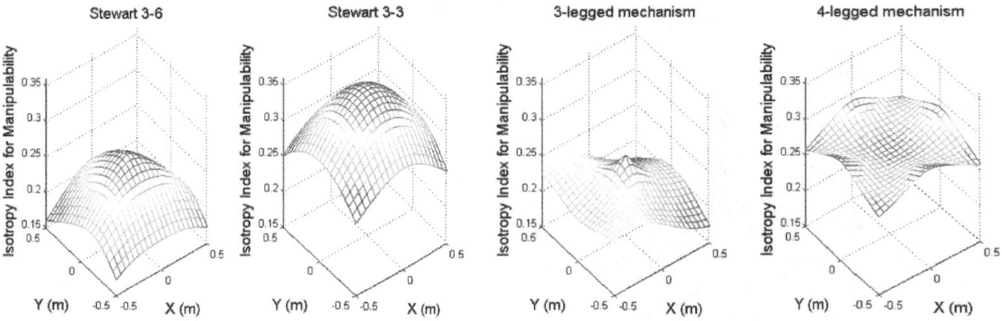

Fig. 7. Isotropy index for manipulability of Stewart 6-6, Stewart 3-6, 3 Legged mechanism and 4 legged mechanism on the plane z= 0.75 m

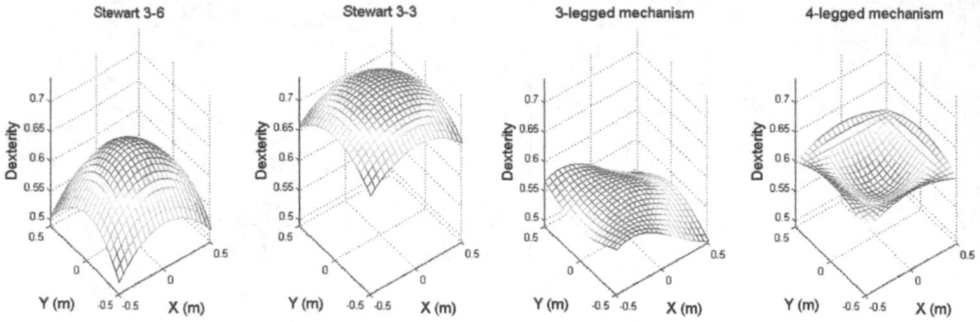

Fig. 8. Dexterity index of Stewart 6-6, Stewart 3-6, 3-Legged mechanism and 4-legged mechanism on the plane z= 0.75 m.

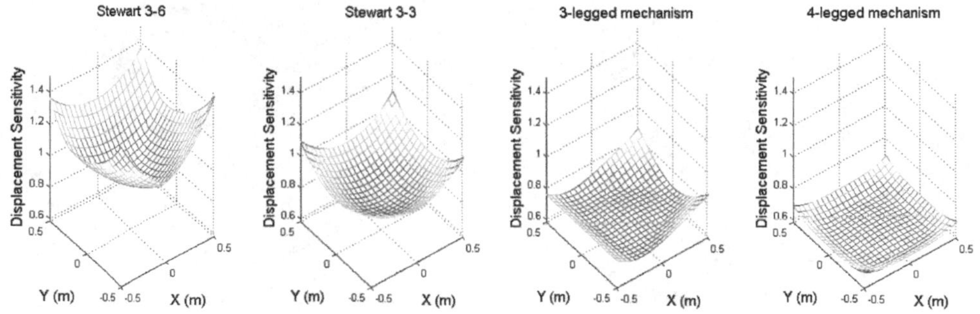

Fig. 9. Displacement Sensitivity index of Stewart 6-6, Stewart 3-6, 3-Legged mechanism and 4-legged mechanism on the plane z= 0.75 m.

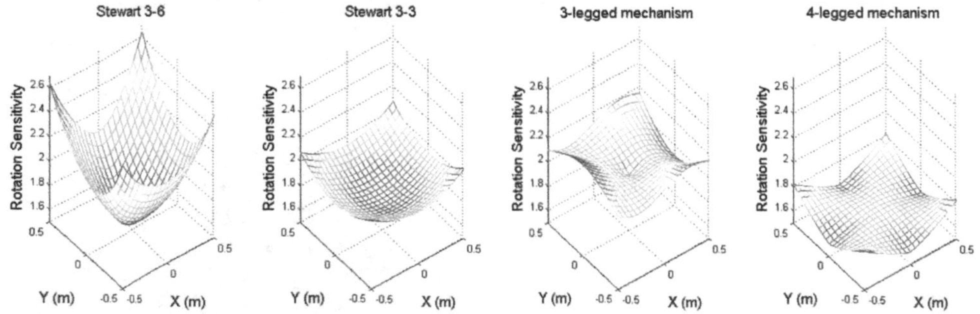

Fig. 10. Rotation Sensitivity index of Stewart 6-6, Stewart 3-6, 3-Legged mechanism and 4-legged mechanism on the plane z= 0.75 m.

(*Z=0.75*). It clearly shows that the 4-legged mechanism has less displacement sensitivity index by far. Fig. 10 depicts the 4-legged mechanism has also less rotation sensitivity compared with other mechanisms.
So far from Figs. 7 to 10, the amounts of performance indices are shown on the planes. To compare the kinetics performance of manipulators over the entire workspace, the global performance Index (GPI) can be evalueated as:

$$GPI = \frac{\int_W PI \;\; dW}{\int_W dW}, \tag{40}$$

which is the average value of local performance index (*PI*) over the workspace (*W*).
The amounts of GPI for Isotropy Index for Manipulability (*IIM*), Dexterity (*DX*), Displacement Sensitivity (*DS*) and Rotation Sensitivity (*RS*) were calculated and obtained results were listed in Table 2.
Table 2 shows that the 4-legged mechanism has a better *IIM* within the selected workspace, which explicitly indicates a better ability for transmitting a certain velocity to its end-effector.
As it is seen from Table 2, the Stewart 3-3 platform has the biggest global *DX* compared with other mechanisms and the 4-legged mechanism has the second (i.e. difference in *DX* is only 5.71% less). It represents that the Stewart 3-3 platform and the proposed 4-legged mechanism have the better kinematics accuracy.
Having the lower sensitivity is a demand for industrial mechanisms. By comparing the values of *DS* and *RS*, which are listed in Table 2, it is obvious the 4-legged mechanism is an appropriate candidate.
Based on the results shown in Table 2, the 4-legged mechanism is recommended for better kinematic performances

	The higher, the better		The lower, the better	
Mechaniasm	*IIM*	*DX*	*DS*	*RS*
Stewart (6-6)	0.0429	0.1589	1.3249	3.5791
Stewart (3-6)	0.1296	0.3969	1.2070	2.5725
Stewart (6-3)	0.1020	0.3449	1.2113	2.8991
Stewart (3-3)	0.1978	**0.5284**	1.0485	2.6077
3-legged mechanism	0.1136	0.3423	0.8538	2.3861
4-legged mechanism	**0.2140**	0.4982	**0.7279**	**1.8441**

Table 2. Performance comparison between 3-legged mechanism, 4-legged mechanism, Stewart 6-6, and Stewart 3-6

6.3 Singularity analysis of 3-legged and 4-legged mechanisms

Several types of workspace can be considered. For example, the 3D constant orientation workspace, which describes all possible locations of an arbitrary point P in the moving system with a constant orientation of the moving platform, the reachable workspace (all the locations that can be reached by P), the orientation workspace (all possible orientations of the end-effector around P for a given position) or the inclusive orientation workspace (all the locations that can be reached by the origin of the end-effector with every orientation in a given set) (Abedinnasab & Vossoughi, 2009).

Out of those types, we have used the inclusive orientation workspace, where for every position in a fixed surface, the moving platform is rotated in every possible orientation to determine if that configuration is singular or not. After trials and errors, we figured out that for a better determination of the singular configurations, the roll-pitch-yaw rotation about the global coordinate is the most critical set of rotations compared to the other rotations such as the reduced Euler rotations.

To illustrate the positive effects of redundancy on eliminating singular configurations, we have done jacobian analysis in planes in different orientations of the workspace as shown in Fig. 11. The results are shown in Figs. 12 and 13. In Figs. 12 and 13, the jacobian determinant of center of the moving platforms of 3-legged and 4-legged mechanisms has been calculated. The platform is rotated simultaneously in three different directions according to the roll-pitch-yaw Euler angles discussed above. Each angle is free to rotate up to ±20°. After the rotations in each position, if the mechanism did not encounter any singular configuration, the color of that position is represented by light gray. If there was any singular configuration inside ±20° region and beyond ±10° region, the color is dark gray. At last, if the singular configuration was encountered in ±10° rotations, the color is black.

As seen from Fig. 12, in the 3-legged mechanism, there exist singular configurations in the most of the X-Y, X-Z and Y-Z planes (black and dark gray regions). However, in the 4-legged mechanism, the singular points do not exist at the most of the plane. Figure 13 shows the same patterns as in the other planes. Figures 12 and 13 simply illustrate the great effect of a simple redundancy; namely, the addition of a leg to the 3-legged mechanism can remove vast singular configurations.

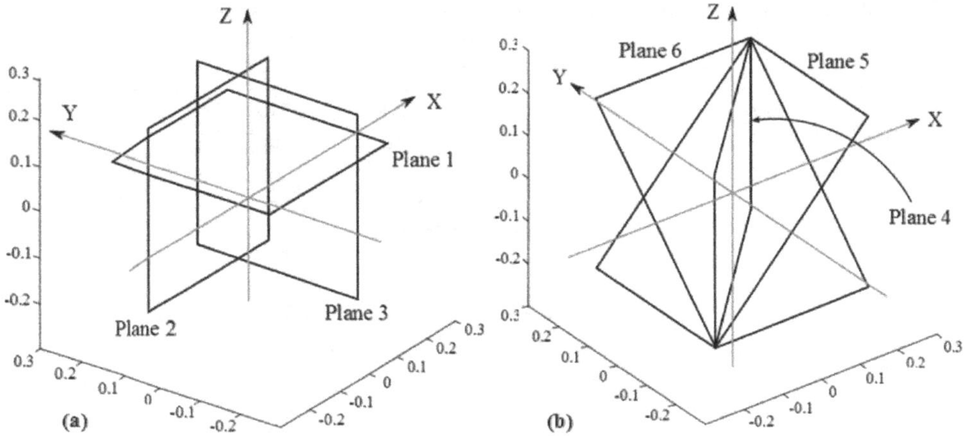

Fig. 11. (a) Schematic view of planes 1, 2, and 3. (b) Schematic view of planes 4, 5, and 6.

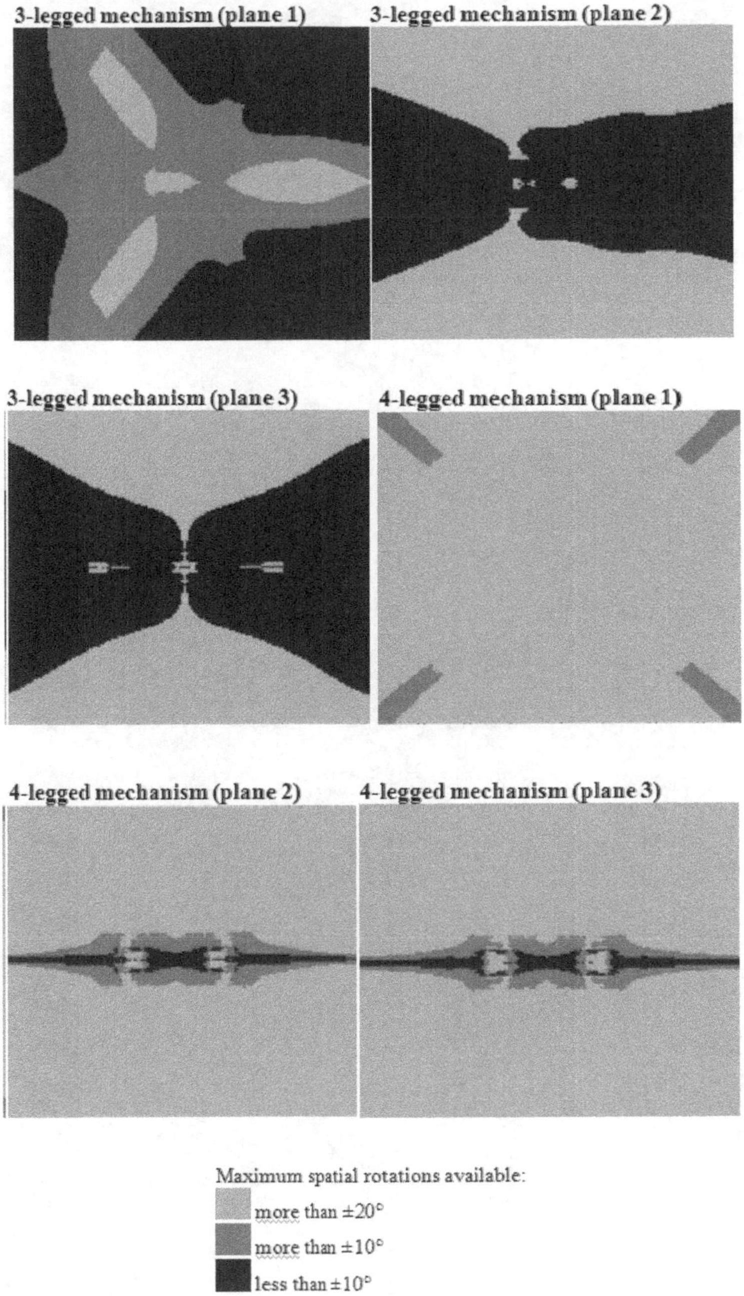

Fig. 12. Singularity analysis in planes 1, 2 and 3 for both 3-legged (non-redundant) and 4-legged (redundant) mechanisms.

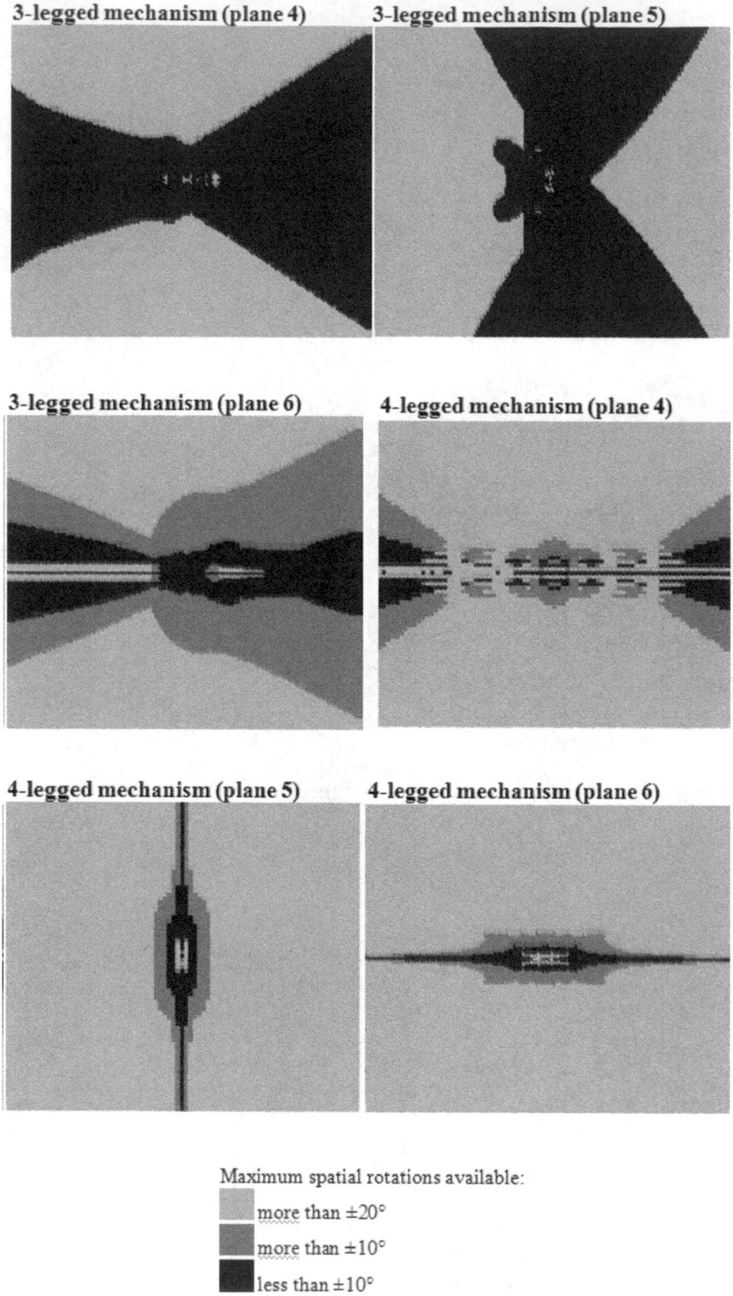

Fig. 13. Singularity analysis in planes 4, 5 and 6 for both 3-legged (non-redundant) and 4-legged (redundant) mechanisms.

7. Conclusion

The effects of redundant actuation are studied. The redundant 4-legged and non-redundant 3-legged parallel mechanisms are compared with 4 well-known architectures of Gough-Stewart platforms. It is shown that the inverse kinematics of the proposed 3-legged and 4-legged mechanisms have a closed-form solution. Also the Jacobian matrix has been determined using the concept of reciprocal screws.

From the design point of view, by replacing the passive universal joints in the Stewart platforms with active joints in the above mentioned mechanisms, the number of legs could be reduced from 6 to 3 or 4. It makes the mechanism to be lighter, since the rotary actuators are resting on the fixed platform, which allows higher accelerations to be available due to smaller inertial effects.

It is illustrated that redundancy improves the ability and performance of the non-redundant parallel manipulator. The redundancy brings some advantages for parallel manipulators such as avoiding kinematic singularities, increasing workspace, improving performance indices, such as dexterity, manipulability, and sensitivity. Finally, we conclude that the redundancy is a key choice to remove singular points, which are common in nearly all parallel mechanisms.

It is worthy to state that the applications of these robots can be found in flight simulators, high precision surgical tools, positioning devices, motion generators, ultra-fast pick-and-place robots, haptic devices, entertainment, multi-axis machine tools, micro manipulators, rehabilitation devices, etc.

8. References

Abedinnasab, M. H. & Vossoughi, G. R. (2009). Analysis of a 6-dof redundantly actuated 4-legged parallel mechanism, Nonlinear Dyn. 58(4): 611–622.

Aghababai, O. (2005). Design, Kinematic and Dynamic Analysis and Optimization of a 6 DOF Redundantly Actuated Parallel Mechanism for Use in Haptic Systems, MSc Thesis, Sharif University of Technology, Tehran, Iran.

Alp, H. & Özkol, I. (2008). Extending the workspace of parallel working mechanisms, Proc. Inst. Mech. Eng., Part C: J. Mech. Eng. Sci. 222(7): 1305–1313.

Angeles, J. & Lopez-Cajun, C. S. (1992). Kinematic isotropy and the conditioning index of serial robotic manipulators, Int. J. Rob. Res. 11(6): 560–571.

Angeles, J. & Rojas, A. A. (1987). Manipulator inverse kinematics via condition-number minimization and continuation, Int. J. Rob. Autom. 2: 61–69.

Arata, J., Kondo, J., Ikedo, N. & Fujimoto, H. (2011). Haptic device using a newly developed redundant parallel mechanism, IEEE Trans. Rob. 27(2): 201–214.

Bai, S. (2010). Optimum design of spherical parallel manipulators for a prescribed workspace, Mech. Mach. Theory 45(2): 200–211.

Beji, L. & Pascal, M. (1999). Kinematics and the full minimal dynamic model of a 6-dof parallel robot manipulator, Nonlinear Dyn. 18: 339–356.

Cardou, P., Bouchard, S. & Gosselin, C. (2010). Kinematic-sensitivity indices for dimensionally nonhomogeneous jacobian matrices, IEEE Trans. Rob. 26(1): 166–173.

Choi, H. B., Konno, A. & Uchiyama, M. (2010). Design, implementation, and performance evaluation of a 4-DOF parallel robot, Robotica 28(1): 107–118.

Deidda, R., Mariani, A. & Ruggiu, M. (2010). On the kinematics of the 3-RRUR spherical parallel manipulator, Robotica 28(6): 821–832.

Gallardo-Alvarado, J., Aguilar-Nájera, C. R., Casique-Rosas, L., Rico-Martínez, J. M. & Islam, M. N. (2008). Kinematics and dynamics of 2(3-RPS) manipulators by means of screw theory and the principle of virtual work, Mech. Mach. Theory 43(10): 1281–1294.

Gallardo-Alvarado, J., Orozco-Mendoza, H. & Rico-Martínez, J. M. (2010). A novel five-degrees-of-freedom decoupled robot, Robotica 28(6): 909–917.

Gallardo-Alvarado, J., Rico-Martínez, J. M. & Alici, G. (2006). Kinematics and singularity analyses of a 4-dof parallel manipulator using screw theory, Mech. Mach. Theory 41(9): 1048–1061.

Gan, D., Liao, Q., Dai, J. & Wei, S. (2010). Design and kinematics analysis of a new 3CCC parallel mechanism, Robotica 28: 1065–1072.

Gao, Z., Zhang, D., Hu, X. & Ge, Y. (2010). Design, analysis, and stiffness optimization of a three degree of freedom parallel manipulator, Robotica 28(3): 349–357.

Gosselin, C. & Angeles, J. (1990). Singularity analysis of closed-loop kinematic chains, IEEE Trans. Rob. Autom. 6(3): 281–290.

Gosselin, C. & Angeles, J. (1991). Global performance index for the kinematic optimization of robotic manipulators, J. Mech. Transm. Autom. Des. 113(3): 220–226.

Gosselin, C. M. (1992). The optimum design of robotic manipulators using dexterity indices, Rob. Autom. Syst. 9(4): 213–226.

Gough, V. E. & Whitehall, S. G. (1962). Universal tyre test machine, in Proc 9th Int. Tech. Congress FISITA pp. 117–137.

Jiang, Q. & Gosselin, C. M. (2009a). Determination of the maximal singularity-free orientation workspace for the gough-stewart platform, Mech. Mach. Theory 44(6): 1281–1293.

Jiang, Q. & Gosselin, C. M. (2009b). Evaluation and Representation of the Theoretical Orientation Workspace of the Gough-Stewart Platform, J. Mech. Rob., Trans. ASME 1(2): 1–9.

Jiang, Q. & Gosselin, C. M. (2009c). Maximal Singularity-Free Total Orientation Workspace of the Gough-Stewart Platform, J. Mech. Rob., Trans. ASME 1(3): 1–4.

Kucuk, S. & Bingul, Z. (2006). Comparative study of performance indices for fundamental robot manipulators, Rob. Autom. Syst. 54(7): 567–573.

Li, H., Gosselin, C. M. & Richard, M. J. (2007). Determination of the maximal singularity-free zones in the six-dimensional workspace of the general, Mech. Mach. Theory 42(4): 497-511.

Li, J.,Wang, S.,Wang, X. & He, C. (2010). Optimization of a novel mechanism for a minimally invasive surgery robot, Int. J. Med. Rob. Comput. Assisted Surg. 6(1): 83–90.

Liu, X., Xie, F., Wang, L. & Wang, J. (2010). Optimal design and development of a decoupled A/B-axis tool head with parallel kinematics, Adv. Mech. Eng. 2010: 1–14.

Lopes, A. M. (2010). Complete dynamic modelling of a moving base 6-dof parallel manipulator, Robotica 28(5): 781–793.

Lu, Y., Hu, B., Li, S.-H. & Tian, X. B. (2008). Kinematics/statics analysis of a novel 2SPS + PRRPR parallel manipulator, Mech. Mach. Theory 43(9): 1099–1111.

Lu, Y., Hu, B. & Yu, J. (2009). Analysis of kinematicsstatics and workspace of a 2(SP+SPR+SPU) serial-parallel manipulator, Multibody Sys. Dyn. 21(4): 361–374.

Lu, Y., Shi, Y. & Yu, J. (2010). Determination of singularities of some 4-dof parallel manipulators by translational/rotational jacobian matrices, Robotica 28(6): 811–819.

Mansouri, I. & Ouali, M. (2011). The power manipulability a new homogeneous performance index of robot manipulators, Rob. Comput. Integr. Manuf. 27(2): 434–449.

Masouleh, M. T., Gosselin, C., Husty, M. & Walter, D. R. (2011). Forward kinematic problem of 5-RPUR parallel mechanisms (3T2R) with identical limb structures, Mech. Mach. Theory 46(7): 945–959.

Merlet, J. P. (2006). Jacobian, manipulability, condition number, and accuracy of parallel robots, J. Mech. Des., Trans. ASME 128(1): 199–206.

Müller, A. (2005). Internal preload control of redundantly actuated parallel manipulators – its application to backlash avoiding control, IEEE Trans. Rob. 21(4): 668–677.

Pendar, H., Mahnama, M. & Zohoor, H. (2011). Singularity analysis of parallel manipulators using constraint plane method, Mech. Mach. Theory 46(1): 33–43.

Piipponen, S. (2009). Singularity analysis of planar linkages, Multibody Sys. Dyn. 22(3): 223–243.

Ropponen, T. & Nakamura, Y. (1990). Singularity-free parameterization and performance analysis of actuation redundancy, Proc. IEEE Int. Conf. Robot. Autom. 2: 806–811.

Sadjadian, H. & Taghirad, H. D. (2006). Kinematic, singularity and stiffness analysis of the hydraulic shoulder: A 3-dof redundant parallel manipulator, Adv. Rob. 20(7): 763–781.

Salisbury, J. K. & Craig, J. J. (1982). Articulated hands - force control and kinematic issues, Int. J. Rob. Res. 1(1): 4–17.

Schwartz, E., Manseur, R. & Doty, K. (2002). Noncommensurate systems in robotics, Int. J. Rob. Autom. 17(2): 86–92.

St-Onge, B. M. & Gosselin, C. M. (2000). Singularity analysis and representation of the general gough-stewart platform, Int. J. Rob. Res. 19(3): 271–288.

Staicu, S. (2009). Dynamics of the spherical 3-U-PS/S parallel mechanism with prismatic actuators, Multibody Sys. Dyn. 22(2): 115–132.

Stewart, D. (1965). A platform with six degrees of freedom, Proc. Inst. Mech. Eng. 180: 371–386.

Tsai, L.-W. (1998). The jacobian analysis of a parallel manipulator using reciprocal screws in Advances in Robot Kinematics: Analysis and Control, J. Lenarcic and M. L. Husty, Eds., Kluwer Academic.

Tsai, L.-W. (1999). Robot analysis: the mechanics of serial and parallel manipulators, Wiley.

Unal, R., Kizltas, G. & Patoglu, V. (2008). A multi-criteria design optimization framework for haptic interfaces, Symp. Haptic Interfaces Virtual Environ. Teleoperator Syst. pp. 231–238.

Vischer, P. & Clavel, R. (1998). Kinematic calibration of the parallel delta robot, Robotica 16(2): 207–218.

Wang, J. & Gosselin, C. M. (2004). Kinematic analysis and design of kinematically redundant parallel mechanisms, J. Mech. Des., Trans. ASME 126(1): 109–118.

Wu, J., Wang, J. & wang, L. (2010). A comparison study of two planar 2-DOF parallel mechanisms : one with 2-RRR and the other with 3-RRR structures, Robotica 28(6): 937–942.

Yi, B.-J., Freeman, R. A. & Tesar, D. (1989). Open-loop stiffness control of overconstrained mechanisms/robotic linkage systems, Proc. IEEE Int. Conf. Robot. Autom. 3: 1340 - 1345.

Yoshikawa, T. (1984). Analysis and control of robot manipulators with redundancy, Int. Symp. Rob. Res. pp. 735-747.

Zhao, J., Feng, Z., Zhou, K. & Dong, J. (2005). Analysis of the singularity of spatial parallel manipulator with terminal constraints, Mech. Mach. Theory 40(3): 275-284.

Zhao, Y. & Gao, F. (2010). The joint velocity, torque, and power capability evaluation of a redundant parallel manipulator. accepted in Robotica.

Zhao, Y., Liu, J. F. & Huang, Z. (2011). A force analysis of a 3-rps parallel mechanism by using screw theory. accepted in Robotica.

2

Inverse Dynamics of RRR Fully Planar Parallel Manipulator Using DH Method

Serdar Küçük
Kocaeli University
Turkey

1. Introduction

Parallel manipulators are mechanisms where all the links are connected to the ground and the moving platform at the same time. They possess high rigidity, load capacity, precision, structural stiffness, velocity and acceleration since the end-effector is linked to the movable plate in several points (Kang et al., 2001; Kang & Mills, 2001; Merlet, J. P. 2000; Tsai, L., 1999; Uchiyama, M., 1994). Parallel manipulators can be classified into two fundamental categories, namely spatial and planar manipulators. The first category composes of the spatial parallel manipulators that can translate and rotate in the three dimensional space. Gough-Stewart platform, one of the most popular spatial manipulator, is extensively preferred in flight simulators. The planar parallel manipulators which composes of second category, translate along the x and y-axes, and rotate around the z-axis, only. Although planar parallel manipulators are increasingly being used in industry for micro-or nano-positioning applications, (Hubbard et al., 2001), the kinematics, especially dynamics analysis of planar parallel manipulators is more difficult than their serial counterparts. Therefore selection of an efficient kinematic modeling convention is very important for simplifying the complexity of the dynamics problems in planar parallel manipulators. In this chapter, the inverse dynamics problem of three-**D**egrees **O**f **F**reedom (**DOF**) RRR **F**ully **P**lanar **P**arallel **M**anipulator (**FPPM**) is derived using DH method (Denavit & Hartenberg, 1955) which is based on 4x4 homogenous transformation matrices. The easy physical interpretation of the rigid body structures of the robotic manipulators is the main benefit of DH method. The inverse dynamics of 3-DOF RRR FPPM is derived using the virtual work principle (Zhang, & Song, 1993; Wu et al., 2010; Wu et al., 2011). In the first step, the inverse kinematics model and Jacobian matrix of 3-DOF RRR FPPM are derived by using DH method. To obtain the inverse dynamics, the partial linear velocity and partial angular velocity matrices are computed in the second step. A pivotal point is selected in order to determine the partial linear velocity matrices. The inertial force and moment of each moving part are obtained in the next step. As a last step, the inverse dynamic equation of 3-DOF RRR FPPM in explicit form is derived. To demonstrate the active joints torques, a butterfly shape Cartesian trajectory is used as a desired end-effector's trajectory.

2. Inverse kinematics and dynamics model of the 3-DOF RRR FPPM

In this section, geometric description, inverse kinematics, Jacobian matrix & Jacobian inversion and inverse dynamics model of the 3-DOF RRR FPPM in explicit form are obtained by applying DH method.

2.1 Geometric descriptions of 3-DOF RRR FPPM

The 3-DOF RRR FPPM shown in Figure 1 has a moving platform linked to the ground by three independent kinematics chains including one active joint each. The symbols θ_i and α_i illustrate the active and passive revolute joints, respectively where i=1, 2 and 3. The link lengths and the orientation of the moving platform are denoted by l_j and ϕ, respectively, j=1, 2, ··· ,6. The points B_1, B_2, B_3 and M_1, M_2, M_3 define the geometry of the base and the moving (Figure 2) platform, respectively. The {XYZ} and {xyz} coordinate systems are attached to the base and the moving platform of the manipulator, respectively. O and M_1 are the origins of the base and moving platforms, respectively. $P(X_B, Y_B)$ and ϕ illustrate the position of the end-effector in terms of the base coordinate system {XYZ} and orientation of the moving platform, respectively.

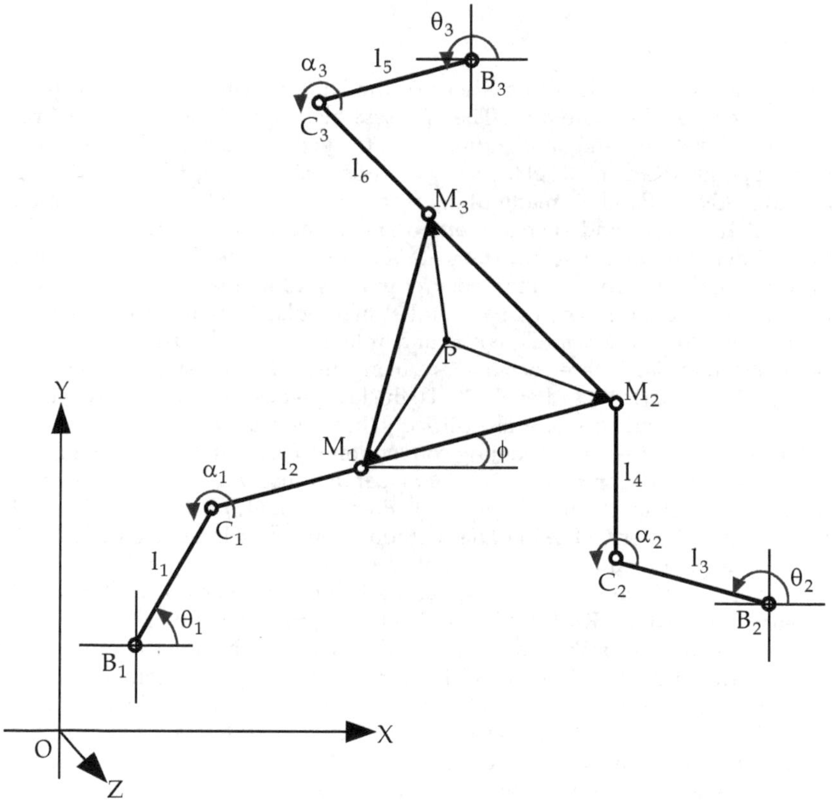

Fig. 1. The 3-DOF RRR FPPM

The lines M_1P, M_2P and M_3P are regarded as n_1, n_2 and n_3, respectively. The γ_1, γ_2 and γ_3 illustrate the angles $B\hat{P}M_1$, $M_2\hat{P}B$, and $B\hat{P}M_3$, respectively. Since two lines AB and M_1M_2 are parallel, the angles $P\hat{M}_1M_2$ and $P\hat{M}_2M_1$ are equal to the angles $A\hat{P}M_1$ and $M_2\hat{P}B$, respectively. $P(x_m, y_m)$ denotes the position of end-effector in terms of {xyz} coordinate systems.

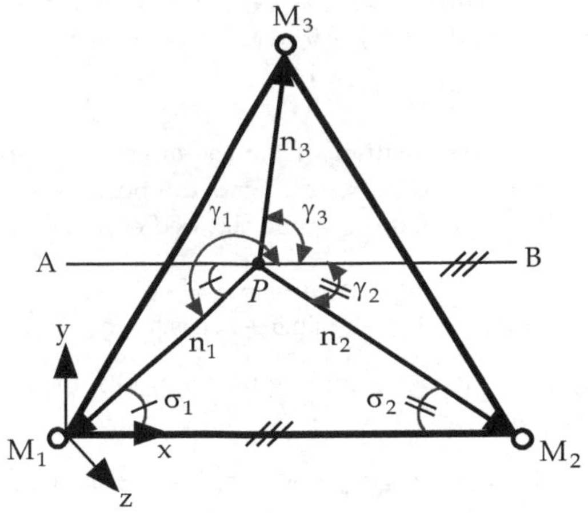

Fig. 2. The moving platform

2.2 Inverse kinematics

The inverse kinematic equations of 3-DOF RRR FPPM are derived using the DH (Denavit & Hartenberg, 1955) method which is based on 4x4 homogenous transformation matrices. The easy physical interpretation of the rigid body structures of the robotic manipulators is the main benefit of DH method which uses a set of parameters ($\alpha_{i-1}, a_{i-1}, \theta_i$ and d_i) to describe the spatial transformation between two consecutive links. To find the inverse kinematics problem, the following equation can be written using the geometric identities on Figure 1.

$$OB_i + B_iM_i = OP + PM_i \tag{1}$$

where i=1, 2 and 3. If the equation 1 is adapted to the manipulator in Figure 1, the ${}^{O_i}_{M_i}T^1$ and ${}^{O_i}_{M_i}T^2$ transformation matrices can be determined as

$$
\begin{aligned}
{}^{O_i}_{M_i}T^1 &= \begin{bmatrix} 1 & 0 & 0 & o_{x_i} \\ 0 & 1 & 0 & o_{y_i} \\ 0 & 0 & 1 & 0 \\ 0 & 0 & 0 & 1 \end{bmatrix} \begin{bmatrix} \cos\theta_i & -\sin\theta_i & 0 & 0 \\ \sin\theta_i & \cos\theta_i & 0 & 0 \\ 0 & 0 & 1 & 0 \\ 0 & 0 & 0 & 1 \end{bmatrix} \begin{bmatrix} \cos\alpha_i & -\sin\alpha_i & 0 & l_{2i-1} \\ \sin\alpha_i & \cos\alpha_i & 0 & 0 \\ 0 & 0 & 1 & 0 \\ 0 & 0 & 0 & 1 \end{bmatrix} \begin{bmatrix} 1 & 0 & 0 & l_{2i} \\ 0 & 1 & 0 & 0 \\ 0 & 0 & 1 & 0 \\ 0 & 0 & 0 & 1 \end{bmatrix} \\
&= \begin{bmatrix} \cos(\theta_i+\alpha_i) & -\sin(\theta_i+\alpha_i) & 0 & o_{x_i} + l_{2i}\cos(\theta_i+\alpha_i) + l_{2i-1}\cos\theta_i \\ \sin(\theta_i+\alpha_i) & \cos(\theta_i+\alpha_i) & 0 & o_{y_i} + l_{2i}\sin(\theta_i+\alpha_i) + l_{2i-1}\sin\theta_i \\ 0 & 0 & 1 & 0 \\ 0 & 0 & 0 & 1 \end{bmatrix}
\end{aligned} \tag{2}
$$

$$
{}^{O_i}_{M_i}T^2 = \begin{bmatrix} 1 & 0 & 0 & P_{X_B} \\ 0 & 1 & 0 & P_{Y_B} \\ 0 & 0 & 1 & 0 \\ 0 & 0 & 0 & 1 \end{bmatrix} \begin{bmatrix} \cos(\gamma_i+\phi) & -\sin(\gamma_i+\phi) & 0 & 0 \\ \sin(\gamma_i+\phi) & \cos(\gamma_i+\phi) & 0 & 0 \\ 0 & 0 & 1 & 0 \\ 0 & 0 & 0 & 1 \end{bmatrix} \begin{bmatrix} 1 & 0 & 0 & n_i \\ 0 & 1 & 0 & 0 \\ 0 & 0 & 1 & 0 \\ 0 & 0 & 0 & 1 \end{bmatrix}
$$

$$= \begin{bmatrix} \cos(\gamma_i+\phi) & -\sin(\gamma_i+\phi) & 0 & P_{X_B}+n_i\cos\gamma_i\cos\phi-n_i\sin\gamma_i\sin\phi \\ \sin(\gamma_i+\phi) & \cos(\gamma_i+\phi) & 0 & P_{Y_B}+n_i\cos\gamma_i\sin\phi+n_i\sin\gamma_i\cos\phi \\ 0 & 0 & 1 & 0 \\ 0 & 0 & 0 & 1 \end{bmatrix} \tag{3}$$

where (P_{X_B}, P_{Y_B}) corresponds the position of the end-effector in terms of the base {XYZ} coordinate systems, $\gamma_1 = \pi + \sigma_1$ and $\gamma_2 = -\sigma_2$. Since the position vectors of ${}^{O_i}_{M_i}T^1$ and ${}^{O_i}_{M_i}T^2$ matrices are equal, the following equation can be obtained easily.

$$\begin{bmatrix} l_{2i}\cos(\theta_i+\alpha_i) \\ l_{2i}\sin(\theta_i+\alpha_i) \end{bmatrix} = \begin{bmatrix} P_{X_B}+b_{x_i}\cos\phi-b_{y_i}\sin\phi-o_{x_i}-l_{2i-1}\cos\theta_i \\ P_{Y_B}+b_{x_i}\sin\phi+b_{y_i}\cos\phi-o_{y_i}-l_{2i-1}\sin\theta_i \end{bmatrix} \tag{4}$$

where $b_{x_i} = n_i\cos\gamma_i$ and $b_{y_i} = n_i\sin\gamma_i$. Summing the squares of the both sides in equation 4, we obtain, after simplification,

$$\begin{aligned} &l_{2i-1}^2 - 2P_{Y_B}o_{y_i} - 2P_{X_B}o_{x_i} + b_{x_i}^2 + b_{y_i}^2 + o_{x_i}^2 + o_{y_i}^2 + P_{X_B}^2 + P_{Y_B}^2 \\ &+2l_{2i-1}b_{y_i}[\sin(\phi-\theta_i)-\cos(\phi-\theta_i)] + 2\cos\phi(P_{X_B}b_{x_i}+P_{Y_B}b_{y_i}-b_{x_i}o_{x_i}-b_{y_i}o_{y_i}) \\ &+2\sin\phi(P_{Y_B}b_{x_i}-P_{X_B}b_{y_i}-b_{x_i}o_{y_i}+b_{y_i}o_{x_i}) + 2l_{2i-1}\cos\theta_i(o_{x_i}-P_{X_B}) \\ &+2l_{2i-1}\sin\theta_i(o_{y_i}-P_{Y_B}) - l_{2i}^2 = 0 \end{aligned} \tag{5}$$

To compute the inverse kinematics, the equation 5 can be rewritten as follows

$$A_i\sin\theta_i + B_i\cos\theta_i = C_i \tag{6}$$

where

$$A_i = 2l_{2i-1}(o_{y_i} - b_{x_i}\sin\phi - b_{y_i}\cos\phi - P_{Y_B})$$

$$B_i = 2l_{2i-1}(o_{x_i} + b_{y_i}\sin\phi - b_{x_i}\cos\phi - P_{X_B})$$

$$\begin{aligned} C_i = -[&l_{2i-1}^2 - 2P_{Y_B}o_{y_i} - 2P_{X_B}o_{x_i} + b_{x_i}^2 + b_{y_i}^2 + o_{x_i}^2 + o_{y_i}^2 + P_{X_B}^2 + P_{Y_B}^2 - l_{2i}^2 \\ &+2\cos\phi(P_{X_B}b_{x_i}+P_{Y_B}b_{y_i}-b_{x_i}o_{x_i}-b_{y_i}o_{y_i}) + 2\sin\phi(P_{Y_B}b_{x_i}-P_{X_B}b_{y_i}-b_{x_i}o_{y_i}+b_{y_i}o_{x_i})] \end{aligned}$$

The inverse kinematics solution for equation 6 is

$$\theta_i = \mathrm{Atan2}(A_i, B_i) \mp \mathrm{Atan2}\left(\sqrt{A_i^2 + B_i^2 - C_i^2}, C_i\right) \tag{7a}$$

Once the active joint variables are determined, the passive joint variables can be computed by using equation 4 as follows.

$$\alpha_i = \mathrm{Atan2}(D_i, E_i) \mp \mathrm{Atan2}\left(\sqrt{D_i^2 + E_i^2 - G_i^2}, G_i\right) \tag{7b}$$

where

$$D_i = -\sin\theta_i, \; E_i = \cos\theta_i$$

and

$$G_i = \left(P_{X_B} + b_{x_i}\cos\phi - b_{y_i}\sin\phi - o_{x_i} - l_{2i-1}\cos\theta_i\right)/l_{2i}$$

Since the equation 7 produce two possible solutions for each kinematic chain according to the selection of plus '+' or mines '-' signs, there are eight possible inverse kinematics solutions for 3-DOF RRR FPPM.

2.3 Jacobian matrix and Jacobian inversion

Differentiating the equation 5 with respect to the time, one can obtain the Jacobian matrices.

$$B\dot{\mathbf{q}} = A\dot{\chi}$$

$$\begin{bmatrix} d_1 & 0 & 0 \\ 0 & d_2 & 0 \\ 0 & 0 & d_3 \end{bmatrix}\begin{bmatrix} \dot{\theta}_1 \\ \dot{\theta}_2 \\ \dot{\theta}_3 \end{bmatrix} = \begin{bmatrix} a_1 & b_1 & c_1 \\ a_2 & b_2 & c_2 \\ a_3 & b_3 & c_3 \end{bmatrix}\begin{bmatrix} \dot{P}_{X_B} \\ \dot{P}_{Y_B} \\ \dot{\phi} \end{bmatrix} \tag{8}$$

where

$$a_i = -2\left(P_{X_B} - o_{x_i} + b_{x_i}\cos\phi - l_{2i-1}\cos\theta_i - b_{y_i}\sin\phi\right)$$

$$b_i = -2\left(P_{Y_B} - o_{y_i} + b_{y_i}\cos\phi - l_{2i-1}\sin\theta_i + b_{x_i}\sin\phi\right)$$

$$c_i = -2\left[l_{2i-1}b_{y_i}\cos(\phi-\theta_i) + l_{2i-1}b_{x_i}\sin(\phi-\theta_i) + \cos\phi\left(P_{Y_B}b_{x_i} - P_{X_B}b_{y_i} - b_{x_i}o_{y_i} + b_{y_i}o_{x_i}\right) + \sin\phi\left(b_{x_i}o_{x_i} + b_{y_i}o_{y_i} - P_{X_B}b_{x_i} - P_{Y_B}b_{y_i}\right)\right]$$

$$d_i = 2\left[l_{2i-1}\cos\theta_i\left(o_{y_i} - P_{Y_B}\right) + l_{2i-1}\sin\theta_i\left(P_{X_B} - o_{x_i}\right) - l_{2i-1}b_{y_i}\cos(\phi-\theta_i) - l_{2i-1}b_{x_i}\sin(\phi-\theta_i)\right]$$

The A and B terms in equation 8 denote two separate Jacobian matrices. Thus the overall Jacobian matrix can be obtained as

$$J = B^{-1}A = \begin{bmatrix} \frac{a_1}{d_1} & \frac{b_1}{d_1} & \frac{c_1}{d_1} \\ \frac{a_2}{d_2} & \frac{b_2}{d_2} & \frac{c_2}{d_2} \\ \frac{a_3}{d_3} & \frac{b_3}{d_3} & \frac{c_3}{d_3} \end{bmatrix} \tag{9}$$

The manipulator Jacobian is used for mapping the velocities from the joint space to the Cartesian space

$$\dot{\theta} = J\dot{\chi} \tag{10}$$

where $\dot{\chi} = [\dot{P}_{X_B} \quad \dot{P}_{Y_B} \quad \dot{\phi}]^T$ and $\dot{\theta} = [\dot{\theta}_1 \quad \dot{\theta}_2 \quad \dot{\theta}_3]^T$ are the vectors of velocity in the Cartesian and joint spaces, respectively.

To compute the inverse dynamics of the manipulator, the acceleration of the end-effector is used as the input signal. Therefore, the relationship between the joint and Cartesian accelerations can be extracted by differentiation of equation 10 with respect to the time.

$$\ddot{\theta} = J\ddot{\chi} + \dot{J}\dot{\chi} \tag{11}$$

where $\ddot{\chi} = [\ddot{P}_{X_B} \quad \ddot{P}_{Y_B} \quad \ddot{\phi}]^T$ and $\ddot{\theta} = [\ddot{\theta}_1 \quad \ddot{\theta}_2 \quad \ddot{\theta}_3]^T$ are the vectors of acceleration in the Cartesian and joint spaces, respectively. In equation 11, the other quantities are assumed to be known from the velocity inversion and the only matrix that has not been defined yet is the time derivative of the Jacobian matrix. Differentiation of equation 9 yields to

$$\dot{J} = \begin{bmatrix} K_1 & L_1 & R_1 \\ K_2 & L_2 & R_2 \\ K_3 & L_3 & R_3 \end{bmatrix} \tag{12}$$

K_i, L_i and R_i in equation 12 can be written as follows.

$$K_i = \frac{\dot{a}_i d_i - a_i \dot{d}_i}{d_i^2} \tag{13}$$

$$L_i = \frac{\dot{b}_i d_i - b_i \dot{d}_i}{d_i^2} \tag{14}$$

$$R_i = \frac{\dot{c}_i d_i - c_i \dot{d}_i}{d_i^2} \tag{15}$$

where

$$\dot{a}_i = -2(\dot{P}_{X_B} - \dot{\phi} b_{x_i} \sin\phi + \dot{\theta}_i l_{2i-1} \sin\theta_i - \dot{\phi} b_{y_i} \cos\phi)$$

$$\dot{b}_i = -2(\dot{P}_{Y_B} - \dot{\phi} b_{y_i} \sin\phi - \dot{\theta}_i l_{2i-1} \cos\theta_i + \dot{\phi} b_{x_i} \cos\phi)$$

$$\begin{aligned} c_i = -2[&-l_{2i-1} b_{y_i} (\dot{\phi} - \dot{\theta}_i) \sin(\phi - \theta_i) + (\dot{\phi} - \dot{\theta}_i) l_{2i-1} b_{x_i} \cos(\phi - \theta_i) \\ &-\dot{\phi} \sin\phi (P_{Y_B} b_{x_i} - P_{X_B} b_{y_i} - b_{x_i} o_{y_i} + b_{y_i} o_{x_i}) + \cos\phi (\dot{P}_{Y_B} b_{x_i} - \dot{P}_{X_B} b_{y_i}) \\ &+\dot{\phi} \cos\phi (b_{x_i} o_{x_i} + b_{y_i} o_{y_i} - P_{X_B} b_{x_i} - P_{Y_B} b_{y_i}) - \sin\phi (\dot{P}_{X_B} b_{x_i} + \dot{P}_{Y_B} b_{y_i})] \end{aligned}$$

$$\begin{aligned} \dot{d}_i = 2[&-l_{2i-1} \dot{\theta}_i \sin\theta_i (o_{y_i} - P_{Y_B}) - l_{2i-1} \cos\theta_i \dot{P}_{Y_B} + l_{2i-1} \dot{\theta}_i \cos\theta_i (P_{X_B} - o_{x_i}) + l_{2i-1} \sin\theta_i \dot{P}_{X_B} \\ &+ l_{2i-1} b_{y_i} (\dot{\phi} - \dot{\theta}_i) \sin(\phi - \theta_i) - l_{2i-1} b_{x_i} (\dot{\phi} - \dot{\theta}_i) \cos(\phi - \theta_i)] \end{aligned}$$

2.4 Inverse dynamics model

The virtual work principle is used to obtain the inverse dynamics model of 3-DOF RRR FPPM. Firstly, the partial linear velocity and partial angular velocity matrices are computed by using homogenous transformation matrices derived in Section 2.2. To find the partial linear velocity matrix, B_{2i-1}, C_{2i-1} and M_3 points are selected as pivotal points of links l_{2i-1}, l_{2i} and moving platform, respectively in the second step. The inertial force and moment of each moving part are determined in the next step. As a last step, the inverse dynamic equations of 3-DOF RRR FPPM in explicit form are derived.

2.4.1 The partial linear velocity and partial angular velocity matrices

Considering the manipulator Jacobian matrix in equation 10, the joint velocities for the link l_{2i-1} can be expressed in terms of Cartesian velocities as follows.

$$\dot{\theta}_i = \begin{bmatrix} \frac{a_i}{d_i} & \frac{b_i}{d_i} & \frac{c_i}{d_i} \end{bmatrix} \begin{bmatrix} \dot{P}_{X_B} \\ \dot{P}_{Y_B} \\ \dot{\phi} \end{bmatrix}, \quad i = 1, 2 \text{ and } 3. \tag{16}$$

The partial angular velocity matrix of the link l_{2i-1} can be derived from the equation 16 as

$$\boldsymbol{\omega}_{2i-1} = \begin{bmatrix} \frac{a_i}{d_i} & \frac{b_i}{d_i} & \frac{c_i}{d_i} \end{bmatrix}, \quad i = 1,2 \text{ and } 3. \tag{17}$$

Since the linear velocity on point B_i is zero, the partial linear velocity matrix of the point B_i is given by

$$\boldsymbol{\upsilon}_{2i-1} = \begin{bmatrix} 0 & 0 & 0 \\ 0 & 0 & 0 \end{bmatrix}, \quad i = 1,2 \text{ and } 3. \tag{18}$$

To find the partial angular velocity matrix of the link l_{2i}, the equation 19 can be written easily using the equality of the position vectors of ${}^{O_i}_{M_i}T^1$ and ${}^{O_i}_{M_i}T^2$ matrices.

$$\begin{bmatrix} o_{x_i} + l_{2i}\cos(\theta_i + \alpha_i) + l_{2i-1}\cos\theta_i \\ o_{y_i} + l_{2i}\sin(\theta_i + \alpha_i) + l_{2i-1}\sin\theta_i \end{bmatrix} = \begin{bmatrix} P_{X_B} + b_{x_i}\cos\phi - b_{y_i}\sin\phi \\ P_{Y_B} + b_{x_i}\sin\phi + b_{y_i}\cos\phi \end{bmatrix} \tag{19}$$

The equation 19 can be rearranged as in equation 20 since the link l_{2i} moves with $\delta_i = \theta_i + \alpha_i$ angular velocity.

$$\begin{bmatrix} o_{x_i} + l_{2i}\cos\delta_i + l_{2i-1}\cos\theta_i \\ o_{y_i} + l_{2i}\sin\delta_i + l_{2i-1}\sin\theta_i \end{bmatrix} = \begin{bmatrix} P_{X_B} + b_{x_i}\cos\phi - b_{y_i}\sin\phi \\ P_{Y_B} + b_{x_i}\sin\phi + b_{y_i}\cos\phi \end{bmatrix} \tag{20}$$

Taking the time derivative of equation 20 yields the following equation.

$$\begin{bmatrix} -l_{2i}\dot{\delta}_i\sin\delta_i - l_{2i-1}\dot{\theta}_i\sin\theta_i \\ l_{2i}\dot{\delta}_i\cos\delta_i + l_{2i-1}\dot{\theta}_i\cos\theta_i \end{bmatrix} = \begin{bmatrix} \dot{P}_{X_B} - \dot{\phi}b_{x_i}\sin\phi - \dot{\phi}b_{y_i}\cos\phi \\ \dot{P}_{Y_B} + \dot{\phi}b_{x_i}\cos\phi - \dot{\phi}b_{y_i}\sin\phi \end{bmatrix} \tag{21}$$

Equation 21 can also be stated as follows.

$$\begin{bmatrix} -\sin\delta_i \\ \cos\delta_i \end{bmatrix} l_{2i}\dot{\delta}_i + \begin{bmatrix} -l_{2i-1}\sin\theta_i \\ l_{2i-1}\cos\theta_i \end{bmatrix} \dot{\theta}_i = \begin{bmatrix} 1 & 0 & -b_{x_i}\sin\phi - b_{y_i}\cos\phi \\ 0 & 1 & b_{x_i}\cos\phi - b_{y_i}\sin\phi \end{bmatrix} \begin{bmatrix} \dot{P}_{X_B} \\ \dot{P}_{Y_B} \\ \dot{\phi} \end{bmatrix} \tag{22}$$

If $\dot{\theta}$ in equation 16 is substituted in equation 22, the following equation will be obtained.

$$\begin{bmatrix} -\sin\delta_i \\ \cos\delta_i \end{bmatrix} l_{2i}\dot{\delta}_i = \left(\begin{bmatrix} 1 & 0 & -b_{x_i}\sin\phi - b_{y_i}\cos\phi \\ 0 & 1 & b_{x_i}\cos\phi - b_{y_i}\sin\phi \end{bmatrix} - \begin{bmatrix} -l_{2i-1}\sin\theta_i \\ l_{2i-1}\cos\theta_i \end{bmatrix} \begin{bmatrix} \frac{a_i}{d_i} & \frac{b_i}{d_i} & \frac{c_i}{d_i} \end{bmatrix} \right) \begin{bmatrix} \dot{P}_{X_B} \\ \dot{P}_{Y_B} \\ \dot{\phi} \end{bmatrix} \tag{23}$$

If the both sides of equation 23 premultiplied by $[-\sin\delta_i \quad \cos\delta_i]$, angular velocity of link l_{2i} is obtained as.

$$\dot{\delta}_i = \begin{bmatrix} -\frac{\sin\delta_i}{l_{2i}} & \frac{\cos\delta_i}{l_{2i}} \end{bmatrix} \left(\begin{bmatrix} 1 & 0 & -b_{x_i}\sin\phi - b_{y_i}\cos\phi \\ 0 & 1 & b_{x_i}\cos\phi - b_{y_i}\sin\phi \end{bmatrix} - \begin{bmatrix} -l_{2i-1}\sin\theta_i \\ l_{2i-1}\cos\theta_i \end{bmatrix} \begin{bmatrix} \frac{a_i}{d_i} & \frac{b_i}{d_i} & \frac{c_i}{d_i} \end{bmatrix} \right) \begin{bmatrix} \dot{P}_{X_B} \\ \dot{P}_{Y_B} \\ \dot{\phi} \end{bmatrix} \tag{24}$$

Finally the angular velocity matrix of l_{2i} is derived from the equation 24 as follows.

$$\omega_{2i} = \begin{bmatrix} -\frac{\sin\delta_i}{l_{2i}} & \frac{\cos\delta_i}{l_{2i}} \end{bmatrix} \left(\begin{bmatrix} 1 & 0 & -b_{x_i}\sin\phi - b_{y_i}\cos\phi \\ 0 & 1 & b_{x_i}\cos\phi - b_{y_i}\sin\phi \end{bmatrix} - \begin{bmatrix} -l_{2i-1}\sin\theta_i \\ l_{2i-1}\cos\theta_i \end{bmatrix} \begin{bmatrix} \frac{a_i}{d_i} & \frac{b_i}{d_i} & \frac{c_i}{d_i} \end{bmatrix} \right) \quad (25)$$

The angular acceleration of the link l_{2i} is found by taking the time derivative of equation 21.

$$\begin{bmatrix} -l_{2i}(\ddot{\delta}_i\sin\delta_i + \dot{\delta}_i^2\cos\delta_i) - l_{2i-1}(\ddot{\theta}_i\sin\theta_i + \dot{\theta}_i^2\cos\theta_i) \\ l_{2i}(\ddot{\delta}_i\cos\delta_i - \dot{\delta}_i^2\sin\delta_i) + l_{2i-1}(\ddot{\theta}_i\cos\theta_i - \dot{\theta}_i^2\sin\theta_i) \end{bmatrix}$$

$$= \begin{bmatrix} \ddot{P}_{X_B} - (\ddot{\phi}b_{x_i}\sin\phi + \dot{\phi}^2 b_{x_i}\cos\phi) - (\ddot{\phi}b_{y_i}\cos\phi - \dot{\phi}^2 b_{y_i}\sin\phi) \\ \ddot{P}_{Y_B} + (\ddot{\phi}b_{x_i}\cos\phi - \dot{\phi}^2 b_{x_i}\sin\phi) - (\ddot{\phi}b_{y_i}\sin\phi + \dot{\phi}^2 b_{y_i}\cos\phi) \end{bmatrix} \quad (26)$$

Equation 26 can be expressed as

$$\begin{bmatrix} -\sin\delta_i \\ \cos\delta_i \end{bmatrix} l_{2i}\ddot{\delta}_i = \begin{bmatrix} s_{i1} \\ s_{i2} \end{bmatrix} \quad (27)$$

where

$$s_{i1} = \ddot{P}_{X_B} - (\ddot{\phi}b_{x_i}\sin\phi + \dot{\phi}^2 b_{x_i}\cos\phi) - (\ddot{\phi}b_{y_i}\cos\phi - \dot{\phi}^2 b_{y_i}\sin\phi) + l_{2i}\dot{\delta}_i^2\cos\delta_i$$
$$+ l_{2i-1}(\ddot{\theta}_i\sin\theta_i + \dot{\theta}_i^2\cos\theta_i)$$

$$s_{i2} = \ddot{P}_{Y_B} + (\ddot{\phi}b_{x_i}\cos\phi - \dot{\phi}^2 b_{x_i}\sin\phi) - (\ddot{\phi}b_{y_i}\sin\phi + \dot{\phi}^2 b_{y_i}\cos\phi) + l_{2i}\dot{\delta}_i^2\sin\delta_i$$
$$- l_{2i-1}(\ddot{\theta}_i\cos\theta_i - \dot{\theta}_i^2\sin\theta_i)$$

If the both sides of equation 27 premultiplied by $[-\sin\delta_i \quad \cos\delta_i]$, angular acceleration of link l_{2i} is obtained as.

$$\ddot{\delta}_i = \begin{bmatrix} -\frac{\sin\delta_i}{l_{2i}} & \frac{\cos\delta_i}{l_{2i}} \end{bmatrix} \begin{bmatrix} s_{i1} \\ s_{i2} \end{bmatrix} \quad (28)$$

where i=1,2 and 3. To find the partial linear velocity matrix of the point C_i, the position vector of ${}^{O_i}_{C_i}T^1$ is obtained in the first step.

$${}^{O_i}_{C_i}T^1 = \begin{bmatrix} 1 & 0 & 0 & o_{x_i} \\ 0 & 1 & 0 & o_{y_i} \\ 0 & 0 & 1 & 0 \\ 0 & 0 & 0 & 1 \end{bmatrix} \begin{bmatrix} \cos\theta_i & -\sin\theta_i & 0 & 0 \\ \sin\theta_i & \cos\theta_i & 0 & 0 \\ 0 & 0 & 1 & 0 \\ 0 & 0 & 0 & 1 \end{bmatrix} \begin{bmatrix} 1 & 0 & 0 & l_{2i-1} \\ 0 & 1 & 0 & 0 \\ 0 & 0 & 1 & 0 \\ 0 & 0 & 0 & 1 \end{bmatrix}$$

$$= \begin{bmatrix} \cos\theta_i & -\sin\theta_i & 0 & o_{x_i} + l_{2i-1}\cos\theta_i \\ \sin\theta_i & \cos\theta_i & 0 & o_{y_i} + l_{2i-1}\sin\theta_i \\ 0 & 0 & 1 & 0 \\ 0 & 0 & 0 & 1 \end{bmatrix} \quad (29)$$

The position vector of ${}^{O_i}_{C_i}T^1$ is obtained from the fourth column of the equation 29 as

$${}^{O_i}_{C_i}T^1_{P(x,y)} = \begin{bmatrix} o_{x_i} + l_{2i-1}\cos\theta_i \\ o_{y_i} + l_{2i-1}\sin\theta_i \end{bmatrix} \quad (30)$$

Taking the time derivative of equation 30 produces the linear velocity of the point C_i.

$$v_{C_i} = \frac{d}{dt}\left({}^{O_i}_{C_i}T^{1}_{P(x,y)}\right) = \begin{bmatrix} -l_{2i-1}\sin\theta_i \\ l_{2i-1}\cos\theta_i \end{bmatrix} \dot{\theta}_i \tag{31}$$

If $\dot{\theta}$ in equation 16 is substituted in equation 31, the linear velocity of the point C_i will be expressed in terms of Cartesian velocities.

$$v_{C_i} = \begin{bmatrix} -l_{2i-1}\sin\theta_i \\ l_{2i-1}\cos\theta_i \end{bmatrix} \begin{bmatrix} \frac{a_i}{d_i} & \frac{b_i}{d_i} & \frac{c_i}{d_i} \end{bmatrix} \begin{bmatrix} \dot{P}_{X_B} \\ \dot{P}_{Y_B} \\ \dot{\phi} \end{bmatrix}$$

$$= \frac{l_{2i-1}}{d_i} \begin{bmatrix} -a_i\sin\theta_i & -b_i\sin\theta_i & -c_i\sin\theta_i \\ a_i\cos\theta_i & b_i\cos\theta_i & c_i\cos\theta_i \end{bmatrix} \begin{bmatrix} \dot{P}_{X_B} \\ \dot{P}_{Y_B} \\ \dot{\phi} \end{bmatrix} \tag{32}$$

Finally the partial linear velocity matrix of l_{2i} is derived from the equation 32 as

$$\boldsymbol{v}_{2i} = \frac{l_{2i-1}}{d_i} \begin{bmatrix} -a_i\sin\theta_i & -b_i\sin\theta_i & -c_i\sin\theta_i \\ a_i\cos\theta_i & b_i\cos\theta_i & c_i\cos\theta_i \end{bmatrix} \tag{33}$$

The angular velocity of the moving platform is given by

$$a_{mp} = \begin{bmatrix} 0 & 0 & 1 \end{bmatrix} \begin{bmatrix} \dot{P}_{X_B} \\ \dot{P}_{Y_B} \\ \dot{\phi} \end{bmatrix} \tag{34}$$

The partial angular velocity matrix of the moving platform is

$$\boldsymbol{\omega}_{mp} = \begin{bmatrix} 0 & 0 & 1 \end{bmatrix} \tag{35}$$

The linear velocity ($l_{v_{mp}}$) of the moving platform is equal to right hand side of the equation 22. Since point M_3 is selected as pivotal point of the moving platform, the b_{x_i} is equal to b_{x_3}.

$$l_{v_{mp}} = \begin{bmatrix} 1 & 0 & -b_{x_3}\sin\phi - b_{y_3}\cos\phi \\ 0 & 1 & b_{x_3}\cos\phi - b_{y_3}\sin\phi \end{bmatrix} \begin{bmatrix} \dot{P}_{X_B} \\ \dot{P}_{Y_B} \\ \dot{\phi} \end{bmatrix} \tag{36}$$

The partial linear velocity matrix of the moving platform is derived from the equation 36 as

$$\boldsymbol{v}_{mp} = \begin{bmatrix} 1 & 0 & -b_{x_3}\sin\phi - b_{y_3}\cos\phi \\ 0 & 1 & b_{x_3}\cos\phi - b_{y_3}\sin\phi \end{bmatrix} \tag{37}$$

2.4.2 The inertia forces and moments of the mobile parts of the manipulator

The Newton-Euler formulation is applied for deriving the inertia forces and moments of links and mobile platform about their mass centers. The m_{2i-1}, m_{2i} and m_{mp} denote the masses of links l_{2i-1} , l_{2i} and moving platform, respectively where i=1,2 and 3. The c_{2i-1} c_{2i} and c_{mp} are the mass centers of the links l_{2i-1}, l_{2i} and moving platform, respectively. Figure 3 denotes dynamics model of 3-DOF RRR FPPM.

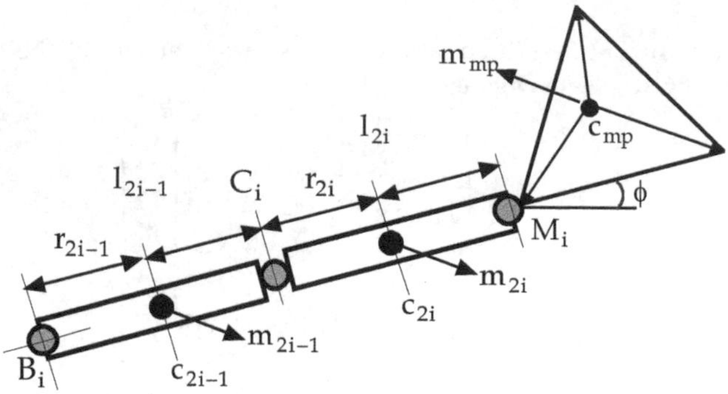

Fig. 3. The dynamics model of 3-DOF RRR FPPM

To find the inertia force of the mass m_{2i-1}, one should determine the acceleration of the link l_{2i-1} about its mass center first. The position vector of the link l_{2i-1} has already been obtained in equation 30. To find the position vector of the center of the link l_{2i-1}, the length r_{2i-1} is used instead of l_{2i-1} in equation 30 as follows

$$ {}^{O_i}_{C_i}T^1_{Pc_{2i-1}} = \begin{bmatrix} o_{x_i} + r_{2i-1}\cos\theta_i \\ o_{y_i} + r_{2i-1}\sin\theta_i \end{bmatrix} \tag{38} $$

The second derivative of the equation 30 with respect to the time yields the acceleration of the link l_{2i-1} about its mass center.

$$ a_{c_{2i-1}} = \frac{d}{dt}\left(\frac{d}{dt}\begin{bmatrix} o_{x_i} + r_{2i-1}\cos\theta_i \\ o_{y_i} + r_{2i-1}\sin\theta_i \end{bmatrix}\right) = r_{2i-1}\begin{bmatrix} -\ddot{\theta}_i\sin\theta_i - \dot{\theta}_i^2\cos\theta_i \\ \ddot{\theta}_i\cos\theta_i - \dot{\theta}_i^2\sin\theta_i \end{bmatrix} \tag{39} $$

The inertia force of the mass m_{2i-1} can be found as

$$ \boldsymbol{F_{2i-1}} = -m_{2i-1}\left(a_{c_{2i-1}} - g\right) $$

$$ = m_{2i-1}r_{2i-1}\begin{bmatrix} \ddot{\theta}_i\sin\theta_i + \dot{\theta}_i^2\cos\theta_i \\ -\ddot{\theta}_i\cos\theta_i + \dot{\theta}_i^2\sin\theta_i \end{bmatrix} \tag{40} $$

where g is the acceleration of the gravity and $\mathbf{g} = [0 \quad 0]^T$ since the manipulator operates in the horizontal plane. The moment of the link l_{2i-1} about pivotal point B_i is

$$ \boldsymbol{M_{2i-1}} = -\left[\ddot{\theta}_i I_{2i-1} + m_{2i-1}\left(\frac{d}{d\theta_i}\,{}^{O_i}_{C_i}T^1_{Pc_{2i-1}}\right)^T a_{B_i}\right] $$

$$ = \ddot{\theta}_i I_{2i-1} \tag{41} $$

where I_{2i-1}, ${}^{O_i}_{C_i}T^{1}_{P_{C_{2i-1}}}$ and a_{B_i}, denote the moment of inertia of the link l_{2i-1}, the position vector of the center of the link l_{2i-1} and the acceleration of the point B_i, respectively. It is noted that $a_{B_i} = 0$.

The acceleration of the link l_{2i} about its mass center is obtained first to find the inertia force of the mass m_{2i}. The position vector of the link l_{2i} has already been given in the left side of the equation 20 in terms of δ_i and θ_i angles. To find the position vector of the center of the link l_{2i} $\left({}^{O_i}_{M_i}T^{1}_{P_{C_{2i}}}\right)$, the length r_{2i} is used instead of l_{2i} in left side of the equation 20.

$$ {}^{O_i}_{M_i}T^{1}_{P_{C_{2i}}} = \begin{bmatrix} o_{x_i} + r_{2i}\cos\delta_i + l_{2i-1}\cos\theta_i \\ o_{y_i} + r_{2i}\sin\delta_i + l_{2i-1}\sin\theta_i \end{bmatrix} \tag{42} $$

The second derivative of the equation 42 with respect to the time produces the acceleration of the link l_{2i} about its mass center.

$$ a_{C_{2i}} = \frac{d}{dt}\left(\frac{d}{dt}\begin{bmatrix} o_{x_i} + r_{2i}\cos\delta_i + l_{2i-1}\cos\theta_i \\ o_{y_i} + r_{2i}\sin\delta_i + l_{2i-1}\sin\theta_i \end{bmatrix}\right) $$

$$ = \begin{bmatrix} -r_{2i}(\ddot{\delta}_i\sin\delta_i + \dot{\delta}_i^2\cos\delta_i) - l_{2i-1}(\ddot{\theta}_i\sin\theta_i + \dot{\theta}_i^2\cos\theta_i) \\ r_{2i}(\ddot{\delta}_i\cos\delta_i - \dot{\delta}_i^2\sin\delta_i) + l_{2i-1}(\ddot{\theta}_i\cos\theta_i - \dot{\theta}_i^2\sin\theta_i) \end{bmatrix} \tag{43} $$

The inertia force of the mass m_{2i} can be found as

$$ \boldsymbol{F}_{2i} = -m_{2i}(a_{C_{2i}} - g) $$

$$ = -m_{2i}\begin{bmatrix} -r_{2i}(\ddot{\delta}_i\sin\delta_i + \dot{\delta}_i^2\cos\delta_i) - l_{2i-1}(\ddot{\theta}_i\sin\theta_i + \dot{\theta}_i^2\cos\theta_i) \\ r_{2i}(\ddot{\delta}_i\cos\delta_i - \dot{\delta}_i^2\sin\delta_i) + l_{2i-1}(\ddot{\theta}_i\cos\theta_i - \dot{\theta}_i^2\sin\theta_i) \end{bmatrix} \tag{44} $$

where $\mathbf{g} = [0 \quad 0]^T$. The moment of the link l_{2i} about pivotal point C_i is

$$ \boldsymbol{M}_{2i} = -\left[\ddot{\delta}_i I_{2i} + m_{2i}\left(\frac{d}{d\delta_i}{}^{O_i}_{M_i}T^{1}_{P_{C_{2i}}}\right)^T a_{C_i}\right] $$

$$ = -\left(\ddot{\delta}_i I_{2i} + m_{2i}r_{2i}l_{2i-1}\left[\sin\delta_i(\ddot{\theta}_i\sin\theta_i + \dot{\theta}_i^2\cos\theta_i) \quad \cos\delta_i(\ddot{\theta}_i\cos\theta_i - \dot{\theta}_i^2\sin\theta_i)\right]\right) \tag{45} $$

where I_{2i}, ${}^{O_i}_{M_i}T^{1}_{P_{C_{2i}}}$ and a_{C_i}, denote the moment of inertia of the link l_{2i}, the position vector of the center of the link l_{2i} in terms of the base coordinate system {XYZ} and the acceleration of the point C_i, respectively. The terms $\frac{d}{d\delta_i}{}^{O_i}_{M_i}T^{1}_{P_{C_{2i}}}$ and a_{C_i} are computed as

$$ \frac{d}{d\delta_i}{}^{O_i}_{M_i}T^{1}_{P_{C_{2i}}} = \frac{d}{d\delta_i}\begin{bmatrix} o_{x_i} + r_{2i}\cos\delta_i + l_{2i-1}\cos\theta_i \\ o_{y_i} + r_{2i}\sin\delta_i + l_{2i-1}\sin\theta_i \end{bmatrix} = r_{2i}\begin{bmatrix} -\sin\delta_i \\ \cos\delta_i \end{bmatrix} \tag{46} $$

$$ a_{C_i} = \frac{d}{dt}\left(\frac{d}{dt}\begin{bmatrix} o_{x_i} + l_{2i-1}\cos\theta_i \\ o_{y_i} + l_{2i-1}\sin\theta_i \end{bmatrix}\right) = -l_{2i-1}\begin{bmatrix} \ddot{\theta}_i\sin\theta_i + \dot{\theta}_i^2\cos\theta_i \\ -\ddot{\theta}_i\cos\theta_i + \dot{\theta}_i^2\sin\theta_i \end{bmatrix} \tag{47} $$

The acceleration of the moving platform about its mass center is obtained in order to find the inertia force of the mass m_{mp}. The position vector of the moving platform has already been given in the right side of the equation 20.

$$ {}^{O_i}_{M_i}T^2 = \begin{bmatrix} P_{X_B} + b_{x_i}\cos\phi - b_{y_i}\sin\phi \\ P_{Y_B} + b_{x_i}\sin\phi + b_{y_i}\cos\phi \end{bmatrix} \tag{48} $$

The second derivative of the equation 48 with respect to the time produces the acceleration of the moving platform about its mass center (c_{mp}).

$$ a_{c_{mp}} = \frac{d}{dt}\left(\frac{d}{dt}\begin{bmatrix} P_{X_B} + b_{x_i}\cos\phi - b_{y_i}\sin\phi \\ P_{Y_B} + b_{x_i}\sin\phi + b_{y_i}\cos\phi \end{bmatrix}\right) $$

$$ = \begin{bmatrix} \ddot{P}_{X_B} - \ddot{\phi}(b_{x_3}\sin\phi + b_{y_3}\cos\phi) - \dot{\phi}^2(b_{x_3}\cos\phi - b_{y_3}\sin\phi) \\ \ddot{P}_{Y_B} + \ddot{\phi}(b_{x_3}\cos\phi - b_{y_3}\sin\phi) - \dot{\phi}^2(b_{x_3}\sin\phi + b_{y_3}\cos\phi) \end{bmatrix} \tag{49} $$

The inertia force of the mass m_{mp} can be found as

$$ \boldsymbol{F_{mp}} = -m_{mp}\left(a_{c_{mp}} - g\right) $$

$$ = -m_{mp}\begin{bmatrix} \ddot{P}_{X_B} - \ddot{\phi}(b_{x_3}\sin\phi + b_{y_3}\cos\phi) - \dot{\phi}^2(b_{x_3}\cos\phi - b_{y_3}\sin\phi) \\ \ddot{P}_{Y_B} + \ddot{\phi}(b_{x_3}\cos\phi - b_{y_3}\sin\phi) - \dot{\phi}^2(b_{x_3}\sin\phi + b_{y_3}\cos\phi) \end{bmatrix} \tag{50} $$

where $\mathbf{g} = [0 \quad 0]^T$. The moment of the moving platform about pivotal point M_3 is

$$ \boldsymbol{M_{mp}} = -\left[\ddot{\phi} I_{mp} + m_{mp}\left(\frac{d}{d\phi}\,{}^{O_i}_{M_3}T^2_{P(x,y)}\right)^T a_{c_{mp}}\right] $$

$$ = -\left(\ddot{\phi} I_{mp} + m_{mp}\left[\ddot{P}_{X_B}(-b_{x_3}\sin\phi - b_{y_3}\cos\phi) + \ddot{P}_{Y_B}(b_{x_3}\cos\phi - b_{y_3}\sin\phi)\right]\right) \tag{51} $$

where I_{mp}, ${}^{O_i}_{M_3}T^2_{P(x,y)}$ and $a_{c_{mp}}$, denote the moment of inertia of the moving platform, the position vector of the moving platform in terms of {XYZ} coordinate system and the acceleration of the point c_{mp}, respectively. The terms $\frac{d}{d\phi}\,{}^{O_i}_{M_3}T^2_{P(x,y)}$ and $a_{c_{mp}}$ are computed as

$$ \frac{d}{d\phi}\,{}^{O_i}_{M_3}T^2_{P(x,y)} = \frac{d}{d\phi}\begin{bmatrix} P_{X_B} + b_{x_3}\cos\phi - b_{y_3}\sin\phi \\ P_{Y_B} + b_{x_3}\sin\phi + b_{y_3}\cos\phi \end{bmatrix} = \begin{bmatrix} -b_{x_3}\sin\phi - b_{y_3}\cos\phi \\ b_{x_3}\cos\phi - b_{y_3}\sin\phi \end{bmatrix} \tag{52} $$

$$ a_{c_{mp}} = \begin{bmatrix} \ddot{P}_{X_B} \\ \ddot{P}_{Y_B} \end{bmatrix} \tag{53} $$

The inverse dynamics of the 3-DOF RRR FPPM based on the virtual work principle is given by

$$ J^T\tau + F = 0 \tag{54} $$

where

$$ F = \sum_{i=1}^{3}\left(\begin{bmatrix}\boldsymbol{v}^T_{2i-1} & \boldsymbol{\omega}^T_{2i-1}\end{bmatrix}\begin{bmatrix}\boldsymbol{F}_{2i-1} \\ \boldsymbol{M}_{2i-1}\end{bmatrix}\right) + \sum_{i=1}^{3}\left(\begin{bmatrix}\boldsymbol{v}^T_{2i} & \boldsymbol{\omega}^T_{2i}\end{bmatrix}\begin{bmatrix}\boldsymbol{F}_{2i} \\ \boldsymbol{M}_{2i}\end{bmatrix}\right) + \begin{bmatrix}\boldsymbol{v}^T_{mp} & \boldsymbol{\omega}^T_{mp}\end{bmatrix}\begin{bmatrix}\boldsymbol{F}_{mp} \\ \boldsymbol{M}_{mp}\end{bmatrix} \tag{55} $$

The driving torques ($\tau_1 \quad \tau_2 \quad \tau_3$) of the 3-DOF RRR FPPM are obtained from equation 54 as

$$ \tau = -(J^T)^{-1}F \tag{56} $$

where $\tau = [\tau_1 \quad \tau_2 \quad \tau_3]^T$.

3. Case study

In this section to demonstrate the active joints torques, a butterfly shape Cartesian trajectory with constant orientation ($\phi = 30^o$) is used as a desired end-effector's trajectory. The time dependent Cartesian trajectory is

$$P_{X_B} = P_{X_0} + a_m \cos(\omega_c \pi t) \quad 0 \leq t \leq 5 \text{ seconds} \tag{57}$$

$$P_{Y_B} = P_{Y_0} + a_m \sin(\omega_s \pi t) \quad 0 \leq t \leq 5 \text{ seconds} \tag{58}$$

A safe Cartesian trajectory is planned such that the manipulator operates a trajectory without any singularity in 5 seconds. The parameters of the trajectory given by 57 and 58 are as follows: $P_{X_{B0}} = P_{Y_{B0}} = 15$, $a_m = 0.5$, $\omega_c = 0.4$ and $\omega_s = 0.8$. The Cartesian trajectory based on the data given above is given by on Figure 4a (position), 4b (velocity) and 4c (acceleration). On Figure 4, the symbols VPBX, VPBY, APBX and APBY illustrate the velocity and acceleration of the moving platform along the X and Y-axes. The first inverse kinematics solution is used for kinematics and dynamics operations. The moving platform is an equilateral triangle with side length of 10. The position of end-effector in terms of {xyz} coordinate systems is $P(x_m, y_m)$=(5, 2.8868) that is the center of the equilateral triangle moving platform. The kinematics and dynamics parameters for 3-DOF RRR FPPM are illustrated in Table 1. Figure 5 illustrates the driving torques (τ_1 τ_2 τ_3) of the 3-DOF RRR FPPM based on the given data in Table 1.

Link lengths		Base coordinates		Masses		Inertias	
l_1	10	o_{x_1}	0	m_1	10	I_1	10
l_2	10	o_{y_1}	0	m_2	10	I_2	10
l_3	10	o_{x_2}	20	m_3	10	I_3	10
l_4	10	o_{y_2}	0	m_4	10	I_4	10
l_5	10	o_{x_3}	10	m_5	10	I_5	10
l_6	10	o_{y_3}	32	m_6, m_{mp}	10	I_6, I_{mp}	10

Table 1. The kinematics and dynamics parameters for 3-DOF RRR FPPM

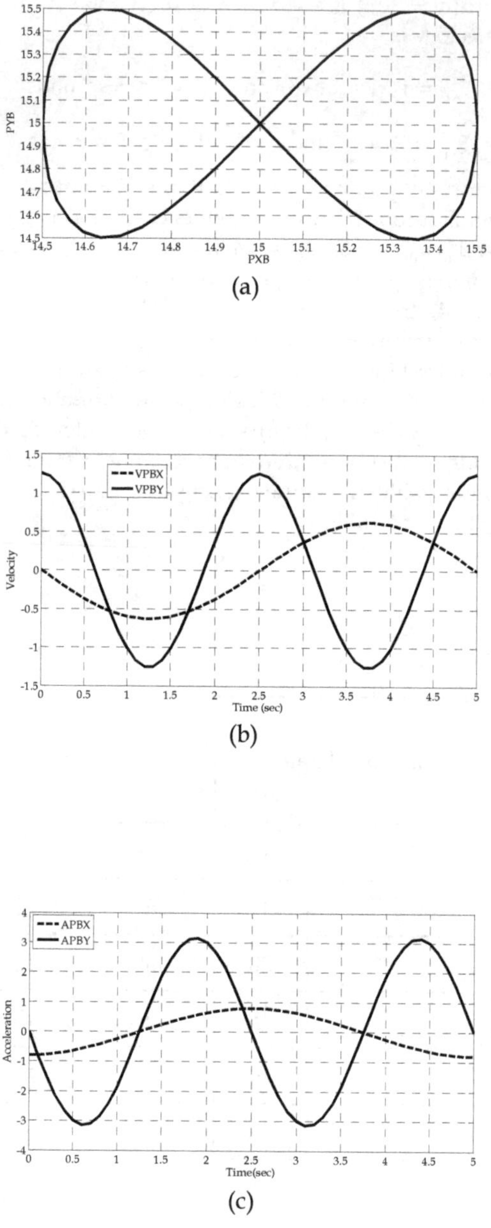

Fig. 4. a) Position, b) velocity and c) acceleration of the moving platform

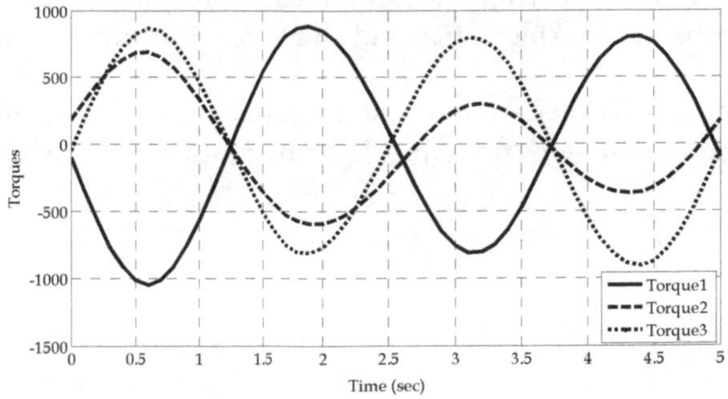

Fig. 5. The driving torques (τ_1 τ_2 τ_3) of the 3-DOF RRR FPPM

4. Conclusion

In this chapter, the inverse dynamics problem of 3-DOF RRR FPPM is derived using virtual work principle. Firstly, the inverse kinematics model and Jacobian matrix of 3-DOF RRR FPPM are determined using DH method. Secondly, the partial linear velocity and partial angular velocity matrices are computed. Pivotal points are selected in order to determine the partial linear velocity matrices. Thirdly, the inertial force and moment of each moving part are obtained. Consequently, the inverse dynamic equations of 3-DOF RRR FPPM in explicit form are derived. A butterfly shape Cartesian trajectory is used as a desired end-effector's trajectory to demonstrate the active joints torques.

5. References

Denavit, J. & Hartenberg, R. S., (1955). A kinematic notation for lower-pair mechanisms based on matrices. Journal of Applied Mechanics, Vol., 22, 1955, pp. 215–221

Hubbard, T.; Kujath, M. R. & Fetting, H. (2001). MicroJoints, Actuators, Grippers, and Mechanisms, CCToMM Symposium on Mechanisms, Machines and Mechatronics, Montreal, Canada

Kang, B.; Chu, J. & Mills, J. K. (2001). Design of high speed planar parallel manipulator and multiple simultaneous specification control, Proceedings of IEEE International Conference on Robotics and Automation, pp. 2723-2728, South Korea

Kang, B. & Mills, J. K. (2001). Dynamic modeling and vibration control of high speed planar parallel manipulator, In Proceedings of IEEE/RJS International Conference on Intelligent Robots and Systems, pp. 1287-1292, Hawaii

Merlet, J. P. (2000) Parallel robots, Kluwer Academic Publishers

Tsai, L. W. (1999). Robot analysis: The mechanics of serial and parallel manipulators, A Wiley-Interscience Publication

Uchiyama, M. (1994). Structures and characteristics of parallel manipulators, Advanced robotics, Vol. 8, no. 6. pp. 545-557

Wu, J.; Wang J.; You, Z. (2011). A comparison study on the dynamics of planar 3-DOF 4-RRR, 3-RRR and 2-RRR parallel manipulators, Robotics and computer-integrated manufacturing, Vol.27, pp. 150–156

Wu, J.; Wang L.; You, Z. (2010). A new method for optimum design of parallel manipulatorbased on kinematics and dynamics, Nonlinear Dyn, Vol. 61, pp. 717–727

Zhang, C. D. & Song, S. M. (1993). An efficient method for inverse dynamics of manipulators based on the virtual work principle, J. Robot. Syst., Vol.10, no.5, pp. 605–627

3

Dynamic Modeling and Simulation of Stewart Platform

Zafer Bingul and Oguzhan Karahan
Mechatronics Engineering, Kocaeli University
Turkey

1. Introduction

Since a parallel structure is a closed kinematics chain, all legs are connected from the origin of the tool point by a parallel connection. This connection allows a higher precision and a higher velocity. Parallel kinematic manipulators have better performance compared to serial kinematic manipulators in terms of a high degree of accuracy, high speeds or accelerations and high stiffness. Therefore, they seem perfectly suitable for industrial high-speed applications, such as pick-and-place or micro and high-speed machining. They are used in many fields such as flight simulation systems, manufacturing and medical applications.

One of the most popular parallel manipulators is the general purpose 6 degree of freedom (DOF) Stewart Platform (SP) proposed by Stewart in 1965 as a flight simulator (Stewart, 1965). It consists of a top plate (moving platform), a base plate (fixed base), and six extensible legs connecting the top plate to the bottom plate. SP employing the same architecture of the Gough mechanism (Merlet, 1999) is the most studied type of parallel manipulators. This is also known as Gough–Stewart platforms in literature.

Complex kinematics and dynamics often lead to model simplifications decreasing the accuracy. In order to overcome this problem, accurate kinematic and dynamic identification is needed. The kinematic and dynamic modeling of SP is extremely complicated in comparison with serial robots. Typically, the robot kinematics can be divided into forward kinematics and inverse kinematics. For a parallel manipulator, inverse kinematics is straight forward and there is no complexity deriving the equations. However, forward kinematics of SP is very complicated and difficult to solve since it requires the solution of many non-linear equations. Moreover, the forward kinematic problem generally has more than one solution. As a result, most research papers concentrated on the forward kinematics of the parallel manipulators (Bonev and Ryu, 2000; Merlet, 2004; Harib and Srinivasan, 2003; Wang, 2007).

For the design and the control of the SP manipulators, the accurate dynamic model is very essential. The dynamic modeling of parallel manipulators is quite complicated because of their closed-loop structure, coupled relationship between system parameters, high nonlinearity in system dynamics and kinematic constraints. Robot dynamic modeling can be also divided into two topics: inverse and forward dynamic model. The inverse dynamic model is important for system control while the forward model is used for system simulation. To obtain the dynamic model of parallel manipulators, there are many valuable studies published by many researches in the literature. The dynamic analysis of parallel manipulators has been traditionally performed through several different methods such as the Newton-Euler method, the Lagrange formulation, the principle of virtual work and the screw theory.

The Newton-Euler approach requires computation of all constraint forces and moments between the links. One of the important studies was presented by Dasgupta and Mruthyunjaya (1998) on dynamic formulation of the SP manipulator. In their study, the closed-form dynamic equations of the 6-UPS SP in the task-space and joint-space were derived using the Newton-Euler approach. The derived dynamic equations were implemented for inverse and forward dynamics of the Stewart Platform manipulator, and the simulation results showed that this formulation provided a complete modeling of the dynamics of SP. Moreover, it demonstrated the strength of the Newton-Euler approach as applied to parallel manipulators and pointed out an efficient way of deriving the dynamic equations through this formulation. This method was also used by Khalil and Ibrahim (2007). They presented a simple and general closed form solution for the inverse and forward dynamic models of parallel robots. The proposed method was applied on two parallel robots with different structures. Harib and Srinivasan (2003) performed kinematic and dynamic analysis of SP based machine structures with inverse and forward kinematics, singularity, inverse and forward dynamics including joint friction and actuator dynamics. The Newton-Euler formulation was used to derive the rigid body dynamic equations. Do and Yang (1988) and Reboulet and Berthomieu, (1991) presented the dynamic modeling of SP using Newton–Euler approach. They introduced some simplifications on the legs models. In addition to these works, others (Guo and Li, 2006; Carvalho and Ceccarelli, 2001; Riebe and Ulbrich, 2003) also used the Newton-Euler approach.

Another method of deriving the dynamics of the SP manipulator is the Lagrange formulation. This method is used to describe the dynamics of a mechanical system from the concepts of work and energy. Abdellatif and Heimann (2009) derived the explicit and detailed six-dimensional set of differential equations describing the inverse dynamics of non-redundant parallel kinematic manipulators with the 6 DOF. They demonstrated that the derivation of the explicit model was possible by using the Lagrangian formalism in a computationally efficient manner and without simplifications. Lee and Shah (1988) derived the inverse dynamic model in joint space of a 3-DOF in parallel actuated manipulator using Lagrangian approach. Moreover, they gave a numerical example of tracing a helical path to demonstrate the influence of the link dynamics on the actuating force required. Guo and Li (2006) derived the explicit compact closed-form dynamic equations of six DOF SP manipulators with prismatic actuators on the basic of the combination of the Newton-Euler method with the Lagrange formulation. In order to validate the proposed formulation, they studied numerical examples used in other references. The simulation results showed that it could be derived explicit dynamic equations in the task space for Stewart Platform manipulators by applying the combination of the Newton-Euler with the Lagrange formulation. Lebret and co-authors (1993) studied the dynamic equations of the Stewart Platform manipulator. The dynamics was given in step by step algorithm. Lin and Chen presented an efficient procedure for computer generation of symbolic modeling equations for the Stewart Platform manipulator. They used the Lagrange formulation for derivation of dynamic equations (Lin and Chen, 2008). The objective of the study was to develop a MATLAB-based efficient algorithm for computation of parallel link robot manipulator dynamic equations. Also, they proposed computer-torque control in order to verify the effectiveness of the dynamic equations. Lagrange's method was also used by others (Gregório and Parenti-Castelli, 2004; Beji and Pascal 1999; Liu et al., 1993).

For dynamic modeling of parallel manipulators, many approaches have been developed such as the principle of virtual work (Tsai, 2000, Wang and Gosselin, 1998; Geike and McPhee, 2003), screw teory (Gallardo et al., 2003), Kane's method (Liu et al., 2000; Meng et al., 2010) and recursive matrix method (Staicu and Zhang, 2008). Although the derived equations for the dynamics of parallel manipulators present different levels of complexity and computational loads, the results of the actuated forces/torques computed by different approaches are equivalent. The main goal of recent proposed approaches is to minimize the number of operations involved in the computation of the manipulator dynamics. It can be concluded that the dynamic equations of parallel manipulators theoretically have no trouble. Moreover, in fact, the focus of attention should be on the accuracy and computation efficiency of the model.

The aim of this paper is to present the work on dynamic formulation of a 6 DOF SP manipulator. The dynamical equations of the manipulator have been formulated by means of the Lagrangian method. The dynamic model included the rigid body dynamics of the mechanism as well as the dynamics of the actuators. The Jacobian matrix was derived in two different ways. Obtaining the accurate Jacobian matrix is very essential for accurate simulation model. Finally, the dynamic equations including rigid body and actuator dynamics were simulated in MATLAB-Simulink and verified on physical system.

This chapter is organized in the following manner. In Section 2, the kinematic analysis and Jacobian matrices are introduced. In Section 3, the dynamic equations of a 6 DOF SP manipulator are presented. In Section 4, dynamic simulations and the experimental results are given in detail. Finally, conclusions of this study are summarized.

2. Structure description and kinematic analysis

2.1 Structure description

The SP manipulator used in this study (Figure 1), is a six DOF parallel mechanism that consists of a rigid body moving plate, connected to a fixed base plate through six independent kinematics legs. These legs are identical kinematics chains, couple the moveable upper and the fixed lower platform by universal joints. Each leg contains a precision ball-screw assembly and a DC- motor. Thus, length of the legs is variable and they can be controlled separately to perform the motion of the moving platform.

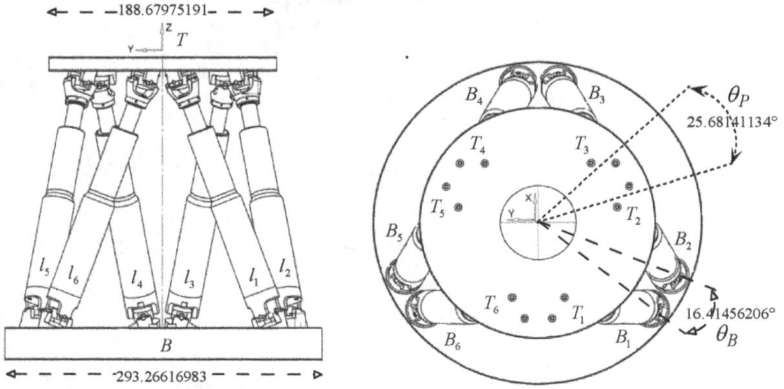

Fig. 1. Solid model of the SP manipulator

2.2 Inverse kinematics

To clearly describe the motion of the moving platform, the coordinate systems are illustrated in Figure 2. The coordinate system (B_{XYZ}) is attached to the fixed base and other coordinate system (T_{xyz}) is located at the center of mass of the moving platform. Points (B_i and T_i) are the connecting points to the base and moving platforms, respectively. These points are placed on fixed and moving platforms (Figure 2.a). Also, the separation angles between points (T_2 and T_3, T_4 and T_5, T_1 and T_6) are denoted by θ_p as shown in Figure 2.b. In a similar way, the separation angles between points (B_1 and B_2, B_3 and B_4, B_5 and B_6) are denoted by θ_b.

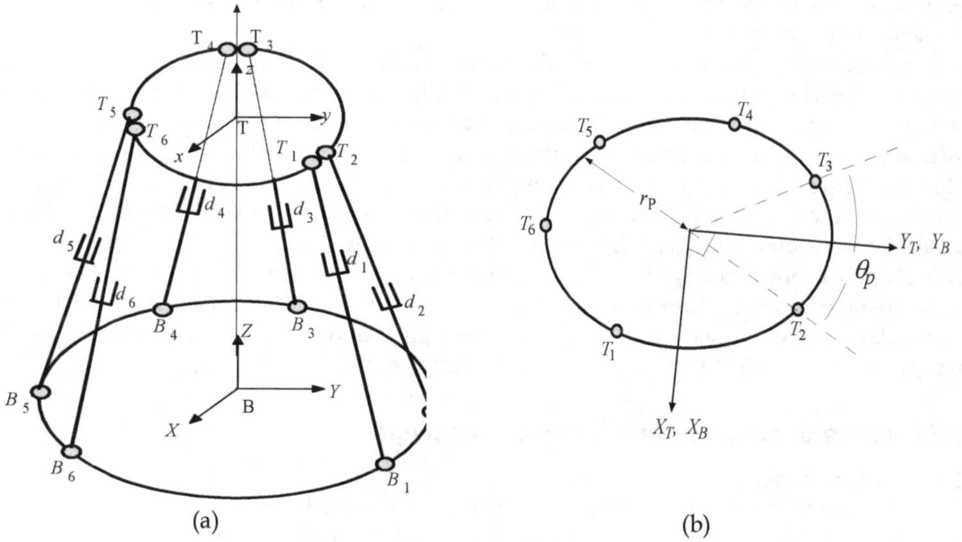

Fig. 2. The schematic diagram of the SP manipulator

From Figure 2.b, the location of the ith attachment point (T_i) on the moving platform can be found (Equation 1). r_p and r_{base} are the radius of the moving platform and fixed base, respectively. By the using the same approach, the location of the ith attachment point (B_i) on the base platform can be also obtained from Equation 2.

$$GT_i = \begin{bmatrix} GT_{xi} \\ GT_{yi} \\ GT_{zi} \end{bmatrix} = \begin{bmatrix} r_p \cos(\lambda_i) \\ r_p \sin(\lambda_i) \\ 0 \end{bmatrix}, \quad \begin{matrix} \lambda_i = \dfrac{i\pi}{3} - \dfrac{\theta_p}{2} & i = 1,3,5 \\ \lambda_i = \lambda_{i-1} + \theta_p & i = 2,4,6 \end{matrix} \tag{1}$$

$$B_i = \begin{bmatrix} B_{xi} \\ B_{yi} \\ B_{zi} \end{bmatrix} = \begin{bmatrix} r_{base} \cos(\upsilon_i) \\ r_{base} \sin(\upsilon_i) \\ 0 \end{bmatrix}, \quad \begin{matrix} \upsilon_i = \dfrac{i\pi}{3} - \dfrac{\theta_b}{2} & i = 1,3,5 \\ \upsilon_i = \upsilon_{i-1} + \theta_b & i = 2,4,6 \end{matrix} \tag{2}$$

The pose of the moving platform can be described by a position vector, P and a rotation matrix, ${}^{B}R_{T}$. The rotation matrix is defined by the roll, pitch and yaw angles, namely, a rotation of α about the fixed *x-axis*, $R_X(\alpha)$, followed by a rotaion of β about the fixed *y-axis*, $R_Y(\beta)$ and a rotaion of γ about the fixed *z-axis*, $R_Z(\gamma)$. In this way, the rotation matrix of the moving platform with respect to the base platform coordinate system is obtained. The position vector *P* denotes the translation vector of the origin of the moving platform with respect to the base platform. Thus, the rotation matrix and the position vector are given as the following.

$$
\begin{aligned}
{}^{B}R_T &= R_Z(\gamma)R_Y(\beta)R_X(\alpha) = \begin{bmatrix} r_{11} & r_{12} & r_{13} \\ r_{21} & r_{22} & r_{23} \\ r_{31} & r_{32} & r_{33} \end{bmatrix} \\
&= \begin{bmatrix} \cos\beta\cos\gamma & \cos\gamma\sin\alpha\sin\beta - \cos\alpha\sin\gamma & \sin\alpha\sin\gamma + \cos\alpha\cos\gamma\sin\beta \\ \cos\beta\sin\gamma & \cos\alpha\cos\gamma + \sin\alpha\sin\beta\sin\gamma & \cos\alpha\sin\beta\sin\gamma - \cos\gamma\sin\alpha \\ -\sin\beta & \cos\beta\sin\alpha & \cos\alpha\cos\beta \end{bmatrix}
\end{aligned} \tag{3}
$$

$$
P = \begin{bmatrix} P_x & P_y & P_z \end{bmatrix}^T \tag{4}
$$

Referring back to Figure 2, the above vectors GT_i and B_i are chosen as the position vector. The vector L_i of the link *i* is simply obtained as

$$
L_i = R_{XYZ}GT_i + P - B_i \qquad \text{i=1,2, … ,6.} \tag{5}
$$

When the position and orientation of the moving platform $X_{p-o} = \begin{bmatrix} P_x & P_y & P_z & \alpha & \beta & \gamma \end{bmatrix}^T$ are given, the length of each leg is computed as the following.

$$
\begin{aligned}
l_i^2 &= \left(P_x - B_{xi} + GT_{xi}r_{11} + GT_{yi}r_{12}\right)^2 \\
&+ \left(P_y - B_{yi} + GT_{xi}r_{21} + GT_{yi}r_{22}\right)^2 + \left(P_z + GT_{xi}r_{31} + GT_{yi}r_{32}\right)^2
\end{aligned} \tag{6}
$$

The actuator length is $l_i = \|L_i\|$.

2.3 Jacobian matrix

The Jacobian matrix relates the velocities of the active joints (actuators) to the generalized velocity of the moving platform. For the parallel manipulators, the commonly used expression of the Jacobian matrix is given as the following.

$$
\dot{L} = J\dot{X} \tag{7}
$$

where $\dot{L}$ and $\dot{X}$ are the velocities of the leg and the moving platform, respectively. In this work, two different derivations of the Jacobian matrix are developed. The first derivation is made using the general expression of the Jacobian matrix given in Equation 7. It can be rewritten to see the relationship between the actuator velocities, $\dot{L}$ and the generalized velocity of the moving platform ($\dot{X}_{p-0}$) as the following

$$
\dot{L} = J_A \dot{X}_{p-0} = J_{IA}\vec{V}_{T_j} \tag{8}
$$

The generalized velocity of the moving platform is below:

$$\vec{V}_{T_j} = J_{IIA}\,\dot{X}_{p-o} \tag{9}$$

where $\vec{V}_{T_j}$ is the velocity of the platform connection point of the leg. Figure 3 shows a schematic view of one of the six legs of the SP manipulator.

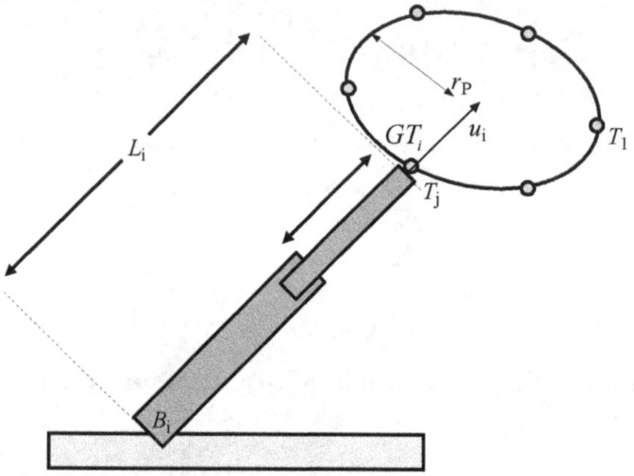

Fig. 3. Schematic view of the ith leg of the parallel manipulator

Now combining Equation 8 and Equation 9 gives

$$\dot{L} = J_A\,\dot{X}_{p-o} = J_{IA}\,J_{IIA}\,\dot{X}_{p-o} \tag{10}$$

The first Jacobian matrix in the equation above is

$$J_{IA} = \begin{bmatrix} \vec{u}_1^T & \cdots & \cdots & \cdots & \cdots & 0 \\ 0 & \vec{u}_2^T & \cdots & \cdots & \cdots & 0 \\ 0 & \cdots & \vec{u}_3^T & \cdots & \cdots & 0 \\ 0 & \cdots & \cdots & \vec{u}_4^T & \cdots & 0 \\ 0 & \cdots & \cdots & \cdots & \vec{u}_5^T & 0 \\ 0 & \cdots & \cdots & \cdots & \cdots & \vec{u}_6^T \end{bmatrix}_{6x18} \tag{11}$$

where u_i is the unit vector along the axis of the prismatic joint of link i (Figure 3). It can be obtained as follows

$$\vec{u}_i = \frac{B_iT_j}{|L_i|} = \frac{L_i}{l_i}, \begin{cases} j = \dfrac{i+1}{2} & if\ i\ is\ odd \\ \\ j = \dfrac{i}{2} & if\ i\ is\ even \end{cases} \tag{12}$$

The second Jacobian matrix in Equation 10 is calculated as the following.

$$J_{IIA} = \begin{bmatrix} I_{3x3} & R_Y(\beta)S(X)R_X(\alpha)R_Z(\gamma)GT_1 & S(Y)R_Y(\beta)R_X(\alpha)R_Z(\gamma)GT_1 & R_Y(\beta)R_X(\alpha)S(Z)R_Z(\gamma)GT_1 \\ \vdots & \vdots & \vdots & \vdots \\ \vdots & \vdots & \vdots & \vdots \\ \vdots & \vdots & \vdots & \vdots \\ \vdots & \vdots & \vdots & \vdots \\ I_{3x3} & R_Y(\beta)S(X)R_X(\alpha)R_Z(\gamma)GT_6 & S(Y)R_Y(\beta)R_X(\alpha)R_Z(\gamma)GT_6 & R_Y(\beta)R_X(\alpha)S(Z)R_Z(\gamma)GT_6 \end{bmatrix}_{18x6} \tag{13}$$

where I_{3x3} denotes the 3x3 identity matrix and S designates the 3x3 screw symmetric matrix associated with the vector $a = \begin{bmatrix} a_x & a_y & a_z \end{bmatrix}^T$,

$$S = \begin{bmatrix} 0 & -a_z & a_y \\ a_z & 0 & -a_x \\ -a_y & a_x & 0 \end{bmatrix} \tag{14}$$

The first proposed Jacobian matrix of the SP manipulator is defined as

$$J_A = J_{IA} \, J_{IIA} \tag{15}$$

The second proposed Jacobian matrix of the SP manipulator is defined as

$$J_B = J_{IB} \, J_{IIB} \tag{16}$$

Given $GT_i = \begin{bmatrix} GT_{xi} & GT_{yi} & GT_{zi} \end{bmatrix}^T$, T_j on the moving platform with reference to the base coordinate system (B_{XYZ}) is obtained as

$$T_j = \begin{bmatrix} P_x & P_y & P_z \end{bmatrix}^T + {}^B R_T GT_i = x + {}^B R_T GT_i \tag{17}$$

The velocity of the attachment point T_j is obtained by differentiating Equation 17 with respect to time

$$\vec{V}_{T_j} = \begin{bmatrix} \dot{P}_x & \dot{P}_y & \dot{P}_z \end{bmatrix}^T + \omega \times {}^B R_T GT_i = \dot{x} + \omega \times {}^B R_T GT_i \tag{18}$$

where $\omega = \left(\omega_x, \omega_y, \omega_z\right)$ is angular velocity of the moving platform with reference to the base platform.

$$\omega = \begin{bmatrix} \omega_x \\ \omega_y \\ \omega_z \end{bmatrix} = \begin{bmatrix} \cos\beta & 0 & 0 \\ 0 & 1 & -\sin\alpha \\ -\sin\beta & 0 & \cos\alpha \end{bmatrix} = \begin{bmatrix} \dot{\alpha} \\ \dot{\beta} \\ \dot{\gamma} \end{bmatrix} \tag{19}$$

Since the projection of the velocity vector ($\dot{T}_j$) on the axis of the prismatic joint of link i produces the extension rate of link i, the velocity of the active joint ($\dot{L}_i$) is computed from

$$\dot{L}_i = \vec{V}_{T_j} \vec{u}_i = \begin{bmatrix} \dot{P}_x & \dot{P}_y & \dot{P}_z \end{bmatrix}^T \cdot \vec{u}_i + \omega \times \left({}^B R_T \, GT_i\right) \cdot \vec{u}_i = \dot{x} \cdot \vec{u}_i + \omega \times \left({}^B R_T \, GT_i\right) \cdot \vec{u}_i \tag{20}$$

Equation 20 is rewritten in matrix format as follows.

$$\dot{L}_i = J_B \begin{bmatrix} \dot{x} \\ \omega \end{bmatrix} = J_B \dot{X}_{p-o} = J_{IB} J_{IIB} \dot{X}_{p-o} \tag{21}$$

The first Jacobian matrix is

$$J_{IB} = \begin{bmatrix} u_{x1} & u_{y1} & u_{z1} & \left({}^{B}R_T GT_1 x\vec{u}_1\right)^T \\ \vdots & \vdots & \vdots & \vdots \\ u_{x6} & u_{y6} & u_{z6} & \left({}^{B}R_T GT_6 x\vec{u}_6\right)^T \end{bmatrix}_{6x6} \tag{22}$$

The second Jacobian matrix is

$$J_{IIB} = \begin{bmatrix} 1 & 0 & 0 & 0 & 0 & 0 \\ 0 & 1 & 0 & 0 & 0 & 0 \\ 0 & 0 & 1 & 0 & 0 & 0 \\ 0 & 0 & 0 & \cos\beta & 0 & 0 \\ 0 & 0 & 0 & 0 & 1 & -\sin\alpha \\ 0 & 0 & 0 & -\sin\beta & 0 & \cos\alpha \end{bmatrix}_{6x6} \tag{23}$$

3. Dynamic modeling

The dynamic analysis of the SP manipulator is always difficult in comparison with the serial manipulator because of the existence of several kinematic chains all connected by the moving platform. Several methods were used to describe the problem and obtain the dynamic modeling of the manipulator. In the literature, there is still no consensus on which formulation is the best to describe the dynamics of the manipulator. Lagrange formulation was used in this work since it provides a well analytical and orderly structure.

In order to derive the dynamic equations of the SP manipulator, the whole system is separated into two parts: the moving platform and the legs. The kinetic and potential energies for both of these parts are computed and then the dynamic equations are derived using these energies.

3.1 Kinetic and potential energies of the moving platform

The kinetic energy of the moving platform is a summation of two motion energies since the moving platform has translation and rotation about three orthogonal axes, *(XYZ)*. The first one is translation energy occurring because of the translation motion of the center of mass of the moving platform. The translation energy is defined by

$$K_{up(trans)} = \frac{1}{2} m_{up} \left(\dot{P}_X^2 + \dot{P}_Y^2 + \dot{P}_Z^2 \right) \tag{24}$$

where m_{up} is the moving platform mass. For rotational motion of the moving platform around its center of mass, rotational kinetic energy can be written as

$$K_{up(rot)} = \frac{1}{2}\vec{\Omega}_{up(mf)}^{T} I_{(mf)} \vec{\Omega}_{up(mf)} \tag{25}$$

where $I_{(mf)}$ and $\vec{\Omega}_{up(mf)}$ are the rotational inertia mass and the angular velocity of the moving platform, respectively. They are given as

$$I_{(mf)} = \begin{bmatrix} I_X & 0 & 0 \\ 0 & I_Y & 0 \\ 0 & 0 & I_Z \end{bmatrix} \tag{26}$$

$$\vec{\Omega}_{up(mf)} = R_Z(\gamma)^T R_X(\alpha)^T R_Y(\beta)^T \vec{\Omega}_{up(ff)} \tag{27}$$

where $\vec{\Omega}_{up(ff)}$ denotes the angular velocity of the moving platform with respect to the base frame. Given the definition of the angles α, β and γ, the angular velocity, $\vec{\Omega}_{up(ff)}$ is

$$\begin{aligned} \vec{\Omega}_{up(ff)} &= \dot{\alpha} R_Y(\beta)\vec{X} + \dot{\beta}\vec{Y} + \dot{\gamma} R_X(\alpha) R_Z(\gamma)\vec{Z} \\ &= \left(\begin{bmatrix} \cos\beta & 0 & \sin\beta \\ 0 & 1 & 0 \\ -\sin\beta & 0 & \cos\beta \end{bmatrix} \begin{bmatrix} 1 & 0 & 0 \\ 0 & 0 & 0 \\ 0 & 0 & 0 \end{bmatrix} + \begin{bmatrix} 0 & 0 & 0 \\ 0 & 1 & 0 \\ 0 & 0 & 0 \end{bmatrix} \right. \\ &\left. + \begin{bmatrix} 1 & 0 & 0 \\ 0 & \cos\alpha & -\sin\alpha \\ 0 & \sin\alpha & \cos\alpha \end{bmatrix} \begin{bmatrix} \cos\gamma & -\sin\gamma & 0 \\ \sin\gamma & \cos\gamma & 0 \\ 0 & 0 & 1 \end{bmatrix} \begin{bmatrix} 0 & 0 & 0 \\ 0 & 0 & 0 \\ 0 & 0 & 1 \end{bmatrix} \right) \begin{bmatrix} \dot{\alpha} \\ \dot{\beta} \\ \dot{\gamma} \end{bmatrix} = \begin{bmatrix} \cos\beta & 0 & 0 \\ 0 & 1 & -\sin\alpha \\ -\sin\beta & 0 & \cos\alpha \end{bmatrix} \begin{bmatrix} \dot{\alpha} \\ \dot{\beta} \\ \dot{\gamma} \end{bmatrix} \end{aligned} \tag{28}$$

In the moving platform coordinate system, the angular velocity of the moving platform given in Equation 27 is calculated as

$$\vec{\Omega}_{up(mf)} = \begin{bmatrix} c\gamma & c\alpha s\gamma & -c\alpha c\gamma s\beta - c\alpha s\alpha s\gamma + c\alpha c\beta s\alpha s\gamma \\ -s\gamma & c\alpha c\gamma & -c\alpha c\gamma s\alpha + c\alpha s\beta s\gamma + c\alpha c\beta s\alpha c\gamma \\ 0 & -s\alpha & s^2\alpha + c^2\alpha c\beta \end{bmatrix} \begin{bmatrix} \dot{\alpha} \\ \dot{\beta} \\ \dot{\gamma} \end{bmatrix} \tag{29}$$

where $s(\cdot) = \sin(\cdot)$ and $c(\cdot) = \cos(\cdot)$. As a result, the total kinetic energy of the moving platform in a compact form is given by

$$\begin{aligned} K_{up} &= K_{up(trans)} + K_{up(rot)} = \frac{1}{2} m_{up}\left(\dot{P}_X^2 + \dot{P}_Y^2 + \dot{P}_Z^2\right) + \frac{1}{2}\vec{\Omega}_{up(mf)}^{T} I_{(mf)} \vec{\Omega}_{up(mf)} \\ &= \frac{1}{2}\dot{X}_{P-O}^{T} \cdot M_{up}\left(X_{P-O}\right) \cdot \dot{X}_{P-O} = \frac{1}{2}\begin{bmatrix} \dot{P}_X & \dot{P}_Y & \dot{P}_Z & \dot{\alpha} & \dot{\beta} & \dot{\gamma} \end{bmatrix} M_{up} \begin{bmatrix} \dot{P}_X \\ \dot{P}_Y \\ \dot{P}_Z \\ \dot{\alpha} \\ \dot{\beta} \\ \dot{\gamma} \end{bmatrix} \end{aligned} \tag{30}$$

where M_{up} is the 6x6 mass diagonal matrix of the moving platform. Also, potential energy of the moving platform is

$$P_{up} = m_{up} g P_Z X_{P-O} = \begin{bmatrix} 0 & 0 & m_{up} g & 0 & 0 & 0 \end{bmatrix} \begin{bmatrix} P_X \\ P_Y \\ P_Z \\ \alpha \\ \beta \\ \gamma \end{bmatrix} \tag{31}$$

where g is the gravity.

3.2 Kinetic and potential energies of the legs

Each leg consists of two parts: the moving part and the fixed part (Figure 4). The lower fixed part of the leg is connected to the base platform through a universal joint, whereas the upper moving part is connected to the moving platform through a universal joint.

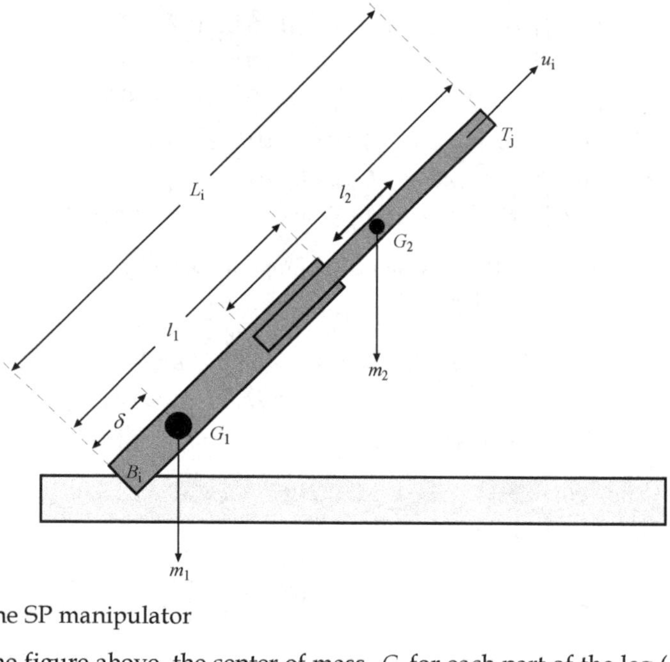

Fig. 4. Leg of the SP manipulator

As shown in the figure above, the center of mass, G_i for each part of the leg ($Leg_i\ i = 1...6$) is considered. G_{1i} denotes the center of mass of the fixed part. l_1 and m_1 are the length and the mass of the fixed part, respectively and δ is the distance between B_i and G_i. For the moving part of the leg, G_{2i} denotes its center of mass. l_2 and m_2 are the length and mass of the part, respectively.

The length of the leg is assumed to be constant. The rotational kinetic energy caused by the rotation around the fixed point B_i as shown in Figure 4 is given by

$$K_{Li(rot)} = \frac{1}{2}(m_1 + m_2)\left[h_i\left(\vec{V}_{T_J}^T \vec{V}_{T_J} - \vec{V}_{T_J}^T \vec{u}_i \vec{u}_i^T \vec{V}_{T_J}\right)\right], \begin{cases} j = \frac{i+1}{2} & \text{if } i \text{ is odd} \\ j = \frac{i}{2} & \text{if } i \text{ is even} \end{cases} \tag{32}$$

where

$$h_i = \left(\frac{\hat{I}}{L_i} + \frac{m_2}{m_1 + m_2}\right)^2, \hat{I} = \frac{1}{m_1 + m_2}\left(\delta m_1 l_1 - \frac{1}{2} m_2 l_2\right) \tag{33}$$

Moreover, the translation kinetic energy due to the translation motion of the leg is computed from

$$K_{Li(trans)} = \frac{1}{2}(m_1 + m_2)\left[\left(\frac{m_2}{m_1 + m_2}\right)^2 \vec{V}_{T_J}^T \vec{u}_i \vec{u}_i^T \vec{V}_{T_J}\right] \tag{34}$$

Therefore, the total kinetic energy of the leg L_i is calculated as the following.

$$K_{Li} = K_{Li(rot)} + K_{Li(trans)} = \frac{1}{2}(m_1 + m_2)\left[\vec{V}_{T_J}^T h_i \vec{V}_{T_J} - \dot{L}_i k_i \dot{L}_i\right] \tag{35}$$

where

$$k_i = \frac{\hat{I}}{L_i}\left(\frac{\hat{I}}{L_i} + \frac{m_2}{m_1 + m_2}\right) = h_i - \left(\frac{m_2}{m_1 + m_2}\right)^2 \tag{36}$$

Remember that $\vec{u}_i$ is the unit vector along the axis of the leg (L_i). By using this vector, the velocity of the leg can be calculated by $\dot{L}_i = \vec{V}_{T_J} \vec{u}_i$.
As a result, the compact expression for the kinetic energy of the six legs can be written as

$$K_{Legs} = \sum_{i=1}^{6} K_{Li} = \frac{1}{2} \dot{X}_{P-O}^T \cdot M_{Legs}(X_{P-O}) \cdot \dot{X}_{P-O} \tag{37}$$

Total potential energy of the legs can be defined as

$$P_{Legs} = (m_1 + m_2) g \sum_{i=1}^{3}\left[\hat{I}\left(\frac{1}{L_{2i}} + \frac{1}{L_{2i-1}}\right) + \frac{2m_2}{m_1 + m_2}\right]\left(p_z + Z_{T_j}\right) \tag{38}$$

where

$$Z_{T_J} = \begin{bmatrix} 0 \\ 0 \\ 1 \end{bmatrix}^T R_Z(\gamma)^T R_X(\alpha)^T R_Y(\beta)^T G T_j, \begin{cases} j = \frac{i+1}{2} & \text{if } i \text{ is odd} \\ j = \frac{i}{2} & \text{if } i \text{ is even} \end{cases} \tag{39}$$

3.3 Dynamic equations

In this subsection, the Lagrange formulation is used to derive the dynamic modeling of the SP manipulator. Considering q and τ as the corresponding generalized coordinates and generalized forces, respectively, the general classical equations of the motion can be obtained from the Lagrange formulation:

$$\frac{d}{dt}\frac{dL}{d\dot{q}}-\frac{\partial L}{\partial q}=\frac{d}{dt}\left(\frac{\partial K(q,\dot{q})}{\partial \dot{q}}\right)-\frac{\partial K(q,\dot{q})}{\partial q}+\frac{\partial P(q)}{\partial q}=\tau \tag{40}$$

where $K(q,\dot{q})$ is the kinetic energy, and $P(q)$ is the potential energy.
Generalized coordinates q is replaced with Cartesian coordinates (X_{p-o}). The dynamic equation derived from Equation 40 can be written as

$$J^{T}(X_{P-O})F=M(X_{P-O})\ddot{X}_{P-O}+V_{m}(X_{P-O},\dot{X}_{P-O})\dot{X}_{P-O}+G(X_{P-O}) \tag{41}$$

where $F=[f_1 \quad f_2 \quad f_3 \quad f_4 \quad f_5 \quad f_6]$, f_i is the force applied by the actuator of leg i in the direction $\vec{u}_i$ and J is the Jacobian matrix. Since the platform is divided into two parts (the moving platform and the legs), inertia, Coriolis-Centrifugal and gravity matrix in Equation 41 are summation of two matrix. Each of these matrices is computed using by two different Jacobian matrices.

$${}^{A}M(X_{p-o})=M_{up}+{}^{A}M_{Legs},\qquad {}^{B}M(X_{p-o})=M_{up}+{}^{B}M_{Legs} \tag{42}$$

$${}^{A}V_{m}(X_{p-o},\dot{X}_{p-o})=V_{mup}+{}^{A}V_{mLegs},\qquad {}^{B}V_{m}(X_{p-o},\dot{X}_{p-o})=V_{mup}+{}^{B}V_{mLegs} \tag{43}$$

$${}^{A}G(X_{p-o})=G_{up}+{}^{A}G_{Legs},\qquad {}^{B}G(X_{p-o})=G_{up}+{}^{B}G_{Legs} \tag{44}$$

where M_{up} obtained from Equation 30, ${}^{A}M_{Legs}$ and ${}^{B}M_{Legs}$ obtained from Equation 37 are the inertia matrix of the moving platform and legs, respectively. V_{mup}, ${}^{A}V_{mLegs}$ and ${}^{B}V_{mLegs}$ are Coriolis-Centrifugal matrix of the moving platform and legs, respectively. G_{up}, ${}^{A}G_{Legs}$ and ${}^{B}G_{Legs}$ are the gravity matrix of the moving platform and legs, respectively. V_{mup}, ${}^{A}V_{mLegs}$ and ${}^{B}V_{mLegs}$ are defined as follows:

$$V_{mup(i,j)}=\frac{1}{2}\sum_{k=1}^{6}\left(\frac{\partial M_{up}(k,j)}{\partial X_{p-o}(i)}+\frac{\partial M_{up}(k,i)}{\partial X_{p-o}(j)}-\frac{\partial M_{up}(i,j)}{\partial X_{p-o}(k)}\right)\dot{X}_{p-o}(k) \tag{45}$$

$${}^{A}V_{Legs(i,j)}=\frac{1}{2}\sum_{k=1}^{6}\left(\frac{\partial {}^{A}M_{Legs}(k,j)}{\partial X_{p-o}(i)}+\frac{\partial {}^{A}M_{Legs}(k,i)}{\partial X_{p-o}(j)}-\frac{\partial {}^{A}M_{Legs}(i,j)}{\partial X_{p-o}(k)}\right)\dot{X}_{p-o}(k) \tag{46}$$

$${}^{B}V_{Legs(i,j)}=\frac{1}{2}\sum_{k=1}^{6}\left(\frac{\partial {}^{B}M_{Legs}(k,j)}{\partial X_{p-o}(i)}+\frac{\partial {}^{B}M_{Legs}(k,i)}{\partial X_{p-o}(j)}-\frac{\partial {}^{B}M_{Legs}(i,j)}{\partial X_{p-o}(k)}\right)\dot{X}_{p-o}(k) \tag{47}$$

Finally, the gravity matrix can be obtained from the equations below.

$$G_{up}(k) = \frac{\partial P_{up}(X_{p-o})}{\partial X_{p-o}(k)} \tag{48}$$

$$\begin{aligned} {}^{A}G_{Legs}(k) &= \frac{\partial\, {}^{A}P_{Legs}(X_{p-o})}{\partial X_{p-o}(k)} \\ &= (m_1+m_2)g\sum_{i=1}^{3}\left[\hat{I}\left[\frac{1}{{}^{A}L_{2i}^{2}}\left(\frac{\partial\, {}^{A}L_{2i}}{\partial X_{p-o}(k)}\right)+\frac{1}{{}^{A}L_{2i-1}^{2}}\left(\frac{\partial\, {}^{A}L_{2i-1}}{\partial X_{p-o}(k)}\right)\right]\right](p_z+Z_{T_i}) \\ &+(m_1+m_2)g\sum_{i=1}^{3}\left[\hat{I}\left(\frac{1}{{}^{A}L_{2i}}+\frac{1}{{}^{A}L_{2i-1}}\right)+\frac{2m_2}{(m_1+m_2)}\right](p_z+Z_{T_i}) \end{aligned} \tag{49}$$

$$\begin{aligned} {}^{B}G_{Legs}(k) &= \frac{\partial\, {}^{B}P_{Legs}(X_{p-o})}{\partial X_{p-o}(k)} \\ &= (m_1+m_2)g\sum_{i=1}^{3}\left[\hat{I}\left[\frac{1}{{}^{B}L_{2i}^{2}}\left(\frac{\partial\, {}^{B}L_{2i}}{\partial X_{p-o}(k)}\right)+\frac{1}{{}^{B}L_{2i-1}^{2}}\left(\frac{\partial\, {}^{B}L_{2i-1}}{\partial X_{p-o}(k)}\right)\right]\right](p_z+Z_{T_i}) \\ &+(m_1+m_2)g\sum_{i=1}^{3}\left[\hat{I}\left(\frac{1}{{}^{B}L_{2i}}+\frac{1}{{}^{B}L_{2i-1}}\right)+\frac{2m_2}{(m_1+m_2)}\right](p_z+Z_{T_i}) \end{aligned} \tag{50}$$

In equation 49 and 50, the expression of $\left(\frac{\partial L_n}{\partial X_{p-o}(k)}\right)$ is needed to compute. This can be obtained using with the Jacobian matrices (J_A and J_B) as follows:

$$\frac{\partial\, {}^{A}L_n}{\partial X_{p-o}(k)} = \sum_{m=1}^{9} J_{IA_{nm}} J_{IIA_{mk}}, \quad \frac{\partial\, {}^{B}L_n}{\partial X_{p-o}(k)} = \sum_{m=1}^{9} J_{IB_{nm}} J_{IIB_{mk}} \tag{51}$$

3.4 Actuator dynamics

6 identical motor-ball-screw drives are used in SP. Dynamic equation of SP with actuator dynamics can be written in matrix form as

$$\tau_m = M_a\ddot{L} + N_a\dot{L} + K_aF \tag{52}$$

$$M_a = \frac{2\pi}{np}\left(J_s + n^2 J_m\right)I_{6x6} \tag{53}$$

$$N_a = \frac{2\pi}{np}\left(b_s + n^2 b_m\right)I_{6x6} \tag{54}$$

$$K_a = \frac{p}{n2\pi}I_{6x6} \tag{55}$$

where M_a, N_a and K_a are the inertia matrix, viscous damping coefficient matrix and gain matrix of the actuator, respectively. Also, J_s and J_m are the mass moment of inertia of the ball-screw and motor, b_s and b_m are the viscous damping coefficient of the ball-screw and motor, p and n are the pitch of the ball-screw and the gear ratio. τ_m and F are the vectors of motor torques and the forces applied by the actuators.
The electrical dynamics of the actuator can be described by the following equations.

$$\tau_m = K_t i \tag{56}$$

$$V = L\frac{di}{dt} + Ri + K_b\dot{\theta}_m \tag{57}$$

where K_t, L, R and K_b are the torque constant, the rotor inductance, terminal resistance and back-emf constant of the actuators, respectively. V and i are the motor voltage and motor current , respectively. The angular velocity of the motor is given as

$$\dot{\theta}_m = \frac{2\pi n}{p}\dot{L} \tag{58}$$

Since the dynamics of the platform is derived in the moving platform coordinates (Cartesian space, X_{p-o}), Equation 52 can be generally expressed in Cartesian space as the follows.

$$\tau_m = \bar{M}_c(X)\ddot{X} + \bar{N}_c(X)\dot{X} + \bar{G}_c(X) \tag{59}$$

The terms in the equation above are obtained from joint space terms and the Jacobian matrix.

$${}^{A}\bar{M}_c(X) = K_a(J_A)^{-T}\,{}^{A}M(X_{p-o}) + M_a J_A \tag{60}$$

$${}^{B}\bar{M}_c(X) = K_a(J_B)^{-T}\,{}^{B}M(X_{p-o}) + M_a J_B \tag{61}$$

$${}^{A}\bar{N}_c(X) = K_a(J_A)^{-T}\,{}^{A}N(X_{p-o}, \dot{X}_{p-o}) + N_a J_A + M_a \dot{J}_A \tag{62}$$

$${}^{B}\bar{N}_c(X) = K_a(J_B)^{-T}\,{}^{B}N(X_{p-o}, \dot{X}_{p-o}) + N_a J_B + M_a \dot{J}_B \tag{63}$$

$${}^{A}\bar{G}_c(X) = K_a(J_A)^{-T}\,{}^{A}G(X_{p-o}) \tag{64}$$

$${}^{B}\bar{G}_c(X) = K_a(J_B)^{-T}\,{}^{B}G(X_{p-o}) \tag{65}$$

Figure 5 shows the simulation block diagram of the Stewart Platform manipulator including the actuator dynamics (Equation 56, 57, 58 and 59). In order to model the platform dynamics without using forward kinematics, the block diagram is developed below.

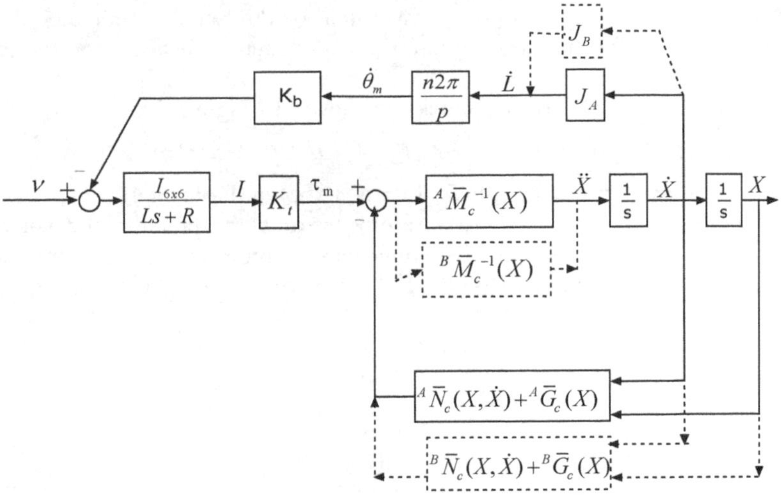

Fig. 5. Simulation block diagram for SP dynamics

4. Results

4.1 Experimental setup and simulation blocks

Figure 6 shows the Stewart Platform manipulator used in the experiments. It is constructed from two main bodies (top and base plates), six linear motors, controller, space mouse, accelerometer, gyroscope, force/torque sensor, power supply, emergency stop circuit and interface board as shown in the figure. Inverse kinematics and control algorithm of SP are embedded in MATLAB/Simulink module. Moreover, controller board like the space DS1103owning real-time interface implementation software to generate and then download the real time code to specific space board is used, and it is fully programmable from MATLAB/Simulink environment. Thus, it is possible for the user to observe the real process and collect the data from encoders for each leg while the experiment is in progress.

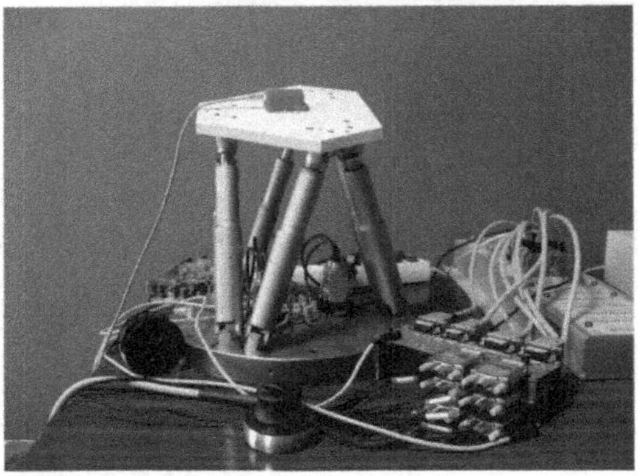

Fig. 6. SP manipulator used in experiments

To demonstrate the effectiveness and the validation of the two dynamic models of the SP manipulator including the actuator dynamics, experimental tests are performed on the manipulator.

The first Simulink model ("*Desired*" block) as shown in Figure 7 is used for the inverse kinematic solution for a given $X_{p-o} = \begin{bmatrix} P_x & P_y & P_z & \alpha & \beta & \gamma \end{bmatrix}^T$. Thus, the reference lengths of the legs are obtained from this block. Figure 8 shows the forward dynamics Simulink block (Jacobian matrix and its derivative, motor torques, the position and velocity of the moving platform, etc.). Figure 9 shows the developed Simulink model for the actual lengths of the legs. This block is designed using the following equation.

$$\ddot{L}_i = \dot{J}\,\dot{X}_{P-O} + J\,\ddot{X}_{P-O} \tag{66}$$

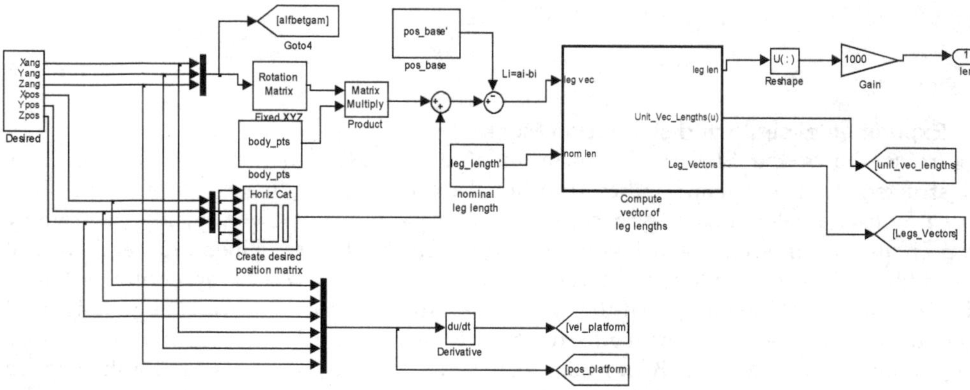

Fig. 7. Inverse kinematic solution Simulink model, "*Leg_Trajectory*"block

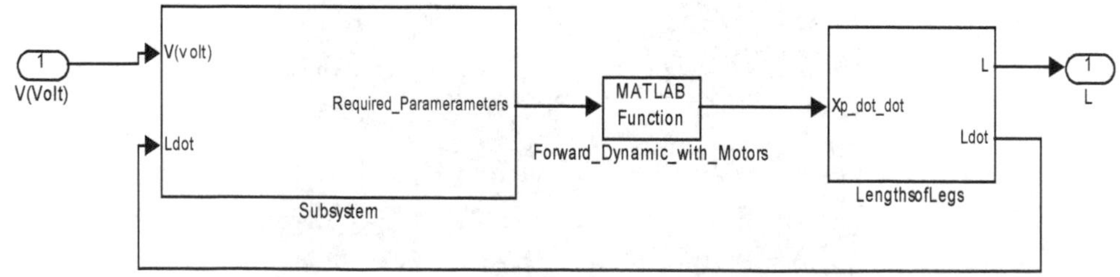

Fig. 8. The Simulink model of the forward dynamics, "*Stewart_Platform_Dynamics*" block

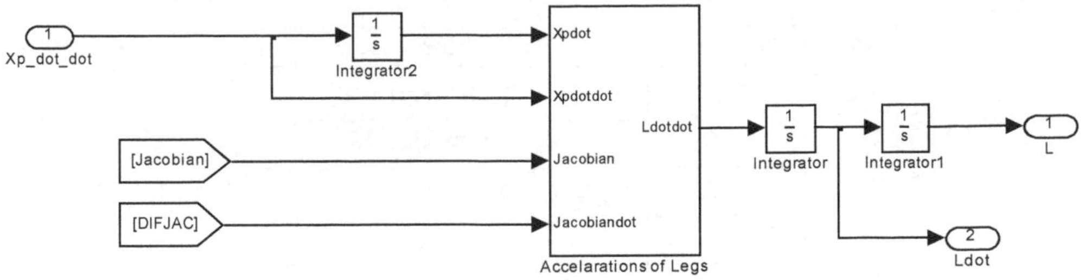

Fig. 9. The leg Simulink model, "*LengthofLegs*"block

To examine trajectory tracking performance of the SP dynamic model, a Simulink model shown in Figure 10 is developed.

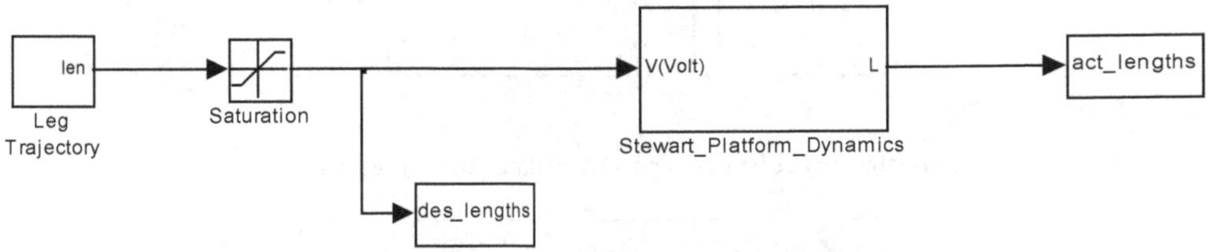

Fig. 10. Simulink model of the SP manipulator

In the figure, the reference lengths of the legs are limited between -25 mm and 25 mm with saturation block. The"*Stewart_Platform_Dynamics*" block computes the dynamic equations of the SP manipulator and outputs the actual lengths of the legs.
The attachment points GT_i and B_i on the moving and base platform, respectively are given in Table 1 and Table 2.

	GT_1 (m)	GT_2 (m)	GT_3 (m)	GT_4 (m)	GT_5 (m)	GT_6 (m)
X	0.0641	0.0641	0.0278	-0.0919	-0.0919	0.0278
Y	-0.0691	0.0691	0.0901	0.0209	-0.0209	-0.0901
Z	0	0	0	0	0	0

Table 1. Attachment point on the moving platform

	B_1 (m)	B_2 (m)	B_3 (m)	B_4 (m)	B_5 (m)	B_6 (m)
x	0.1451	0.1451	-0.0544	-0.0906	-0.0906	-0.0544
y	-0.0209	0.0209	0.1361	0.1152	-0.1152	-0.1361
z	0	0	0	0	0	0

Table 2. Attachment point on the base platform

Also, the system constants are given in Table 3.

Parameter	Value
m_{up}	1.1324 (Kg)
m_1	0.4279 (Kg)
m_2	0.1228 (Kg)
l_1	0.22 (m)
l_2	0.05 (m)
r_p	0.18867975191 (m)
r_{base}	0.29326616983 (m)
$I_{(mf)}$	$\begin{bmatrix} 0.0025 & 0 & 0 \\ 0 & 0.0025 & 0 \\ 0 & 0 & 0.005 \end{bmatrix} Kg.m^2$

Table 3. SP constants

The constants of the motor used in the SP manipulator are given in Table 4.

Parameter	Value
R	7.10 (ohm)
L	265e-6 (H)
K_b	2.730e-3 (V/rpm)
K_t	26.10e-3 (Nm/A)
n	1 (Rad/rad)
J_m	0.58e-6 (Kg.m^2)
J_s	0.002091e-3 (Kg.m^2)
b_m	0.0016430e-3 (N.s/rad)
b_s	0.11796e-3 (N.s/rad)
P	0.001 (m)

Table 4. SP motor constants

4.2 Dynamic simulations and experimental results

In this subsection, the effectiveness and the validation of the dynamic models of the SP manipulator with the actuator dynamics were investigated. In order to compare the experimental results with the simulation results, three trajectory tracking experiments were conducted. Also, the dynamic models obtained with two different Jacobian matrices (J_A and J_B) are examined. In all experiments, SP was worked in open-loop. In the first experiments, the translation motion along z-axis was applied to the SP system.

$$x(t) = 0 \qquad \alpha(t) = 0$$

$$y(t) = 0 \qquad \beta(t) = 0 \tag{67}$$

$$z(t) = 0.25 + 10\sin(\pi t) \qquad \gamma(t) = 0$$

Figure 11 shows the reference lengths of the legs, the actual lengths from the encoder and the lengths predicted by the dynamic equations of the SP manipulator with two different Jacobian J_A (Sim-A) and J_B (Sim-B).

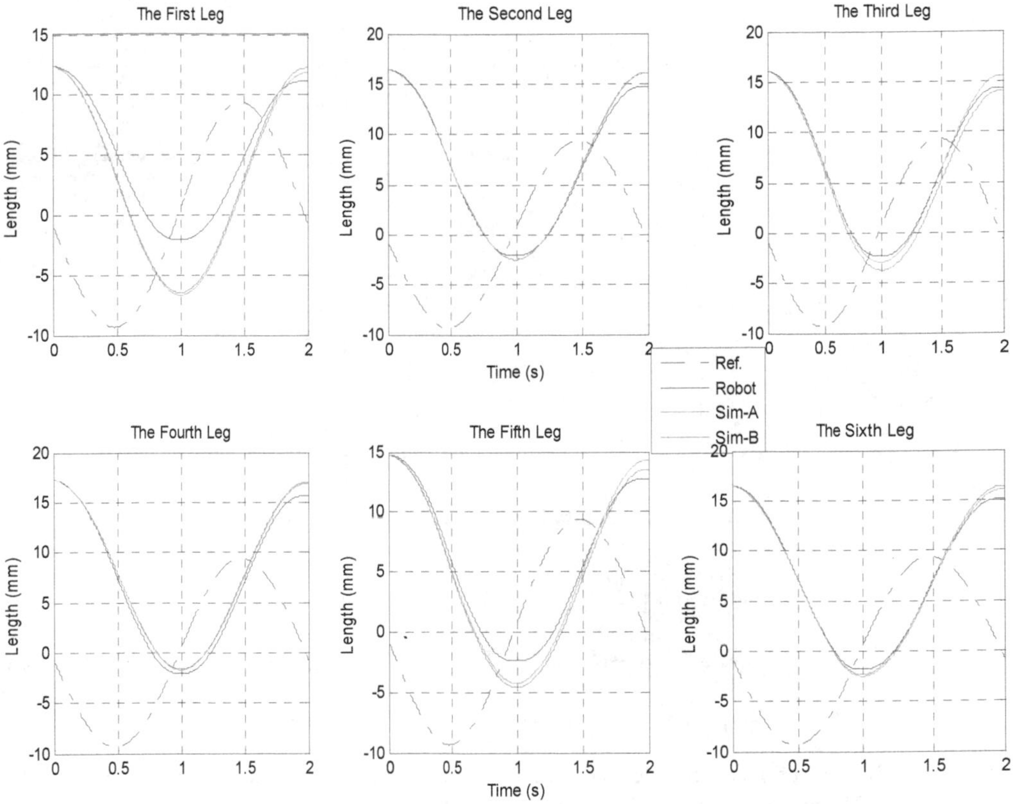

Fig. 11. Comparison of simulation results (red and blue) and experimental results (black) for the first experiment

Response of two dynamic simulation models (Sim-A and Sim-B) is almost same. But, it is observed that the *Sim-A* shows better performance than the *Sim-B* for this trajectory. Also, simulation and experimental results are very close to each other.

The second experiment, both translational and rotational motion along all axes (x,y,z) was applied to the SP system. The trajectory is defined in Equation 68. The experimental and simulation results for this trajectory are shown in Figure 12.

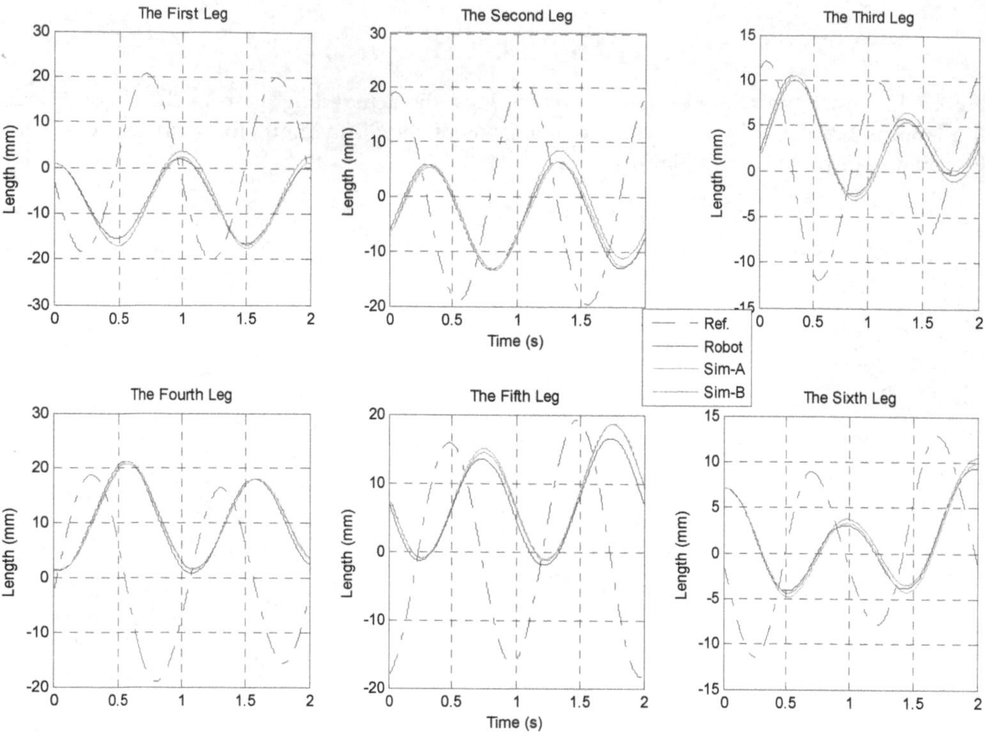

Fig. 12. Comparison of simulation results (red and blue) and experimental results (black) for the second experiment

$$x(t) = 5\sin(\pi t) \qquad \alpha(t) = 10\cos(2\pi t)$$
$$y(t) = 5\cos(\pi t) \qquad \beta(t) = 10\sin(2\pi t) \qquad (68)$$
$$z(t) = 0.25 + \sin(\pi t) \qquad \gamma(t) = 10\cos(2\pi t)$$

As can be shown in the figure, the dynamic model (red) has better performance compared to other (blue). There is good match between the simulation and experimental results.

In the last experiment, the fast translational and rotational motion along all axes (x,y,z) was conducted. The trajectory is given below.

$$x(t) = 10\sin(2\pi t) \qquad \alpha(t) = 5\cos(4\pi t)$$
$$y(t) = 10\cos(3\pi t) \qquad \beta(t) = 5\sin(5\pi t) \qquad (69)$$
$$z(t) = 0.25 + 3\sin(\pi t) \qquad \gamma(t) = 5\cos(3\pi t)$$

Figure 13 illustrates the results obtained from dynamic models and experimental results for this trajectory. In accordance with the results shown in Figure 13, the relative small deviations between the models and the experimental data are occurred. However, the obtained dynamic models can track the high frequency reference trajectory.

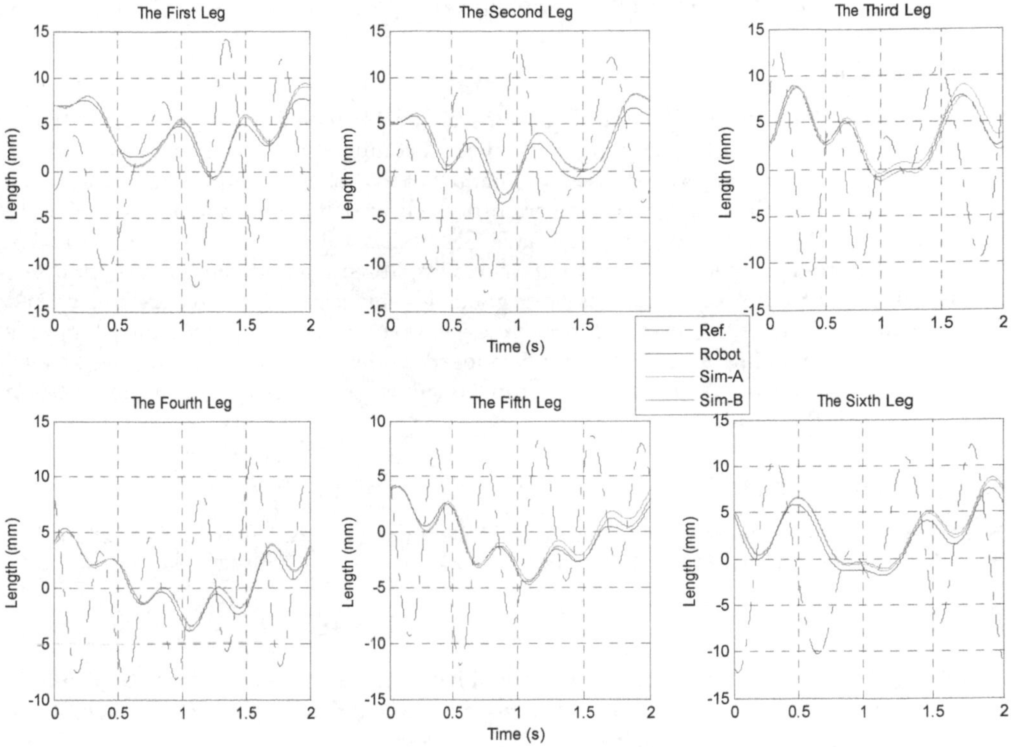

Fig. 13. Comparison of simulation results (red and blue) and experimental results (black) for the third experiment

The errors between simulation and experimental results were computed for all experiments. The cost function for the modeling error is defined as

$$E = \frac{1}{N}\sum_{i=1}^{N} |e_1(i)| + |e_2(i)| + |e_3(i)| + |e_4(i)| + |e_5(i)| + |e_6(i)| \tag{70}$$

where $e_1(i)...e_6(i)$ are the trajectory errors of ith sample obtained from difference between the experimental and the simulation results and N is the number of sample.

Table 5 gives the cost function values obtained from the two different dynamic models.

The two dynamic models of the SP manipulator using J_A and J_B exhibit very good performance in terms of model accuracy. Performance of the dynamic model using J_A is better than which of other model.

Experiment	Sim-A	Sim-B
1-EXP(z)	4.4337	5.0121
2-EXP(x,y,z,α,β,γ)	4.5389	4.5877
3-EXP(x,y,z,α,β,γ)	3.6470	3.3697

Table 5. The cost function values for the two different dynamic models

5. Conclusion

In this paper, closed-form dynamic equations of the SP manipulator with the actuator dynamics were derived using Lagrangian method. A computational highly efficient method was developed for the explicit dynamic equations. Besides, two simple methods for the calculation of the Jacobian matrix of SP were proposed. Two dynamic models of the SP were obtained using these Jacobian matrices. Two SP models were simulated in a MATLAB-Simulink. In order to verify the simulation results, three experiments were conducted. Considering all of the results, there is very good agreement between the experiments and the simulations. Modeling errors for each experiment were computed. Based on the modeling error, modeling accuracy of the developed models is very high. Thus, the verified model of the SP can be used for control and design purposes. Especially, a model based controller needs the verified model.

6. Acknowledgment

This work is supported by The Scientific and Technological Research Council of Turkey (TUBITAK) under the Grant No. 107M148.

7. References

Abdellatif, H. & Heimann, B. (2009). Computational efficient inverse dynamics of 6-DOF fully parallel manipulators by using the Lagrangian formalism. *Mechanism and Machine Theory*, 44, pp. (192-207)

Beji, L. & Pascal, M. (1999). The Kinematics and the Full Minimal Dynamic Model of a 6-DOF Parallel Robot Manipulator. *Nonlinear Dynamics*, 18, pp. (339-356)

Bonev, I. A., & Ryu, J. A. (2000). New method for solving the direct kinematics of general 6-6 Stewart platforms using three linear extra sensors. *Mechanism and Machine Theory*, 35, pp. (423–436)

Carvalho, J. & Ceccarelli, M. (2001). A Closed-Form Formulation for the Inverse Dynamics of a Cassino Parallel Manipulator. *Multibody System Dynamics*, Vol. 5, pp. (185–210)

Dasgupta, B. & Mruthyunjaya, T. S. (1998). Closed-Form Dynamic Equations Of The General Stewart Platform Through The Newton-Euler. *Mechanism and Machine Theory*, 33 (7), pp. (993-1012)

Do, W. & Yang, D. (1988). Inverse Dynamic Analysis and Simulation of a Platform Type of Robot.*Journal of Robotic Systems*, Vol. 5, pp. (209-227)

Gallardo, J.; Rico, J.M.; Frisoli, A.; Checcacci, D.; Bergamasco, M. (2003). Dynamics of parallel manipulators by means of screw theory. *Mechanism and Machine Theory*, 38, pp. (1113–1131)

Geike, T. & McPhee J. (2003). Inverse dynamic analysis of parallel manipulators with full mobility. *Mechanism andMachine Theory*, 38, pp. (549–562)

Gregório, R. & Parenti-Castelli, V. (2004). Dynamics of a Class of Parallel Wrists. *Journal of Mechanical Design*, Vol. 126, pp. (436-441)

Guo, H. & Li, H. (2006). Dynamic analysis and simulation of a six degree of freedom Stewart platform manipulator. Proceedings of the Institution of Mechanical Engineers, Part C: *Journal of Mechanical Engineering Science*, Vol. 220, pp. (61-72)

Harib , K.; Srinivasan, K. (2003). Kinematic and dynamic analysis of Stewart platform-based machine tool structures, *Robotica*, 21(5), pp. (541–554)

Khalil W, Ibrahim O. (2007). General solution for the dynamic modelling of parallel robots.*J Intell Robot Systems*, 49, pp. (19-37)

Lebret G.; Liu K.; Lewis F. L. (1993). Dynamic analysis and control of a Stewart platform manipulator. *Journal of Robotic System*, 10, pp. (629-655)

Lee, K & Shah, D.K. (1988). Dynamical Analysis of a Three-Degrees-of-Freedom In-Parallel Actuated Manipulator. *IEEE Journal of Robotics and Automation*, 4(3).

Lin, J. & Chen, C.W. (2008). Computer-aided-symbolic dynamic modeling for Stewart-platform manipulator.*Robitica*, 27 (3), pp. (331-341)

Liu, K.; Lewis, F. L.; Lebret, G.; Taylor, D. (1993). The singularities and dynamics of a Stewart platform manipulator. *J. Intell. Robot.* Syst., 8, pp. (287–308)

Liu, M.J.; Li, C.X.; Li, C.N., 2000. Dynamics analysis of the Gough-Stewart platform manipulator. *IEEE Trans. Robot. Automat.*, 16 (1), pp. (94-98)

Meng, Q.; Zhang, T.; He, J-F.; Song, J-Y.; Han, J-W.(2010). Dynamic modeling of a 6-degree-of-freedom Stewart platform driven by a permanent magnet synchronous motor. *Journal of Zhejiang University-SCIENCE C (Computers & Electronics)*, 11(10), pp. (751-761)

Merlet, J.P. (1999). *Parallel Robots*, Kluwer Academic Publishers.

Merlet, J. P. (2004). Solving the forward kinematics of a Gough-type parallel manipulator with internal analysis. *International Journal of Robotics Research*, 23(3), pp. (221–235)

Reboulet, C. & Berthomieu, T. (1991). Dynamic Models of a Six Degree of Freedom Parallel Manipulators. *Proceedings of the IEEE International Conference on Robotics and Automation*, pp. (1153-1157)

Riebe, S. & Ulbrich, H. (2003). Modelling and online computation of the dynamics of a parallel kinematic with six degrees-of-freedom. *Arch ApplMech*, 72, pp. (817–29)

Staicu, S. & Zhang, D. (2008). A novel dynamic modelling approach for parallel mechanisms analysis. *Robot ComputIntegrManufact*, 24, pp. (167–172)

Stewart, D. (1965). A Platform with Six Degrees of Freedom, *Proceedings of the Institute of Mechanical Engineering*, Vol. 180, Part 1, No. 5, pp. (371-386)

Tsai L-W. (2000). Solving the inverse dynamics of Stewart–Gough manipulator by the principle of virtual work. *J Mech Des*, 122, pp. (3–9)

Wang, J. & Gosselin, C. (1998). A New Approach for the Dynamic Analysis of Parallel Manipulators. *Multibody System Dynamics*, 2, pp. (317-334)

Wang , Y. (2007). A direct numerical solution to forward kinematics of general Stewart-Gough platforms, *Robotica*, 25(1), pp. (121-128)

4

Parallel, Serial and Hybrid Machine Tools and Robotics Structures: Comparative Study on Optimum Kinematic Designs

Khalifa H. Harib[1], Kamal A.F. Moustafa[1],
A.M.M. Sharif Ullah[2] and Salah Zenieh[3]
[1]*UAE University*
[2]*Kitami Institute of Technology*
[3]*Tawazun Holding*
[1,3]*United Arab Emirates*
[2]*Japan*

1. Introduction

After their inception in the past two decades as possible alternatives to conventional Computer Numerical Controlled (CNC) machine tools structures that dominantly adapt serial structures, Parallel Kinematic Machines (PKM) were anticipated to form a basis for a new generation of future machining centers. However this hope quickly faded out as most problems associated with this type of structures still persist and could not be completely solved satisfactorily. This especially becomes more apparent in machining applications where accuracy, rigidity, dexterity and large workspace are important requirements. Although the PKMs possess superior mechanical characteristics to serial structures, particularly in terms of high rigidity, accuracy and dynamic response, however the PKMs have their own drawbacks including singularity problems, inconsistent dexterity, irregular workspace, and limited range of motion, particularly rotational motion.

To alleviate the PKMs' limitations, considerable research efforts were directed to solve these problems. Optimum design methods are among the various methods that are attempted to improve the dexterity as well as to maximize the workspace (Stoughton and Arai, 1993, Huang et al., 2000). Various methods to evaluate the workspace were suggested (Gosselin, 1990; Luh et al., 1996; Conti et al., 1998; Tsai et al., 2006). Workspace optimization is also addressed (Wang and Hsieh, 1998). A new shift in tackling the aforementioned problems came when researchers start to look at hybrid structures, consisting of parallel and serial linkages as a compromise to exploit the advantageous characteristics of the serial and parallel structures. This shift creates new research and development needs and founded new ideas.

Among the early hybrid kinematic designs, the Tricept was considered as the first commercially successful hybrid machine tools. This hybrid machine which was developed by Neos Robotics, has a three-degrees-of-freedom parallel kinematic structure and a standard two-degrees-of-freedom wrest end-effector holding joint. The constraining passive leg of the machine has to bear the transmitted torque and moment between the moving platform and the base (Zhang and Gosselin, 2002). Recently the Exechon machine is introduced as an improvement over the Tricept design. The Exechon adopts a unique overconstrained structure, and it has been improved based on the success of the Tricept (Zoppi, et al., 2010, Bi and Jin, 2011). Nonetheless, regardless of the seemingly promising prospect of the hybrid kinematic structures, comprehensive study and understanding of the involved kinematics, dynamics and design of these structures are lacking. This paper is attempting to provide a comparative study and a formulation for the kinematic design of hybrid kinematic machines. The remainder of this paper is as follows: Section 2 provides a discussion on the mobility of serial, parallel and hybrid kinematic structures and the involved effects of overconstrain on the mobility of the mechanism. Section 3 provides a discussion on kinematic design for hybrid machines and the implication of the presented method. Concluding remarks are presented in Section 4.

2. Mobility of robotic structures

Mobility is a significant structural attribute of mechanisms assembled from a number of links and joints. It is also one of the most fundamental concepts in the kinematic and the dynamic modeling of mechanisms and robotic manipulators. IFToMM defines the mobility or the degree of freedom as the number of independent co-ordinates needed to define the configuration of a kinematic chain or mechanism (Gogu, 2005, Ionescu, 2003). Mobility, M, is used to verify the existence of a mechanism (M > 0), to indicate the number of independent parameters in the kinematic and the dynamic models and to determine the number of inputs needed to drive the mechanism.

The various methods proposed in the literature for mobility calculation of the closed loop mechanisms can be grouped in two categories (Ionescu, 2003): (a) approaches for mobility calculation based on setting up the kinematic constraint equations and their rank calculation for a given position of the mechanism with specific joint location, and (b) formulas for a quick calculation of mobility without need to develop the set of constraint equations. The approaches for mobility calculation based on setting up the kinematic constraint equations and their rank calculation are valid without exception. The major drawback of these approaches is that the mobility cannot be determined quickly without setting up the kinematic model of the mechanism. Usually this model is expressed by the closure equations that must be analyzed for dependency. There is no way to derive information about mechanism mobility without performing kinematic analysis by using analytical tools. For this reason, the real and practical value of these approaches is very limited in spite of their valuable theoretical foundations.

Many formulas based on approach (b) above have been proposed in the literature for the calculation of mechanisms' mobility. Many of these methods are reducible to the Cebychev-Grubler-Kutzbach's mobility formula given by Equation 1 below (Gogu, 2005). Using this formula, the mobility M of a linkage composed of L links connected with j joints can be determined from the following equation.

$$M = 6(L - 1 - j) + \sum_{i=1}^{j} f_i \quad (1)$$

where f_i is the DOF associated with joint i. Equation 1 is used to calculate the mobility of spatial robotic mechanisms as most industrial robots and machine tools structures are serial structures with open kinematics chains.

2.1 Mobility of planner mechanisms

To gain an insight into the effect of mobility on the kinematic analysis and design of serial, parallel and hybrid kinematic structures, we will also look at the mobility of planner mechanisms, which can be obtained from the following planner Kutzbach-Gruebler's equation (Gogu , 2005, Norton, 2004).

$$M = 3(L-1-j) + \sum_{i=1}^{j} f_i \quad (2)$$

where M, L, j and f_i are as defined before in Equation 1. As shown in Figure 1, the robotic structures are arranged in serial, parallel and hybrid kinematic chains, and thus have different number of links and joints. Using Equation 2, all the three structures in Figure 1 have three degrees of freedom, or mobility three. This gives the end-effector two translational degrees-of-freedom to position it arbitrarily in the x-y plane, and one rotational degree-of-freedom to orient it about the z-axis. In the serial kinematic structure all three joints are actuated, whereas for the parallel and hybrid structures only the three prismatic joints are actuated whereas the revolute joints are passive. The parallel kinematic part of the hybrid structure in Figure 1.c, has two degrees of freedom, which is achieved by reducing the number of legs to two and eliminating one of the passive revolute joints.

Figure 2 shows an alternative way to reduce the degrees of freedom of the parallel kinematic mechanism, and hence to reduce the number of actuated prismatic joints. In this example this is done by eliminating one of the revolute joints which connect the legs to the platform. The corresponding leg has a passive prismatic joint to constraint one of the degrees-of-freedom. By removing the revolute joint though the leg becomes a three-force member and hence it will be carrying bending moments necessitating considerable design attention to maintain desired stiffness levels and accuracy. This concept of reducing the degrees of freedom is adopted in the spatial Tricept mechanism to reduce the degrees-of-freedom from six to three. Compared to the mechanism in 1.c, this mechanism has more joints and links, which is not desirable from the design point of view.

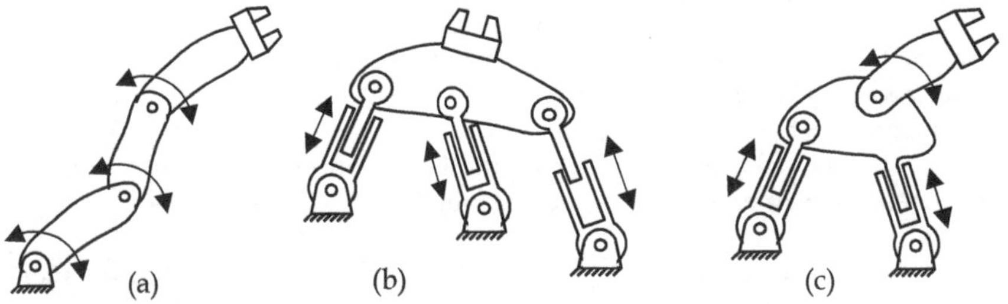

Fig. 1. Schematics of planner 3 degrees-of-freedom robotic structures with a) fully serial (L=4, j=3, Σf_i =3), b) fully parallel (L=8, j=9, Σf_i =9), and c) hybrid topologies (L=6, j=6, Σf_i =6)

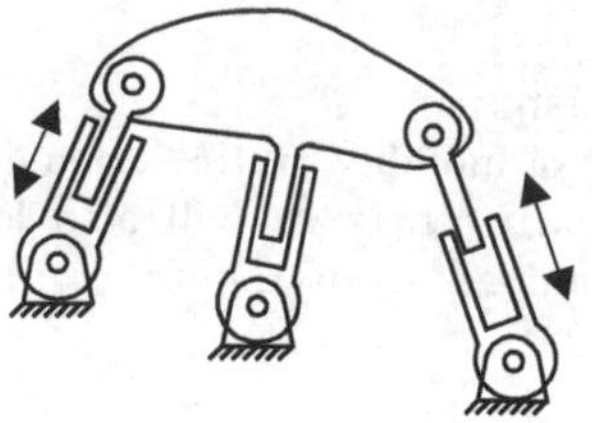

Fig. 2. Schematics of a 2-degrees-of-freedom planner parallel robotic structure, L=7, j=8, Σf_i =8. The prismatic joint of the middle leg is passive (unactuated)

It should be noted here that the planner mechanisms are realized by necessitating that the involved revolute joints to be perpendicular to the plane and the prismatic joints to be confined to stay in the plane. As such these mechanisms can also be viewed as special cases of spatial mechanisms that are confined to work in a plane through overconstrains and thus Equation 1, with proper modification, rather than Equation 2 could be use, as discussed in the next section,

2.2 Over-constrained mechanisms

Formula for a quick calculation of mobility is an explicit relationship between structural parameters of the mechanism: the number of links and joints, the motion/constraint parameters of the joints and of the mechanism. Usually, these structural parameters are easily determined by inspection without a need to develop a set of kinematic constraint equations. However, not all known formulas for a quick calculation of mobility fit for many classical mechanisms and in particular parallel robotic manipulators (Ionescu, 2003). Special geometric conditions play a significant role in the determination of mobility of such mechanisms, which are called paradoxical mechanisms, or overconstrained, yet mobile linkage (Waldron and Kinzel, 1999). However, as mentioned above, there are overconstrained mechanisms that have full range mobility and therefore they are mechanisms even though they should be considered as rigid structures according to the mobility criterion (i.e. the mobility $M < 1$ as calculated from Equation 1. The mobility of such mechanisms is due to the existence of a particular set of geometric conditions between the mechanism joint axes that are called overconstraint conditions.

Overconstrained mechanisms have many appealing characteristics. Most of them are spatial mechanisms whose spatial kinematic characteristics make them good candidates in modern linkage designs where spatial motion is needed. Another advantage of overconstrained mechanisms is that they are mobile using fewer links and joints than it is expected.

In fact, the planner mechanisms in Figures 1 and 2 can also be viewed as overconstrained spatial mechanisms, and thus the spatial version of Kutzbach-Gruebler's equation (Equation 1), does not work for some of these planner mechanisms. In particular, for the parallel and hybrid kinematic planner mechanisms, Equation 1 will result in negative mobility values suggesting that these mechanisms are rigid structures, although they are not. Since this is not true, it should be concluded that Equation 1 cannot be used for these over-constraint mechanisms (Mavroidis and Roth, 1995). The overconstraint in planner parallel and hybrid kinematic mechanisms is due to the geometrical requirement on the involved joint-axes in relation to each other. To solve the problem when using the spatial version of the Kutzbach-Gruebler's equation for planner mechanisms or for over-constraint mechanisms in general, Equation 1 has to be modified by adding a parameter reflecting the number of overconstaints existing in the mechanism (Cretu, 2007). The resulting equation is called the universal Somo-Malushev's mobility equation. For the case of mechanisms that do not involve any passive degrees of freedom it is written as

$$M = 6(L-1-j) + \sum_{i=1}^{j} f_i + s \tag{3}$$

where s is the number of overconstraint (geometrical) conditions. For example, the parallel kinematic mechanism in Figure 1.b has L = 8, j= 9, and Σf_i = 9. Using these parameters in Equation 1 gives M = -3. However using Equation (3) and observing that there are 6 overconstraints in this mechanism, the mobility will amount to M = 3. The overconstraints in this mechanism are due to the necessity for confining the axes of the three prismatic joints to form a plane or parallel planes (two overcosntraints), and for the axes of the three revolute joints of the moving platform (two overconstraints), and the three revolute joints of the base to be perpendicular to the plane formed by the prismatic joints.

3. Kinematic designs of robotic structures

A widely used kinematic design strategy for serial kinematic robotic structures to optimize the workspace is to use the first group of links and joints to position the end-effector and the remaining links and joints to orient the end-effector, and thus breaking the design problem into two main tasks. For the 6-DOF Puma robot schematic shown in Figure 3, the first three links and joint are responsible for positioning the end-effector at the desired position, while the last three joints and links form a 3-DOF concurrent wrest joint that orient the end-effector.

Conventional five axis machining centers achieve similar decoupling by splitting the five axes (three translational axes and two rotational axes) into two groups of axes. One group of serially connected axes is responsible for positioning/orienting the worktable which is holding the workpiece, while the other group of axes moves/orients the spindle (Bohez, 2002).

Unfortunately this strategy cannot be adopted for parallel kinematic structures due to the similarity of the legs and their way of working in parallel. As such decoupling of the two functions (positioning and orienting the end-effector) is not straightforward to do for parallel kinematic structures if not impossible. Partial decoupling has been attempted by Harib and Sharif Ullah (2008) using the axiomatic design approach.

On the other hand, it should be noted here that parallel structures, and to some extent hybrid structures, can be built from identical parts and modules, and thus lend themselves well to adaptation as reconfigurable machines (Zhang, 2006). This attribute is not strongly relevant to serial structures which consist of axes that are stacked on each other making the links and joints differ considerably in terms of size and shape.

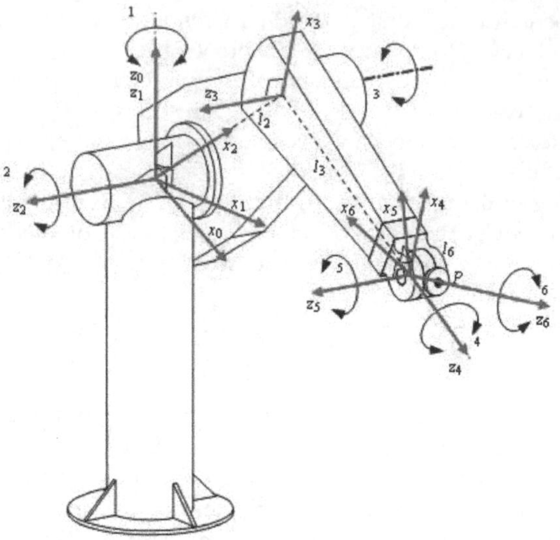

Fig. 3. A schematic of a 6-DOF Puma robot (serial kinematic structure)

3.1 Parallel kinematic designs

A main objective of the optimal design of parallel kinematic machines is to maintain consistent dexterity within the workable space of the machine. Dexterity of the mechanism is a measure of its ability to change its position and orientation arbitrarily, or to apply forces and torques in arbitrary directions. As such the Jacobian matrix of the mechanism is widely used in formulating the dexterity measure. For a six degrees-of-freedom hexapod mechanism (Harib and Srinivasan, 2003), shown in Figure 4, the Jacobian matrix J relates the translational and rotational velocity vectors of the moving platform to the extension rate of the legs as indicated below (Harib and Sharif Ullah, 2008).

$$\left[\dot{l}_1 \cdots \dot{l}_6 \right]^T = \mathbf{J}_1 \dot{\mathbf{c}} + \mathbf{J}_2 \boldsymbol{\omega} = \mathbf{J} \left[\dot{\mathbf{c}}^T \quad \boldsymbol{\omega}^T \right]^T \tag{4}$$

where $\mathbf{J} = [\mathbf{J}_1\ \mathbf{J}_2]$ is the Jacobian matrix of the hexapod, which consists of two 6×3 sub-matrices $\mathbf{J}_1$ and $\mathbf{J}_2$ that are given as

$$\mathbf{J}_1 = [\mathbf{u}_1 \cdots \mathbf{u}_6]^T \tag{5}$$

$$\mathbf{J}_2 = [{}^M\mathbf{R}_C\, {}^C\mathbf{a}_1 \times \mathbf{u}_1 \cdots {}^M\mathbf{R}_C\, {}^C\mathbf{a}_6 \times \mathbf{u}_6]^T \tag{6}$$

where $\mathbf{u}_i$ and ${}^C\mathbf{a}_i$ are respectively a unit vector along the ith leg and the position vector of its attachment point to the moving platform in the platform coordinate frame C, and ${}^M\mathbf{R}_C$ is the rotational matrix of the moving platform. The Jacobian matrix J relates also the external task space forces and torques and the joint space forces as indicated below.

$$[\mathbf{F}\ \ \mathbf{T}]^T = \mathbf{J}^T [f_1 \cdots f_6]^T \tag{7}$$

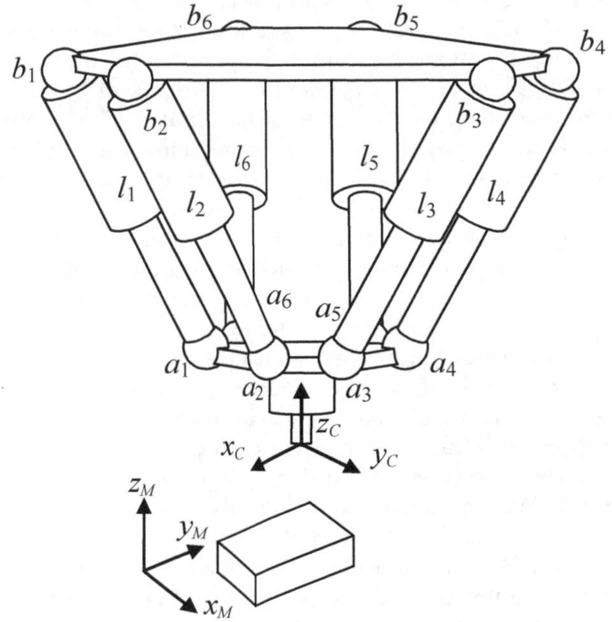

Fig. 4. Typical Construction of a hexapodic machine tools.

where **F** and **T** are respectively the resultants 3-D external force and torque systems applied to the movable platform. This result suggests that to support external force and torque along arbitrary directions, $\mathbf{J}_1$ and $\mathbf{J}_2$ must both have a rank three. Now to support these external force and torque resultants using bounded joint space forces, the condition numbers of $\mathbf{J}_1$ and $\mathbf{J}_2$ must be both as close to unity as possible.
An overall local performance measure *PM* can be obtained from the following relation

$$PM = w\,PM_1 + (1-w)\,PM_2 \tag{8}$$

where w is a weighing factor in the range $[0\cdots1]$ which signify how much emphases is given to translational and rotational dexterities, and PM_1 and PM_2 are respectively performance measures for the translational motion and the rotational motion of the structure, and are defined as (Harib and Sharif Ullah, 2008, Stoughton and Arai, 1993).

$$PM_1 = V' \Big/ \int_{V'} \kappa(\mathbf{J}_1)\,dV \tag{9}$$

$$PM_2 = V' \Big/ \int_{V'} \kappa(\mathbf{J}_2)\,dV \tag{10}$$

In the previous equations, $\kappa(\cdot)$ is the condition number function and V' is the workspace which is a subset of the total reachable space of the mechanism V. PM will then be in the range $[0\cdots1]$, with 1.0 corresponding to the best possible performance, which in turn corresponds to a perfectly conditioned Jacobian matrix.

The workspace of PKMs is another design issue that needs careful attention due to the computational complexity involved. Algorithms proposed in the literature to determine the workspace of PKM structures use the geometric constraints of the structures, including maximum/minimum leg lengths, passive joint limits. The complexity of these computational methods varies depending on the constraints imposed. For example if the cross sectional variation hexapod legs is also considered as a factor to avoid leg collisions considerable computational requirement will be necessary (Conti et. al, 1998). If the considered design would ensure that the operation of the machine is far enough from possibility of leg collisions in the first place considerable design efforts could be saved.

Harib and Sharif Ullah (2008) used the axiomatic design methodology (Suh, 1990) to analyze the kinematic design of PKM structures. In terms of the kinematic functions of PKM structures and based on the aforementioned contemplation, the following basic Functional Requirements (FRs) were identified: (1) The mechanism should be able to support arbitrary 3-D system of forces i.e. PM1 should be as close to unity as possible. (2) The mechanism should be able to support arbitrary 3-D systems of torques i.e. PM2 should be as close to unity as possible. (3) The mechanism should be able to move the cutting tool through a desired workspace. (4) The mechanism should be able to orient the spindle at a desired range within the desired workspace. On the other hand, to achieve the FRs the following two Design Parameters (DPs) are often used: (1) Determine the lengths and strokes of the legs. (2) Determine the orientation of the legs relative to the fixed base and to the moving platform in the home position. From the perspective of AD this implies that the kinematic design of hexapodic machine tools is sort of coupled design. Therefore, gradual decomposition of FRs and DPs are needed to make the system consistent with the AD.

Figure 5 shows a 2-DOF planar parallel kinematics structure. The structure includes two extendable legs with controllable leg lengths l_1 and l_2 and three revolute joints a_1, a_2, and c. The controlled extension of the two legs places the end-effectors point c at an arbitrary position (x, y) in the x-y plane.

The way the function requirements are fulfilled is this design is by assembling the mechanism such that the two legs are orthogonal to each other at the central position of the workspace as shown in Figure 5. This result is coherent with the isotropic configuration that could be obtained for this mechanism (Huang et al., 2004). Away from that position the mechanism is not expected to deviate much from this condition for practical configuration if the limits of the leg lengths are appropriately selected. It is clear that arbitrary strokes and average lengths of the two legs can be selected while maintaining leg orthogonally condition by adjusting the position of b_1 and b_2.

The reachable space of the 2-DOF PKM of Figure 5 is bounded by four circular arc segments with radii $l_{1\text{-}max}$, $l_{1\text{-}min}$, $l_{2\text{-}max}$ and $l_{2\text{-}min}$ and centers b_1 and b_2. With the two legs normal to each other the workspace can be modified along any of the two orthogonal directions independent from the other.

An extension of the previous design method to three DOF planner PKM structures is shown in Figure 6. Selecting the reference point of the mechanism to be the concurrent attachment point of the two orthogonal legs serves the purpose of showing the validity of the previously established result of uncoupled design in terms of the previously defined FRs and DPs. As indicated on Figure 6, with this choice of reference point, the same workspace of the 2-DOF structure is obtained.

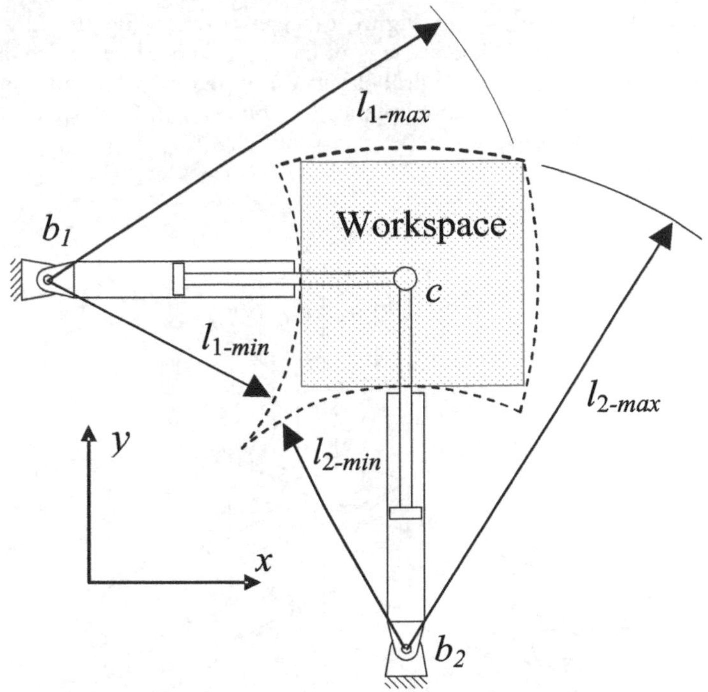

Fig. 5. A 2-DOF Planner PKM System (Harib and Sharif Ullah, 2008)

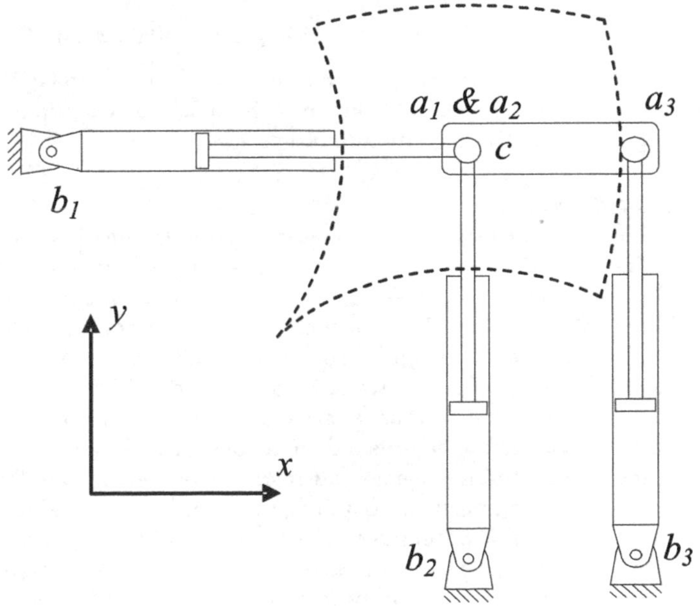

Fig. 6. A 3-DOF Planner PKM System (Harib and Sharif Ullah, 2008)

The previous 3-DOF PKM design of Figure 6 suggests extending the idea to a 6-DOF structure, as shown in Figure 7. The six legs of the suggested structure are arranged such that the idea remains the same (two parallel legs connected by a link and one orthogonal leg) in each of three mutually orthogonal planes. The purpose of the design is to support an arbitrary 6-DOF force and torque system.

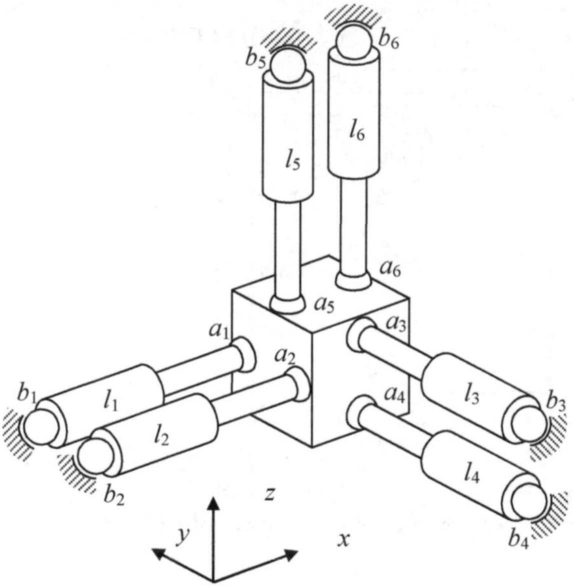

Fig. 7. A schematic of a 6-DOF spatial PKM (Harib and Sharif Ullah, 2008)

While the FRs' and DPs' of the axiomatic design methods are difficult to be decoupled here, this design of the 6 DOF mechanism is shown to be a logical extension from planner mechanisms designed with such design methodology.

3.2 Hybrid kinematic designs

Similar to the serial kinematic robotic design strategy, hybrid kinematic structures could be designed such the first three links and joints, forming the parallel structure, handle the gross positioning of the end-effector. The rest of joints and links could be made to form a concurrent serial kinematic structure that is responsible for orienting the end-effector. Thus this strategy decoupled two main functional requirements (FRs) of the mechanism and their design parameters (DPs). Now, while the serial kinematic part, which is responsible for the orientation of the end-effector, could be a standard wrest joint consisting concurrent revolute joints, the focus could bow be directed on the design of the parallel kinematic part which still requires considerable design attention. The decoupling of the design requirements reduces the design problem to a design of a three-degrees-of-freedom parallel kinematic spatial structure that position the concurrent wrest joint along the x, y and z axes. Although the design requirements on the orientation are not part of the design requirements of the parallel part of the mechanism, ability to support a system of transmitted torque is still part of the design requirements. This is in addition to the requirements of having ability to provide arbitrary motion along three directions and to support associated force system along these directions.

3.2.1 The Exechon mechanism

The Exechon machining center is based on a hybrid five degrees-of-freedom mechanism that consists of parallel and serial kinematic linkages (Zoppi et al., 2010). The parallel kinematic structure of the Exechon is an overconstrained mechanism with eight links and a total of nine joints; three prismatic joints with connectivity one, three revolute joints with connectivity one, and three universal joints with connectivity two. This mechanism is shown schematically in Figure 8.

The number of overconstraint (geometrical) conditions s is 3. These conditions require that the two prismatic joints l_1 and l_2 form a plane, and that the two axes of the joints a_1 and a_2 to be perpendicular to this plane, and the axis of joint a_3 be perpendicular to the axes of joints a_1 and a_2. The parameters of the underlying mechanism can be identified as: $L = 8$, $j = 9$, $\sum f_i$ =12 for all the nine revolute, prismatic and universal joints. The mobility of this mechanism is erroneously calculated by Equation 1 as $M = 0$, which indicates that the mechanism is a structure. Nevertheless, if the geometrical constraints involved in this mechanism are considered and Equation 3 is applied, the mobility is correctly calculated as $M = 3$. These three degrees of freedom obviously correspond to the three actuating linear motors. The overconstraints in this mechanism considerably reduce the required joints, which obviously improves the rigidity of the mechanism. However, the geometric constraints that result in reducing the mobility to three require structural design for the joints to bear the transmitted bending moments and torque components. This requirement is more stringent in the case of the prismatic joints of the three legs. These legs will not be two-force members as in the six DOF hexapodic mechanism and have to be designed to hold bending moments.

The parallel kinematic part can be viewed as a 2-DOF planner mechanism formed by the two struts l_1 and l_2 and the platform, which could be revolved about an axis (the axes of the base joints b_1 and b_2, shown as dashed line in Figure 8) via the actuation of the third strut l_3. To achieve 2-DOF in the planner mechanism, three overconstraints are required. As indicated before these overconstraints come as requirements on the axes of the revolute joints a_1 and a_2 to be normal to the plane formed by l_1 and l_2, and on the third revolute joint a_3 to be normal to the other two joints. Thus the projection of this strut onto the plane is constraining the rotational degree-of-freedom of the moving platform in the plane. This situation resembles the 2-DOF planner mechanism of Figure 2. When this projection onto the plane vanishes (i.e. when the angle between the third strut and the plane made by other two struts is 90 degree), the mechanism becomes singular (attains additional degree-of-freedom).

3.2.2 Alternative hybrid kinematic mechanism

In this section we demonstrate employoing the Axiomatic Design to evaluate a potential design of a 5-axes alternative hybrid kinematic machine tools mechanism consisting of a 3-DOF parallel kinematic structure and a 2-DOF wrest joint. Axiomatic design is a structured design methodology which is developed to improve design activities by establishing criteria on which potential designs may be evaluated and enhanced (Suh, 1990). The general function requirements (FRs) for the proposed hybrid mechanism can be listed as follows. The mechanism should 1) provide required positioning and orientation capabilities, 2) have adequate and consistent dexterity throughout the workspace, 3) have good structural rigidity, and 4) have a large and well shaped workspace. The design parameters (DPs) that could be used to achieve the function requirements concerning the parallel kinematic part of the mechanism include 1) the configuration of the wrest joint, 2) the configuration of the parallel kinematic mechanism, 3) the types of the end joints, and 4) the strokes and average lengths of the legs.

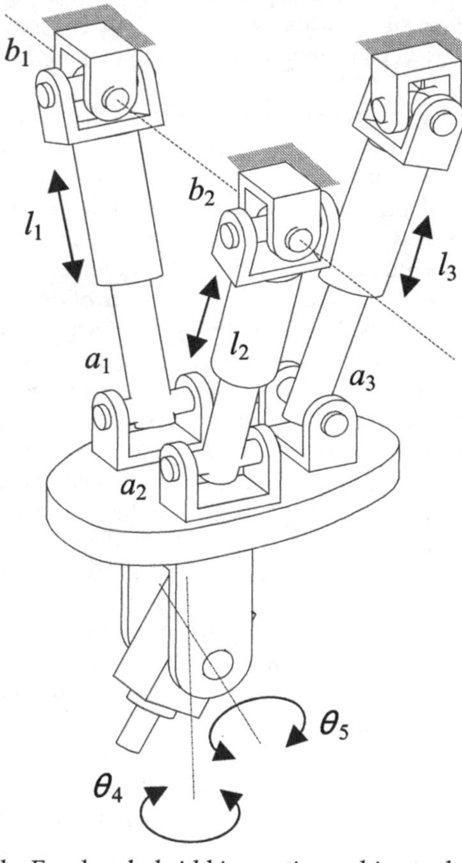

Fig. 8. A schematic of the Exechon hybrid kinematic machine tools mechanism

Based on the discussion in the previous sections and the axiomatic design formulation previously used for planner parallel kinematic structures (Harib and Sharif Ullah, 2008) a kinematic design of an alternative design for a hybrid kinematic machine tools mechanism is proposed. A schematic of the proposed mechanism is depicted in Figure 9 below. The parallel kinematic part has three perpendicular struts when the mechanism is at the center of the workspace, and consists of movable platform and three extendable struts. As shown in Figure 1, the first strut is rigidly connected to the platform, which in turn is connected to other two struts via revolute and universal joints. The struts are connected respectively to the machine frame via universal joints and a spherical joint with connectivity three. The number of overconstraint (geometrical) conditions s is 2. These conditions require that the two prismatic joints l_1 and l_2 form a plane and that the axis of joint a_2 to be perpendicular to this plane. Calculating the mobility using Equation 1 yields $M = 1$. However considering the overconstraints ($s = 2$), the mobility of the mechanism, as calculated by Equation 3, will be $M = 3$.

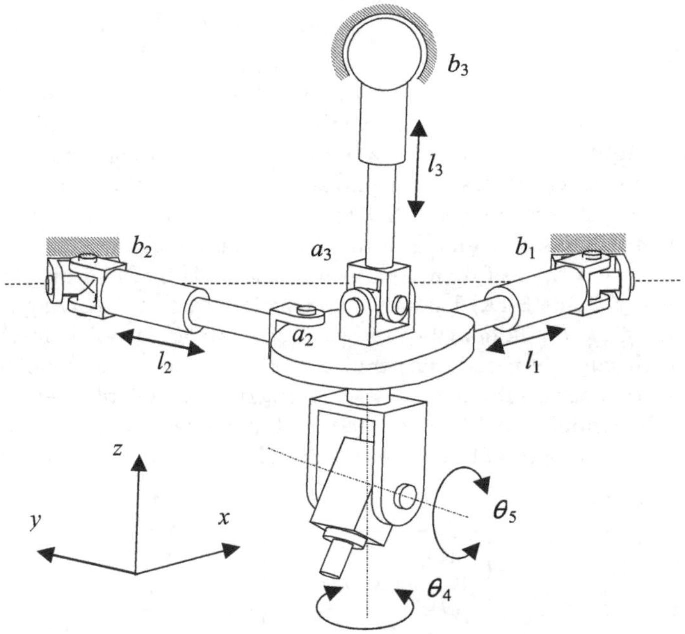

Fig. 9. A schematic of a proposed hybrid machine tools mechanism

In order to reach an optimum design, the Axiomatic Design FRs and DPs are grouped hierarchically. The design problem is also formulated such that the FRs are independent from each other (to fulfill the Independence Axiom), and the DPs are uncoupled at least partially (to fulfill the Information Axiom). Thus, the design strategy is directed to fulfill the FRs using uncoupled DPs first. Figure 10 shows main FRs for a hybrid kinematic mechanism design arranged hierarchically. The fundamental function requirement (FR1 = positioning and orientation capabilities) is split into two independent function requirements (FR11 and FR12) which can be addressed using independent design parameters. FR12 is split into three function requirements (FR121, FR122, FR123). For a given configuration of the parallel kinematic mechanism, the function requirements (FR121, FR122, FR123) can be addressed using the following design parameters:

DP121i: the type of the ith platform end joint a_i

DP122i: the type of the ith base end joint b_i

DP123i: the stroke of ith leg ($l_{i\text{-}max} - l_{i\text{-}min}$)

DP124i: average lengths of the ith leg $(l_{i\text{-}max} + l_{i\text{-}min})/2$

It is worth mentioning here that the joint axes resemble the five axes of the machine tools at the center of the workspace, and could be maintained to be close to this situation by proper design and choice of the leg strokes and mean lengths. Also as an alternative configuration, the 2-DOF wrest joint that hold the spindle could also be replaced by a 2-DOF rotary table, transferring the relative rotational motion to the workpiece. A redundant hybrid structure consisting of a hexapod machine tools and a 2-DOF rotary table is suggested and analyzed by Harib et al. (2007).

4. Conclusions

The considerable interest that is shifted to hybrid kinematic structures to exploit the advantageous features of the serial and parallel kinematic structures and avoiding their drawbacks has brought about some interest in overconstrained hybrid mechanisms. A study on the mobility of the three classes of mechanisms is presented and focuses on the mobility of overconstrained structures in view of their application in parallel and hybrid structures to reduce the number of passive joints. The mobility of the Exechon mechanism is analyzed and discussed as an example of a successful machine tools mechanism. The study of this mechanism reveals that its 3-DOF parallel kinematic part is a revolving 2-DOF planner mechanism. Strategies for kinematic designs of planner parallel mechanisms were developed and discussed based on the axiomatic design methodology. Optimum configurations for planner mechanisms were presented for 2-DOF planner mechanisms and were shown to be extendable to 3-DOF planner and spatial mechanisms by proper choice of joints and constraints. An alternative optimum parallel and hybrid mechanism is discussed and analyzed.

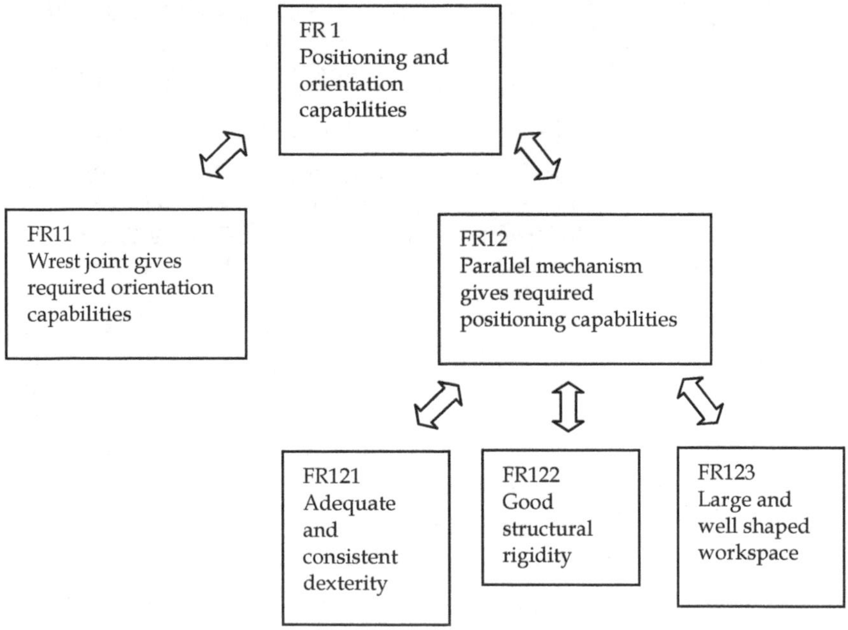

Fig. 10. Main Function Requirements of a hybrid design

5. Acknowledgment

The authors gratefully acknowledge the support of the Emirates Foundation (EF) under Grant No. 2008/052.

6. References

Bi, Z.M., Jin, Y., (2011) Kinematic modeling of Exechon parallel kinematic machine, Journal of Robotics and Computer-Integrated Manufacturing, Vol. 27, Issue 1, February 2011, 186-193

Bohez, E.L.J. (2002) Five-axis milling machine tool kinematic chain design and analysis Int. Journal of Machine Tools & Manufacture 42 Pages 505–520

Cretu, S.M., (2007) TRIZ applied to establish mobility of some mechanisms," Proceedings of the 12th IFToMM World Congress, Besancon, France, 18-21 June

Conti, J.P. Clinton, C.M., Zhang, G. and A. J. Wavering, A.J. (1998). Workspace Variation of a Hexapod Machine Tool, NISTIR 6135, National Institute of Standards and Technology, Gaithersburg, MD, March 1998.

Gogu, G., (2005) Mobility of mechanisms: a critical review," Mechanism and Machine Theory, Vol. 40, p. 1608-1097.

Gosselin, C. (1990). Determination of the Workspace of 6-DOF Parallel Manipulator, Journal of Applied Mechanical Design, 112, 331-336Luh, C., Adkins, F. Haung, E. and Qiu, C. (1996). Working Capapbility Analysis of Stewart Platform Manipulator, ASME Journal of Mechanical Design 118, 220-227.

Harib, K.H., Sharif Ullah, A.M.M. (2008), Axiomatic Design of Hexapod-based Machine Tool Structures, Reliability and Robust Design in Automotive Engineering, SAE/SP-2170: 331-337.

Harib, K.H, Sharif Ullah, A.M.M. and Hammami, H. (2007). A Hexapod-Based Machine Tool with Hybrid Structure: Kinematic Analysis and Trajectory Planning, International Journal of Machine Tools and Manufacture, 47, 1426-1432

Harib, K., Srinivasan, K. (2003) Kinematic and Dynamic Analysis of Stewart Platform-Based Machine Tool Structures, Robotica, 21, (5), 541-554

Huang, T., Li, Li, A., Chetwynd, D.G. and Whitehouse, D.J. (2004). Optimal Kinematic Design of 2-DOF Parallel Manipulators with Well-Shaped Workspace Bounded by a Specified Conditioning Index, IEEE Transactions on Robotics and Automation, 20(3), 538-543

Huang, T., Wang, J., Gosselin, C.M. and Whitehouse, D.J. (2000). Kinematic Synthesis of Hexapods with Specified Orientation Capability and Well-Conditioned Dexterity, Journal of Manufacturing Processes, 2 (1), 36-47.

Ionescu T.G., (2003) Terminology for mechanisms and machine science," Mechanism and Machine Theory, Vol. 38, p. 7-10

Mavroidis C. and Roth B. (1995) Analysis of Overconstrained Mechanisms, Journal of Mechanical Design, Transactions of the ASME, Vol. 117, pp. 69-74.

Norton, R. L., (2004), Design of Machinery, McGraw Hill

Stoughton, R. & Arai, T. (1993). A Modified Stewart Platform Manipulator with Improved Dexterity, IEEE Trans Robotics and Automation, 9 (2), pp. 166-173.

Suh, N.P. (1990). The Principles of Design, New York: Oxford University Press.

Tsai, K.Y. Lee, T.K. and Huang, K.D. (2006). Determining the workspace boundary of 6-DOF parallel manipulators, Robotica, 24(5), 605-611.

Waldron, K.J. and Kinzel, G.L. (1999) Kinematics, Dynamics, and Design of Machinery, John Wiley & Sons

Wang, L.-C.T. and Hsieh, J.-H. (1998). Extreme reaches and reachable workspace analysis of general parallel robotic manipulators, Journal of Robotic Systems 15(3), 145-159.

Zhang, D., Gosselin, C.M.(2002) Kinetostatic Analysis and Design Optimization of the Tricept Machine Tool Family, J. Manufacturing. Science and Engineering., August, Vol. 124, Issue 3, 725-733

Zoppi, M., Zlatanov, D., and Molfino, R. (2010) Kinematics Analysis of the Exechon Tripod, ASME 2010 International Design Engineering Technical Conferences and Computers and Information in Engineering Conference (IDETC/CIE2010), August 15–18, Montreal, Quebec, Canada

5

Kinematic and Dynamic Modelling of Serial Robotic Manipulators Using Dual Number Algebra

R. Tapia Herrera, Samuel M. Alcántara,
J.A. Meda-Campaña and Alejandro S. Velázquez
Instituto Politécnico Nacional
México

1. Introduction

The kinematic and dynamic modelling of robotic manipulators has, as a specific field of robotics, represented a complex problem. To deal with this, the researchers have based their works on a great variety of mathematical theories (Seiling, 1999). One of these tools is the Dual Algebra, which is a concept originally introduced in 1893 by William Kingdon Clifford (Fisher, 1998; Funda, 1988). A dual number is a compact form that can be used to represent the rigid body motion in the space (Keller, 2000; Pennestrí & Stefanelli, 2007), it has therefore, a natural application in the analysis of spatial mechanisms specifically mechanical manipulators (Bandyopadhyay, 2004, 2006; Bayro-Corrochano & Kähler, 2000, 2001).

Several works related to dual-number in kinematics, dynamics and synthesis of mechanisms have been developed (Cheng, 1994; Fisher, 1998, 1995, 2003) in (Moon & Kota, 2001) is presented a methodology for combining basic building blocks to generate alternative mechanism concepts. The methodology is based on a mathematical framework for carrying out systematic conceptual design of mechanisms using dual vector algebra. The dual vector representation enables separation of kinematic function from mechanism topology, allowing a decomposition of a desired task into subtask, in order to meet either kinematic function or spatial constraints. (Ying et al, 2004) use dual angles as an alternative approach to quantify general spatial human joint motion. Ying proposes that dual Euler angles method provides a way to combine rotational and translational joint motions in Cartesian Coordinate systems, which can avoid the problems caused by the use of the joint coordinate system due to non-orthogonality. Hence the dual angles method is suitable for analyzing the motion characteristics of the ankle joint. The motion at the ankle joint complex involves rotations about and translations along three axes. In the same field of biomechanics (KiatTeu et al, 2006) present a method that provides a convenient assessment of golf-swing effectiveness. The method can also be applied to other sports to examine segmental rotations. In general, this method facilitates the study of human motion with relative ease. The use of a biomechanical model, in conjunction with dual-number coordinate transformation for motion analysis, was shown to provide accurate and reliable results. In particular, the advantage of using the dual Euler angles based on the dual-number coordinate transformation approach, is that it allows, for a complete 3D motion, to use only six parameters for each anatomical joint. KiatTeu infers that the method has proved to be an effective means to examine high-speed movement in 3D space. It also provides an option in assessing the contributions of the individual segmental rotations in production of the relevant velocity of the end-effector.

(Page et al, 2007) present the location of the instantaneous screw axis (ISA) in order to obtain useful kinematic models of the human body for applications such as prosthesis and orthoses design or even to help in disease diagnosis techniques. Dual vectors are used to represent and operate with kinematic screws with the purpose of locating the instantaneous screw axes which characterize this instantaneous motion. A photogrammetry system based on markers is used to obtain the experimental data from which the kinematic magnitudes are obtained. A comprehensive analysis of the errors in the measurement of kinematic parameters was developed, obtaining explicit expressions for them based on the number of markers and their distribution.

1.1 Dual-number representation in robotics

The dual-number representation has been extended to other fields of mechanics; rigid body mechanics is an area where the dual number formulation has been applied, especially in the kinematics and dynamics of mechanisms.

The homogeneous transformation is a point transformation; in contrast, a line transformation can also naturally be defined in 3D Cartesian space, in which the transformed element is a line in 3D space instead of a point. In robotic kinematics and dynamics, the velocity and acceleration vectors are often the direct targets of analysis. The line transformation will have advantages over the ordinary point transformation, since the combination of the linear and angular quantities can be represented by lines in 3D space. Since a line in 3D space is determined by four independent parameters. (Gu, 1988) presents a procedure that, offers an algorithm which deals with the symbolic analysis for both rotation and translation. In particular, the aforementioned is effective for the direct determination of Jacobian matrices and their derivatives. The dual-number transformation procedure, based on these properties and the principle of transference, can be used for finding Jacobian matrices in robotic kinematics and their derivatives in robotic dynamics and control modeling. A related work was performed in (Pennock & Mattson, 1996) where the forward position problem of two PUMA robots manipulating a planar four bar linkage payload is solved using closed-form solutions for the remaining unknown angular displacements based in orthogonal transformation matrices with dual-number. (Brosky & Shoham, 1998; Sai-Kai, 2000) introduce the generalized Jacobian matrix which consists of the complete dual transformation matrices. The generalized Jacobian matrix relates force and moment at the end-effector to force and moment at the joints for each axe. Furthermore, the generalized Jacobian matrix also relates motion in all directions at the joints to the motion of the end-effector, an essential relation required at the design stage of robot manipulators. An extension of these kinematics and statics schemes into dynamics is possible by applying the dual inertia operator. (Brodsky, 1999) formulated the representation of rigid body dynamic equations introducing the dual inertia operator. Brodsky gives a general expression for the three-dimensional dynamic equation of a rigid body with respect to an arbitrary point. Then the dual Lagrange equation is established by developing derivative rules of a real function with respect to dual variables.

(Bandyopadhyay & Ghosal, 2004) performs a study in order to determinate principal twist of the end-effector of a multi-degree of freedom manipulator, which plays a central role in the analysis, design, motion planning and determination of singularities.
(Yang & Wang, 2010) solve the direct and inverse kinematics of a SCARA robot. They proved that the dual number allows compact formulations considerably facilitating the analysis of robot kinematics. To deal with coordinate transformation in three dimensional Cartesian space, the homogeneous transformation is usually applied. It is defined in the four-dimensional space and its matrix multiplication performs the simultaneous rotation and translation.

2. Mathematical preliminaries

Let us consider a transformation of coordinates between the Cartesian Coordinate system (x,y) and the oblique coordinate system (u,v) given by the equations:

$$x = au + bv; \quad y = 0u + av \tag{1}$$

With a, b real numbers. The geometry is depicted in Figure 1.
It is well known that the point (u,v) is localized by the vector $\vec{r} = x\hat{\imath} + y$ from the origin of the Cartesian coordinate. From the transformation (1):

$$\vec{r} = (au + bv)\hat{\imath} + avj \tag{2}$$

The tangent vectors to u and v are:

$$\frac{\partial \vec{r}}{\partial u} = ai; \quad \frac{\partial \vec{r}}{\partial v} = bi + aj \tag{3}$$

From the obvious $\hat{e}_u \cdot \hat{e}_v = \frac{b}{\sqrt{b^2 + a^2}}$ it is clear that the coordinates (u, v) are not orthogonal.
The unit vectors of the Cartesian frame can be written in the form of column vectors:

$$i = \begin{pmatrix} 1 \\ 0 \end{pmatrix}; \quad j = \begin{pmatrix} 0 \\ 1 \end{pmatrix} \tag{4}$$

We can describe the oblique frame (u,v) in terms of the tangent vectors ai and bi+aj written as a column vectors:

$$ai = \begin{pmatrix} a \\ 0 \end{pmatrix} \quad ; \quad bi + aj = \begin{pmatrix} b \\ a \end{pmatrix} \tag{5}$$

The column vectors $\begin{bmatrix} 1 & 0 \end{bmatrix}^T$, $\begin{bmatrix} 0 & 1 \end{bmatrix}^T$ can be combined into a single matrix describing the Cartesian Frame:

$$\text{Cartesian Frame: } \begin{pmatrix} 1 & 0 \\ 0 & 1 \end{pmatrix}$$

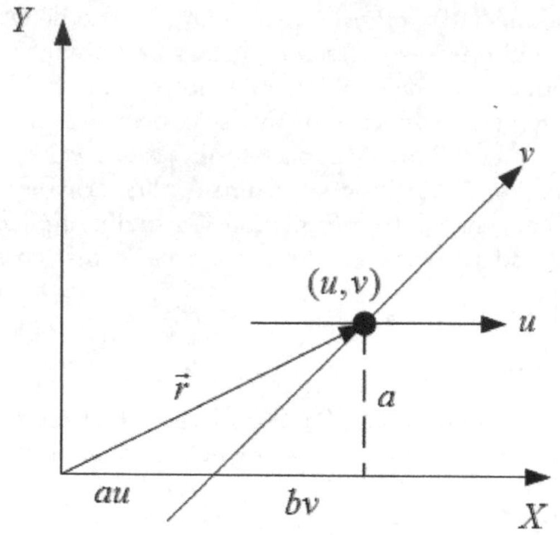

Fig. 1. Oblique plane

The column vectors $[a \quad 0]^T \ [b \quad a]^T$ can be combined into a single matrix describing the oblique frame:

$$\text{Oblique Frame: } \begin{pmatrix} a & b \\ 0 & a \end{pmatrix}$$

The matrix $\begin{pmatrix} a & b \\ a & a \end{pmatrix}$ is the transformation matrix that describes the Oblique Frame relative to the Cartesian Frame.

The matrix $\begin{pmatrix} a & b \\ a & a \end{pmatrix}$ can be decomposed as:

$$\begin{pmatrix} a & b \\ 0 & a \end{pmatrix} = \begin{pmatrix} 1 & 0 \\ 0 & 1 \end{pmatrix} a + b \begin{pmatrix} 0 & 1 \\ 0 & 0 \end{pmatrix} \tag{6}$$

Where $\begin{pmatrix} 1 & 0 \\ 0 & 1 \end{pmatrix} = 1$ is the unitary matrix and $\begin{pmatrix} 0 & 1 \\ 0 & 0 \end{pmatrix} = \varepsilon$ is a matrix with the following properties:

a. ε is nilpotent:

$$\varepsilon^2 = \begin{pmatrix} 0 & 1 \\ 0 & 0 \end{pmatrix}^2 = \begin{pmatrix} 0 & 1 \\ 0 & 0 \end{pmatrix}\begin{pmatrix} 0 & 1 \\ 0 & 0 \end{pmatrix} = \begin{pmatrix} 0 & 0 \\ 0 & 0 \end{pmatrix} = 0$$

b. ε is a ninety degree rotation operator

$$\varepsilon j = \begin{pmatrix} 0 & 1 \\ 0 & 0 \end{pmatrix}\begin{pmatrix} 1 \\ 0 \end{pmatrix} = i$$

Finally expression (6) is written in the form:

$$\begin{pmatrix} a & b \\ 0 & a \end{pmatrix} = a + \varepsilon b \tag{7}$$

Equation (7) is easily extended to 3D:

$$1 = \begin{pmatrix} 1 & 0 & 0 \\ 0 & 1 & 0 \\ 0 & 0 & 1 \end{pmatrix}; \quad \varepsilon = \begin{pmatrix} 0 & 0 & 1 \\ 0 & 0 & 0 \\ 0 & 0 & 0 \end{pmatrix}$$

Study in 1903 called the expression $a + \varepsilon b$ "dual number" because it is constructed from the pair of real numbers (a, b). A dual number is usually denoted in the form:

$$\hat{a} = a + \varepsilon b \tag{8}$$

The algebra of dual numbers has been originally conceived by William Kingdon Clifford. Elementary operation of addition is defined as: $\left(\hat{a} = a + \varepsilon a_0; \quad \hat{b} = b + \varepsilon b_0\right)$

$$\hat{a} + \hat{b} = (a + b) + \varepsilon(a_0 + b_0) \tag{9}$$

Multiplication is defined as:

$$\hat{a}\hat{b} = (a + \varepsilon a_0)(b + \varepsilon b_0) = ab + \varepsilon(ab_0 + a_0 b) \tag{10}$$

Division is defined as:

$$\frac{\hat{a}}{\hat{b}} = \frac{a + \varepsilon a_0}{b + \varepsilon b_0} = \left(\frac{a + \varepsilon a_0}{b + \varepsilon b_0}\right)\left(\frac{b + \varepsilon b_0}{b + \varepsilon b_0}\right) = \frac{a}{b} + \varepsilon \frac{a_0 b - a b_0}{b^2} \tag{11}$$

Division by a pure dual number (εa_0) is no defined. It immediately follows that:

$$\hat{a}^n = (a + \varepsilon a_0)^n = a^n + \varepsilon n a_0 a^{n-1} \tag{12}$$

$$\sqrt{\hat{a}} = \sqrt{a} + \varepsilon \frac{a_0}{a\sqrt{a}} \tag{13}$$

A function F of a dual variable $\hat{x} = x + \varepsilon x_0$ can be represented in the form:

$$F(\hat{x}) = f(x, x_0) + \varepsilon g(x, x_0) \tag{14}$$

Where f and g are real functions of real variables x & x_0. The necessary and sufficient conditions in order that F be analytic are:

$$\frac{\partial f}{\partial x_0} = 0; \quad \frac{\partial f}{\partial x} = \frac{\partial g}{\partial x_0} \tag{15}$$

From these it immediately follows the Taylor Series expansion:

$$f(\hat{x}) = f(x + \varepsilon x_0) = f(x) + \varepsilon x_0 \frac{\partial f(x)}{\partial x} \tag{16}$$

Because $\varepsilon^2 = 0$, $\varepsilon^3 = 0$, $\varepsilon^4 = 0$ and so on, all formal operation of dual number are the same as those of ordinary algebra.

2.1 The dual angle

The dual angle represents the relative displacement and orientation between two lines L_1 and L_2 in the 3D space (Figure 2).

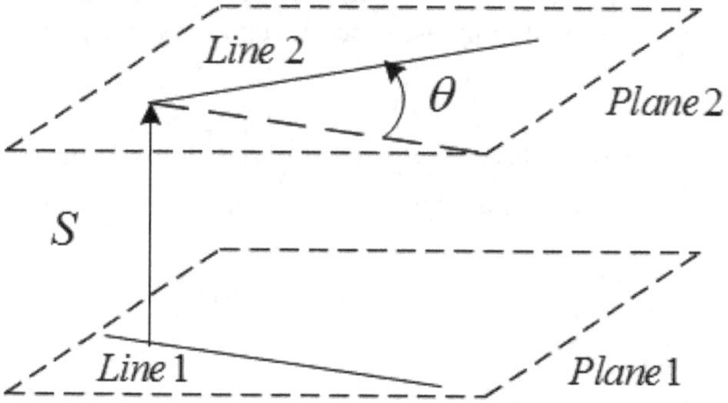

Fig. 2. Geometric representation of a dual angle.

The dual angle is defined as:

$$\hat{\theta} = \theta + \varepsilon S \tag{17}$$

This concept was introduced by Study in 1903. θ the primary component of $\hat{\theta}$ is the projected angle between L_1 and L_2. S the dual component of $\hat{\theta}$ is the shortest distance between lines L_1 & L_2 (as is obvious S is the length of common perpendicular to plane 1 and plane 2. Table 1 summarizes some properties:

$\hat{c} = \hat{a} + \hat{b}$	$\hat{c} = (a+b) + \varepsilon(a_0 + b_0)$
$\hat{c} = \hat{a}\hat{b}$	$\hat{a}\hat{b} = (a + \varepsilon a_0)(b + \varepsilon b_0)$
$\hat{a}^n$	$\hat{a}^n = a^n + \varepsilon n a_0 a^{n-1}$
$\sqrt{\hat{a}}$	$\sqrt{\hat{a}} = \sqrt{a} + \varepsilon \frac{a_0}{2\sqrt{a}}$
$\hat{a}/\hat{b}$	$\hat{c} = \frac{a}{b} + \varepsilon \frac{a_0 b - a b_0}{b^2}$
$f(\hat{a})$	$f(\hat{a}) = f(a) + \varepsilon a_0 \frac{df(a)}{da}$

Table 1. Basic dual algebra operations

In particular a dual angle is an advantageous tool to represent the coordinates of a rigid body in the space relative to other rigid body, if two planes are parallel and exists a line in each plane, dual angle will be the distance between the planes and the angle produced by the projection of one of the lines onto plane, thus a dual angle is used to describe each one of a robot's joints as a cylindrical one, which means that the entire topology is formulated as a set of dual angles (Fisher, 1995; Cecchini et al, 2004).

2.2 Dual vectors

A dual vector $\hat{V} = \vec{V} + \varepsilon(\vec{r} \times \vec{V})$ is a vector constrained to lie upon a given line L in 3D space. The primary component $\vec{V}$ is called the "resultant vector" and comprises the magnitude and direction of dual vector $\hat{V}$. It is independent of the location frame origin. The dual component $\vec{r} \times \vec{V}$ is called the "moment vector". The vector $\vec{r}$ is the position vector from the frame origin to any point on the line L of dual vector $\vec{V}$. $\vec{r} \times \vec{V}$ is invariant for any choice of point on line L, it depends on the choice of the frame origin. (Figure 3).

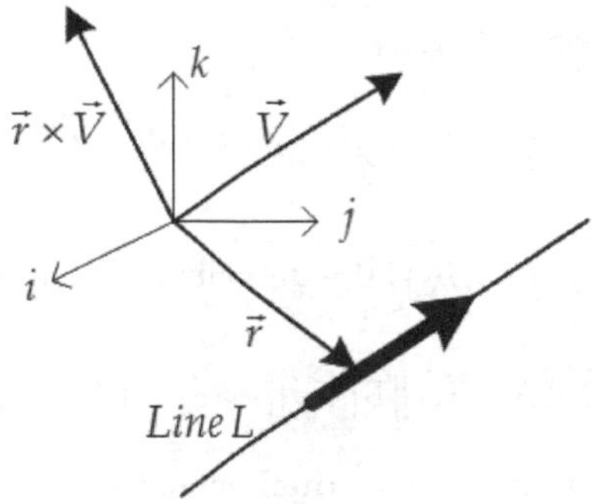

Fig. 3. Dual vector

Among the important dual vectors are:

1. Velocity defined as: $\hat{V} = (\omega + \varepsilon v)\hat{u}$
2. Linear momentum: $\hat{p} = \vec{p} + \varepsilon \vec{r} \times \vec{p}$
3. Force: $\hat{F} = \vec{F} + \varepsilon \vec{r} \times \vec{F}$
4. Angular momentum: $\hat{L} = \vec{L} + \varepsilon \vec{r} \times \vec{L}$

Important dual rotations around and along z, y, x axis are (Figure 4):

$$\hat{R}_{x,\hat{\alpha}} = \begin{bmatrix} 1 & 0 & 0 \\ 0 & \cos\hat{\alpha} & -\sin\hat{\alpha} \\ 0 & \sin\hat{\alpha} & \cos\hat{\alpha} \end{bmatrix}$$

$$\hat{R}_{y,\hat{\phi}} = \begin{bmatrix} \cos\hat{\phi} & 0 & sen\hat{\phi} \\ 0 & 1 & 0 \\ -sen\hat{\phi} & 0 & \cos\hat{\phi} \end{bmatrix} \qquad \hat{R}_{z,\hat{\theta}} = \begin{bmatrix} \cos\hat{\theta} & -\sin\hat{\theta} & 0 \\ sen\hat{\theta} & \cos\hat{\theta} & 0 \\ 0 & 0 & 1 \end{bmatrix}$$

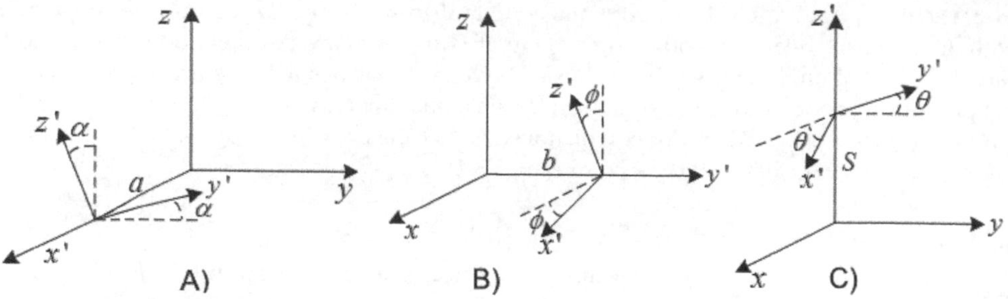

Fig. 4. Dual rotations: A) around and along x, B) around and along y, C) around and along z.

2.3 Algebra of dual vectors and matrices

Let: $\hat{A} = \vec{a} + \varepsilon(\vec{r}_1 \times \vec{a})$ and $\hat{B} = \vec{b} + \varepsilon(\vec{r}_2 \times \vec{b})$:

$$\hat{A} + \vec{B} = \vec{a} + \vec{b} + \varepsilon(\vec{r}_1 \times \vec{a} + \vec{r}_2 \times \vec{b})$$

Definition of dot and cross products are:

$$\hat{A} \cdot \hat{B} = ab\cos\hat{\theta}; \quad \hat{A} \times \hat{B} = ab\sin\hat{\theta}\hat{S}$$

Product of two dual matrices:

Let $\left[\hat{A}\right] = [A] + \varepsilon[A_0]$ & $\left[\hat{B}\right] = [B] + \varepsilon[B_0]$, the definition of their dual product is:

$$\left[\hat{A}\right]\left[\hat{B}\right] = [A][B] + \varepsilon\{[A][B_0] + [A_0][B]\}$$

It is similar with the multiplication rule for dual numbers. The inverse of a square matrix is defined as:

$$\left[\hat{A}\right]\left[\hat{A}\right]^{-1} = [I]$$

$$\left[\hat{A}\right]^{-1} = [A]^{-1} - \varepsilon\left\{[A]^{-1}[A_0][A]^{-1}\right\}$$

3. Denavit – Hartenberg parameters

Mechanisms analysis is facilitated by fixing a coordinate system on each link in a specific manner. With reference to Figure 5, a coordinate frame $\{i+1\}$ is fixed on the distal end of a link i joining joints i and $i+1$ such that:

$\hat{k}_{i+1}$ axis coincident with axis of joint $i+1$

$\hat{\imath}_{i+1}$ axis coincident with shortest distance between axes $\hat{k}_i$ & $\hat{k}_{i+1}$

$\hat{\imath}_{i+1}$ axis perpendicular to both axes $\hat{\imath}_{i+1}$ & $\hat{k}_{i+1}$

The origin of frame $\{i+1\}$ is located at the intersection of axes $\hat{k}_{i+1}$ and i_{i+1}. Frame $\{i\}$ is fixed on the previous link $i-1$. Translation S_i is the distance from point i, the origin of frame $\{i\}$ to the line segment whose length a_i is the shortest distance between joint axes $\hat{k}_i$ and $\hat{k}_{i+1}$. That line segment of shortest distance between join axes intersects axis $\hat{k}_{i+1}$ at point $i+1$, the origin of frame $\{i+1\}$ fixed on the distal end of link i. The projected angle between axes $\hat{k}_i$ and $\hat{k}_{i+1}$ represent the twist α_i of link i. The values α, θ, a, S are the well-known Denavit-Hartenberg parameters.

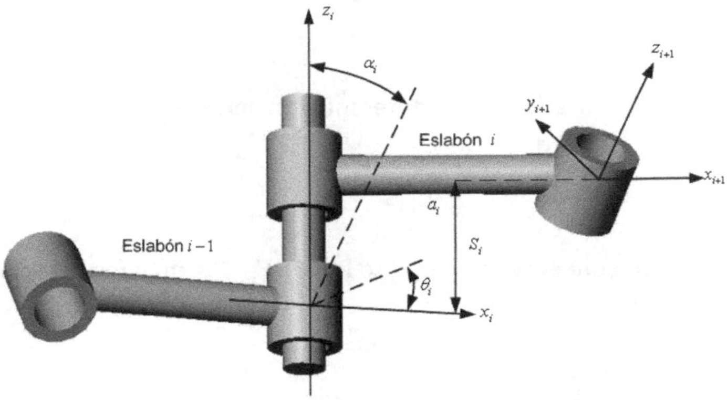

Fig. 5. Denavit and Hartenberg parameters (Pennestrí & Stefanelli, 2007)

A dual matrix rotation that represent the necessary motions of frame $\{i\}$ in terms of an attached $\{i-1\}$ frame is the composition of rotation $R_x(\hat{\alpha}_i)$ and rotation $R_z(\hat{\theta}_i)$, i.e.

$$^{i-1}\hat{M}_i = \left(\hat{R}_{z,\hat{\theta}}\right)\left(\hat{R}_{x,\hat{\alpha}}\right) = \begin{bmatrix} \cos\hat{\theta}_i & -\cos\hat{\alpha}_i \sin\hat{\theta}_i & \sin\hat{\alpha}_i \sin\hat{\theta}_i \\ \sin\hat{\theta}_i & \cos\hat{\alpha}_i \cos\hat{\theta}_i & -\sin\hat{\alpha}_i \cos\hat{\theta}_i \\ 0 & \sin\hat{\alpha}_i & \cos\hat{\alpha}_i \end{bmatrix} \tag{18}$$

So the open chain dual equation is:

$$^{0}\hat{M}_n = \prod_{i=1}^{n} {}^{i-1}\hat{M}_i \tag{19}$$

According with Funda the dual rotation matrix $^{0}\hat{M}_n = {}^{0}T_n + \varepsilon {}^{0}D_n$

The above expression is useful for modeling prismatic, rotational and cylindrical joints, this represents a main advantage respecting to real numbers, to represent the relative position of a point respecting an inertia frame an alternative is establishing the representation of Denavit and Hartenberg trough the dual angles θ, α:

4. Dual Jacobian matrix

If a point P on a body j is moving with respect to a body I (Fig. 6), the velocity can be expressed in terms of inertial frame R $\left({}^{R}\hat{V}^{P}_{j,i}\right)$.

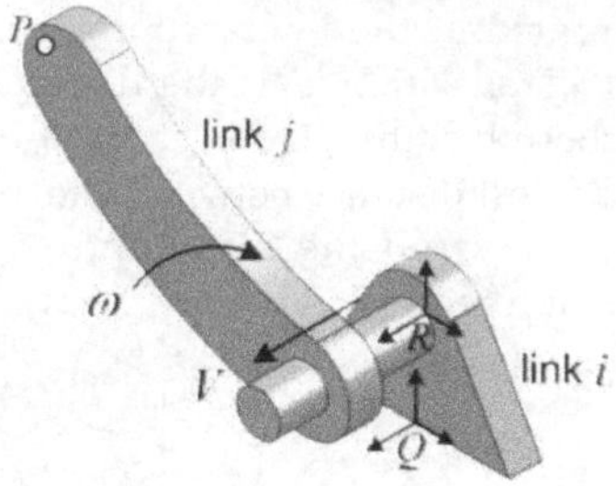

Fig. 6. Dual velocity scheme.

When the dual velocity needs to be represented in terms of frame Q, a rotation from frame R is done:

$$^{Q}\hat{V}_{j,i}^{P} = {}^{Q}T_{p}\,{}^{P}\hat{V}_{j,i}^{P} \tag{20}$$

The relative velocity of a link k with respect to link i $\left(\hat{V}_{k,i}\right)$ in dual form is established as:

$$\hat{V}_{k,i} = \hat{V}_{j,i} + \hat{V}_{k,j} \tag{21}$$

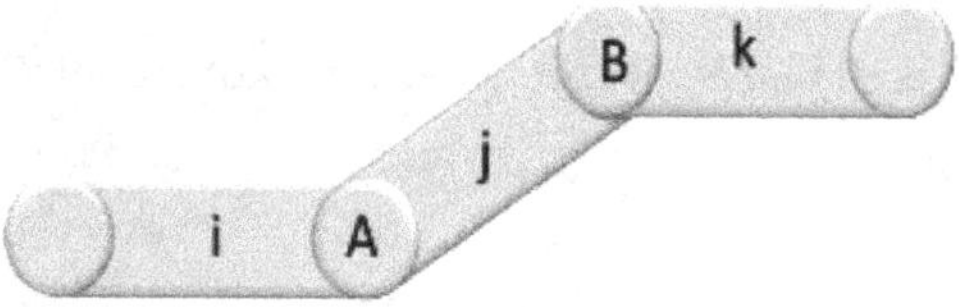

Fig. 7. Relative dual velocity theorem in a kinematic chain.

From dual velocities theorem, the vector of dual velocities in the end of n link in terms of the n frame can be found as:

$$^{0}\hat{V}_{n,0}^{n} = {}^{0}T_{n}\sum_{i=1}^{n}{}^{n}\hat{M}_{i-1}\,{}^{i-1}\hat{V}_{i,i-1}^{i-1} \tag{22}$$

Where $^{0}T_{n}$ is the primary component of the dual matrix(19).
The generalized dual Jacobian matrix is obtained by applying the relative velocity theorem in dual form. The differential motions, whether axial or radial, are expressed in a matrix formed by the dual homogenous matrices, in contrast with the conventional Jacobian matrix that is obtained from specific columns of homogeneous transformation matrix (Sai-Kai, 2000).

$$^{0}\hat{V}_{n,0}^{n} = \begin{bmatrix} \omega_x + \varepsilon V_x \\ \omega_y + \varepsilon V_y \\ \omega_z + \varepsilon V_z \end{bmatrix} = \left[{}^{0}T_{n}\right]\left[{}^{n}\hat{M}_{0} \quad {}^{n}\hat{M}_{1} \quad {}^{n}\hat{M}_{2}\ldots \quad {}^{n}\hat{M}_{n-1}\right]\begin{bmatrix} {}^{0}\hat{V}_{1,0}^{0} \\ \hline {}^{1}\hat{V}_{2,1}^{1} \\ \hline \vdots \\ {}^{n-1}\hat{V}_{n,n-1}^{n-1} \end{bmatrix} \tag{23}$$

The block matrix $\left[{}^n\hat{M}_0 \quad {}^n\hat{M}_1 \quad {}^n\hat{M}_2 \ldots \quad {}^n\hat{M}_{n-1}\right]$ is called Dual Jacobian matrix (Brodsky).

5. Dynamic analysis: Dual force

One of the most important features of dual number formulation is the capability of generalization for a great variety of robot topologies, without modifying the main program, this is an advantage when compared to typical homogenous matrices wherein is required to specify in dynamical model whether a joint is rotational or prismatic.
In dual algebra, if a force and a momentum act with respect a coordinate system, they can be represented in an expression called dual force:

$$\hat{F} = \vec{F} + \varepsilon\vec{\tau} \tag{24}$$

A clear example would be a screwdriver where is necessary to apply a force axially and around to screw.

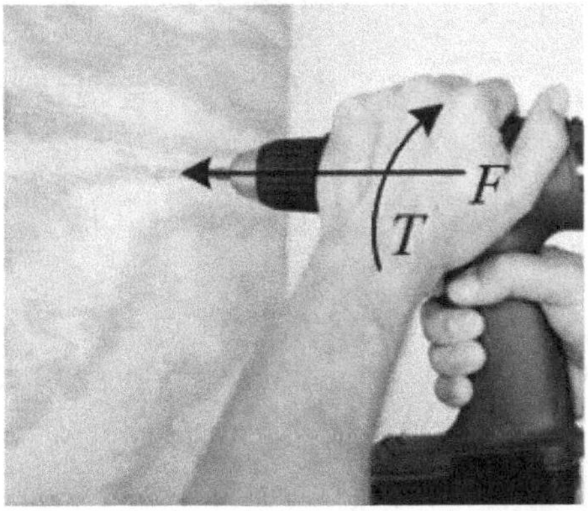

Fig. 8. Example of dual force

If a dual force is applied on a point "B" different to the origin point "A", the effect on the point "B" will be determined by a coordinate transformation. Then a dual force applied on "A" in terms of the frame "B" is given by:

$${}^B\hat{F}_A = {}^BT_A\,{}^A\vec{F}_A + \varepsilon\,{}^BT_A\,{}^A\vec{\tau}_A \tag{25}$$

5.1 Dual momentum

Dual momentum concept is introduced due to the acceleration is a dual pseudo-vector, that means that it can not be established as a dual vector.

$${}^B\hat{H}_A = \int_A \left({}^B\vec{P}_p + \varepsilon\,{}^B\vec{H}_p\right) \tag{26}$$

The terms ${}^B\vec{P}_p$ & ${}^B\vec{H}_p$ are the linear and angular momentum of a particle "p" on a body "A" respectively, in terms of frame "B".

$$ {}^{B}P_A = m_A\left[{}^{B}V_A^B\right] - \left[{}^{B}S_A^B\right]\left[{}^{B}\omega_A^B\right] $$

$$ {}^{B}H_A = \left[{}^{B}S_A^B\right]\left[{}^{B}V_A^B\right] + \left[{}^{B}J_A^B\right]\left[{}^{B}\omega_A^B\right] $$

5.2 Dual inertial force

According with (Pennock & Meehan, 2000) the dual inertial forces on a rigid body are the derivative of the dual momentum:

$$ {}^{B}\hat{f}_A = \frac{d}{dt}\left({}^{B}\hat{H}_A\right) \tag{27} $$

$$ {}^{B}\hat{f}_A = \begin{bmatrix} {}^{B}\dot{\hat{H}}_{Ai} \\ {}^{B}\dot{\hat{H}}_{Aj} \\ {}^{B}\dot{\hat{H}}_{Ak} \end{bmatrix} + \begin{bmatrix} 0 & -{}^{B}\hat{V}_{Ak}^{B} & {}^{B}\hat{V}_{Aj}^{B} \\ {}^{B}\hat{V}_{Ak}^{B} & 0 & -{}^{B}\hat{V}_{Ai}^{B} \\ -{}^{B}\hat{V}_{Aj}^{B} & {}^{B}\hat{V}_{Ai}^{B} & 0 \end{bmatrix} \begin{bmatrix} {}^{B}\hat{H}_{Ai} \\ {}^{B}\hat{H}_{Aj} \\ {}^{B}\hat{H}_{Ak} \end{bmatrix} $$

$$ \begin{bmatrix} {}^{B}\hat{f}_{Ai} \\ {}^{B}\hat{f}_{Aj} \\ {}^{B}\hat{f}_{Ak} \end{bmatrix} = \begin{bmatrix} {}^{B}\dot{\hat{H}}_{Ai} - {}^{B}\hat{V}_{Ak}^{B}\,{}^{B}\hat{H}_{Aj} + {}^{B}\hat{V}_{Aj}^{B}\,{}^{B}\hat{H}_{Ak} \\ {}^{B}\dot{\hat{H}}_{Aj} - {}^{B}\hat{V}_{Ai}^{B}\,{}^{B}\hat{H}_{Ak} + {}^{B}\hat{V}_{Ak}^{B}\,{}^{B}\hat{H}_{Ai} \\ {}^{B}\dot{\hat{H}}_{Ak} - {}^{B}\hat{V}_{Aj}^{B}\,{}^{B}\hat{H}_{Ai} + {}^{B}\hat{V}_{Ai}^{B}\,{}^{B}\hat{H}_{Aj} \end{bmatrix} $$

5.3 Dynamic equilibrium

Extending the D'Alembert principle to dual numbers for dynamic equilibrium

$$ {}^{B}\hat{M}_A\,{}^{A}\hat{F}_A - {}^{B}F_B = {}^{B}\hat{f}_A \tag{28} $$

6. Example: Robot with cylindrical joints

The robot shown in the Figure 9 is a clear example where the dual numbers can be employed:

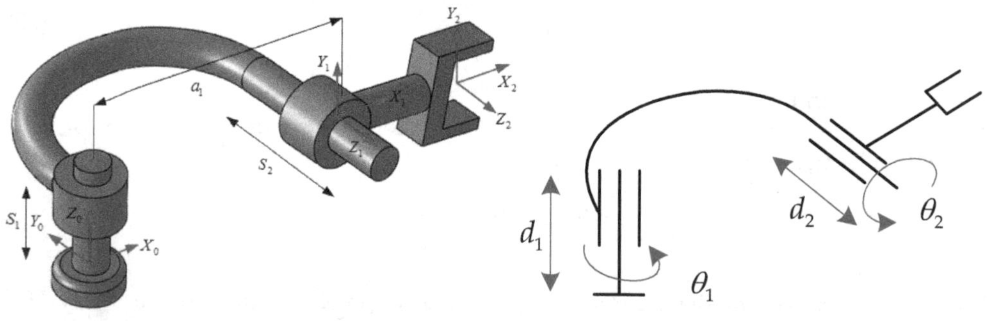

Fig. 9. Assignment of reference systems and Denavit-Hartenberg parameters.

i	θ_i	S_i	a_i	α_i
1	θ_1	d_1	l_1	90°
2	θ_2	d_2	l_2	0°

Table 2. Denavith and Hartemberg parameters of 2C robotic arm.

From Table 2, the dual angles $\hat{\theta}$ & $\hat{\alpha}$ are constructed as:

$$\hat{\theta}_1 = \theta_1 + \varepsilon d_1 \qquad \hat{\theta}_2 = \theta_2 + \varepsilon d_2 \qquad \hat{\alpha}_1 = \alpha_1 + \varepsilon a_1 \qquad \hat{\alpha}_2 = \alpha_2 + \varepsilon a_2$$

It is observed that different topologies can be solved from the assigned values to $\theta_1, d_1, \theta_2, d_2$; for example if θ_1 is 0 the robot will change the original topology CC to PC then nine different robots can be solved from the same aforementioned equations.

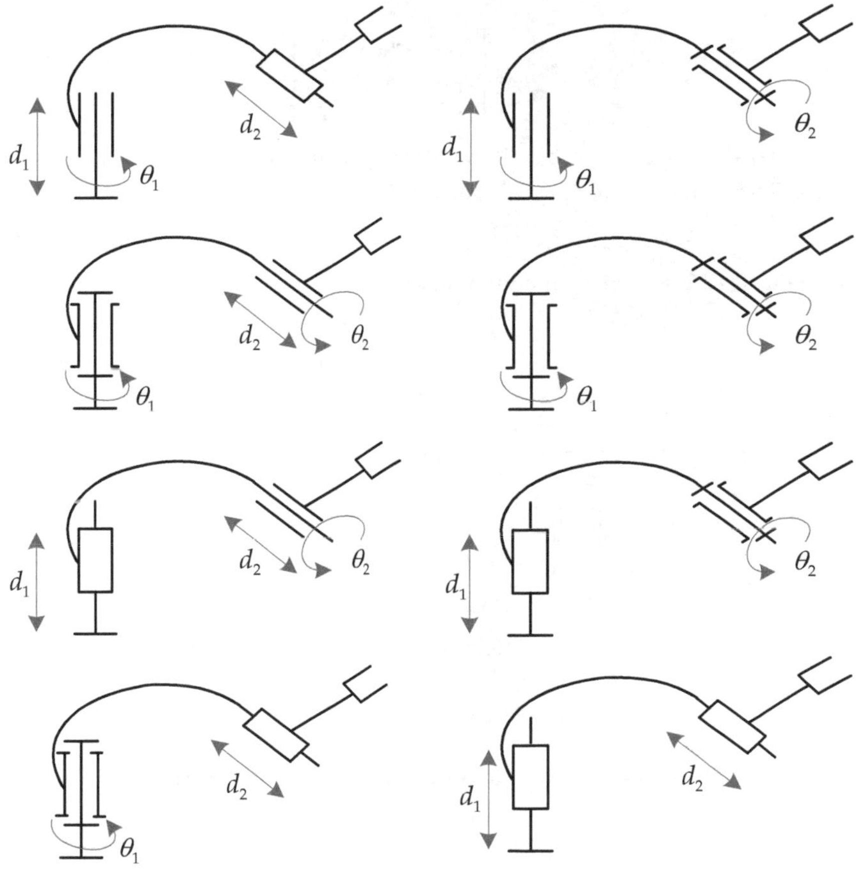

Fig. 10. Possible topologies for different values of θ_1 & θ_2.

$$ {}^{0}\hat{M}_2 = \begin{bmatrix} \cos\hat{\theta}_1 & -\cos\hat{\alpha}_1\sin\hat{\theta}_1 & \sin\hat{\alpha}_1\sin\hat{\theta}_1 \\ \sin\hat{\theta}_1 & \cos\hat{\alpha}_1\cos\hat{\theta}_1 & -\sin\hat{\alpha}_1\cos\hat{\theta}_1 \\ 0 & \sin\hat{\alpha}_1 & \cos\hat{\alpha}_1 \end{bmatrix} \begin{bmatrix} \cos\hat{\theta}_2 & -\cos\hat{\alpha}_2\sin\hat{\theta}_2 & \sin\hat{\alpha}_2\sin\hat{\theta}_2 \\ \sin\hat{\theta}_2 & \cos\hat{\alpha}_2\cos\hat{\theta}_2 & -\sin\hat{\alpha}_2\cos\hat{\theta}_2 \\ 0 & \sin\hat{\alpha}_2 & \cos\hat{\alpha}_2 \end{bmatrix} $$

For the inverse solution:

$$ {}^{0}\hat{M}_2 = \begin{bmatrix} n_x & o_x & a_x \\ n_y & o_y & a_y \\ n_z & o_z & a_z \end{bmatrix} + \varepsilon \begin{bmatrix} -p_z n_y + p_y n_z & -p_z o_y + p_y o_z & -p_z a_y + p_y a_z \\ p_z n_x - p_x n_z & p_z o_x - p_x o_z & p_z a_x - p_x a_z \\ -p_y n_x + p_x n_y & -p_y o_x + p_x o_y & -p_y a_x + p_x a_y \end{bmatrix} $$

$$ {}^{0}\hat{M}_2 = \begin{bmatrix} c_1 c_2 & -c_1 s_2 & s_1 \\ s_1 c_2 & -s_1 s_2 & -c_1 \\ s_2 & c_2 & 0 \end{bmatrix} + \varepsilon \begin{bmatrix} 100.5 s_1 s_2 - d_1 s_1 c_2 - d_2 c_1 s_2 & 70 s_1 + 100.5 s_1 c_2 - d_2 c_1 c_2 + d_1 s_1 s_2 & c_1 (d_1 + 70 s_2) \\ 100.5 c_1 s_2 + d_1 c_1 c_2 - d_2 s_1 s_2 & 70 c_1 - 100.5 c_1 c_2 - d_1 c_1 s_2 - d_2 s_1 c_2 & s_1 (d_1 + 70 s_2) \\ d_2 c_2 & -d_2 s_2 & -70 c_2 - 100.5 \end{bmatrix} $$

Dividing elements (2,1) and (1,1) in both matrices:

$$ \theta_1 = tg^{-1}\left(\frac{n_y}{n_x}\right) \; ; \; \theta_2 = tg^{-1}\left(\frac{\pm\sqrt{1-o_z^2}}{o_z^2}\right) $$

The velocities in the cylindrical joints are given by:

$$ {}^{0}\hat{V}_{1,0}^{0} = \begin{bmatrix} 0 \\ 0 \\ \omega_1 \end{bmatrix} + \varepsilon \begin{bmatrix} 0 \\ 0 \\ v_1 \end{bmatrix} \qquad {}^{1}\hat{V}_{2,1}^{1} = \begin{bmatrix} 0 \\ 0 \\ \omega_2 \end{bmatrix} + \varepsilon \begin{bmatrix} 0 \\ 0 \\ v_2 \end{bmatrix} $$

Computing the velocities on the end-effector:

$$ {}^{0}\hat{V}_{2,0}^{2} = {}^{0}T_2 \left\{ {}^{2}\hat{M}_0 \, {}^{0}\hat{V}_{1,0}^{0} + {}^{2}\hat{M}_1 \, {}^{1}\hat{V}_{2,1}^{1} \right\} $$

The above expression can be rewritten in terms of dual Jacobian matrix.

$$ {}^{0}\hat{V}_{2,0}^{2} = \begin{bmatrix} \omega_x + \varepsilon V_x \\ \omega_y + \varepsilon V_y \\ \omega_z + \varepsilon V_z \end{bmatrix} = \left[{}^{0}T_2 \right] \left[{}^{2}\hat{M}_0 \quad | \quad {}^{2}\hat{M}_1 \right] \begin{bmatrix} {}^{0}\hat{V}_{1,0}^{0} \\ \hline {}^{1}\hat{V}_{2,1}^{1} \end{bmatrix} $$

$$ {}^{0}\hat{V}_{2,0}^{2} = \begin{bmatrix} c_1c_2 & -c_1s_2 & s_1 \\ s_1c_2 & -s_1s_2 & -c_1 \\ s_2 & c_2 & 0 \end{bmatrix} \begin{bmatrix} {}^{0}\hat{M}_2^T & {}^{1}\hat{M}_2^T \end{bmatrix} \begin{bmatrix} 0 \\ 0 \\ \omega_1 + \varepsilon v_1 \\ 0 \\ 0 \\ \omega_2 + \varepsilon v_2 \end{bmatrix} $$

Dual velocities:

$$ {}^{1}\hat{V}_{1,0}^{1} = {}^{1}\hat{M}_0 \, {}^{0}\hat{V}_{1,0}^{0} = \begin{bmatrix} 0 \\ \sin\hat{\alpha}_1(\omega_1 + \varepsilon v_1) \\ \cos\hat{\alpha}_1(\omega_1 + \varepsilon v_1) \end{bmatrix} \qquad {}^{2}\hat{V}_{2,0}^{2} = {}^{2}\hat{M}_1 \, {}^{1}\hat{V}_{2,0}^{1} = \begin{bmatrix} 0 \\ \sin\hat{\alpha}_2(\omega_2 + \varepsilon v_2) \\ \cos\hat{\alpha}_2(\omega_2 + \varepsilon v_2) \end{bmatrix} $$

6.1 Dynamic analysis

Once obtained the velocities, the next step for solving the dynamic equations is, according with Fisher, to compute the dual momentum, being for each link:

$$ {}^{1}\hat{H}_0 = m_1\left[{}^{1}\vec{V}_{1,0}^{1}\right] - \left[{}^{1}S_1\right]\left[{}^{1}\vec{\omega}_{1,0}^{1}\right] + \varepsilon\left\{\left[{}^{1}S_1\right]\left[{}^{1}\vec{V}_{1,0}^{1}\right] + \left[{}^{1}I_1\right]\left[{}^{1}\vec{\omega}_{1,0}^{1}\right]\right\} $$

$$ {}^{2}\hat{H}_1 = m_2\left[{}^{2}\vec{V}_{2,0}^{2}\right] - \left[{}^{2}S_2\right]\left[{}^{2}\vec{\omega}_{2,0}^{2}\right] + \varepsilon\left\{\left[{}^{2}S_2\right]\left[{}^{2}\vec{V}_{2,0}^{2}\right] + \left[{}^{2}I_2\right]\left[{}^{2}\vec{\omega}_{2,0}^{2}\right]\right\} $$

Derivating the above expressions:

$$ {}^{1}\dot{\hat{H}}_0 = m_1\left[\frac{d}{dt}\left({}^{1}\vec{V}_{1,0}^{1}\right)\right] - \left[{}^{1}S_1\right]\left[\frac{d}{dt}\left({}^{1}\vec{\omega}_{1,0}^{1}\right)\right] $$

$$ +\varepsilon\left\{\left[{}^{1}S_1\right]\left[\frac{d}{dt}\left({}^{1}\vec{V}_{1,0}^{1}\right)\right] + \left[{}^{1}I_1\right]\left[\frac{d}{dt}\left({}^{1}\vec{\omega}_{1,0}^{1}\right)\right]\right\} $$

$$ \begin{bmatrix} {}^{1}f_{0i} + \varepsilon\,{}^{1}t_{0i} \\ {}^{1}f_{0j} + \varepsilon\,{}^{1}t_{0j} \\ {}^{1}f_{0k} + \varepsilon\,{}^{1}t_{0k} \end{bmatrix} $$

$$ = \begin{bmatrix} {}^{1}\dot{P}_{0i} - {}^{1}\omega_{1,0k}^{1}\,{}^{1}P_{0j} + {}^{1}\omega_{1,0j}^{1}\,{}^{1}P_{0k} \\ {}^{1}\dot{P}_{0j} - {}^{1}\omega_{1,0i}^{1}\,{}^{1}P_{0k} + {}^{1}\omega_{1,0k}^{1}\,{}^{1}P_{0i} \\ {}^{1}\dot{P}_{0k} - {}^{1}\omega_{1,0j}^{1}\,{}^{1}P_{0i} + {}^{1}\omega_{1,0i}^{1}\,{}^{1}P_{0j} \end{bmatrix} $$

$$ +\varepsilon \begin{bmatrix} {}^{1}\dot{H}_{0i} - {}^{1}\omega_{1,0k}^{1}\,{}^{1}H_{0j} - {}^{1}V_{1,0k}^{1}\,{}^{1}P_{0j} + {}^{1}\omega_{1,0j}^{1}\,{}^{1}H_{0k} + {}^{1}V_{1,0j}^{1}\,{}^{1}P_{0k} \\ {}^{1}\dot{H}_{0j} - {}^{1}\omega_{1,0i}^{1}\,{}^{1}H_{0k} - {}^{1}V_{1,0i}^{1}\,{}^{1}P_{0k} + {}^{1}\omega_{1,0k}^{1}\,{}^{1}H_{0i} + {}^{1}V_{1,0k}^{1}\,{}^{1}P_{0i} \\ {}^{1}\dot{H}_{0k} - {}^{1}\omega_{1,0j}^{1}\,{}^{1}H_{0i} - {}^{1}V_{1,0j}^{1}\,{}^{1}P_{0i} + {}^{1}\omega_{1,0i}^{1}\,{}^{1}H_{0j} + {}^{1}V_{1,0i}^{1}\,{}^{1}P_{0j} \end{bmatrix} $$

$$\begin{bmatrix} {}^2f_{1i}+\varepsilon\,{}^2t_{1i} \\ {}^2f_{1j}+\varepsilon\,{}^2t_{1j} \\ {}^2f_{1k}+\varepsilon\,{}^2t_{1k} \end{bmatrix} = \begin{bmatrix} {}^2\dot{P}_{1i} - {}^2\omega_{2,0k}^2\,{}^2P_{1j} + {}^2\omega_{2,0j}^2\,{}^2P_{1k} \\ {}^2\dot{P}_{1j} - {}^2\omega_{2,0i}^2\,{}^2P_{1k} + {}^2\omega_{2,0k}^2\,{}^2P_{1i} \\ {}^2\dot{P}_{1k} - {}^2\omega_{2,0j}^2\,{}^2P_{1i} + {}^2\omega_{2,0i}^2\,{}^2P_{1j} \end{bmatrix}$$

$$+\varepsilon\begin{bmatrix} {}^2\dot{H}_{1i} - {}^2\omega_{2,0k}^2\,{}^2H_{1j} - {}^2V_{2,0k}^2\,{}^2P_{1j} + {}^2\omega_{2,0j}^2\,{}^2H_{1k} + {}^2V_{2,0j}^2\,{}^2P_{1k} \\ {}^2\dot{H}_{1j} - {}^2\omega_{2,0i}^2\,{}^2H_{1k} - {}^2V_{2,0i}^2\,{}^2P_{1k} + {}^2\omega_{2,0k}^2\,{}^2H_{1i} + {}^2V_{2,0k}^2\,{}^2P_{1i} \\ {}^2\dot{H}_{1k} - {}^2\omega_{2,0j}^2\,{}^2H_{1i} - {}^2V_{2,0j}^2\,{}^2P_{1i} + {}^2\omega_{2,0i}^2\,{}^2H_{1j} + {}^2V_{2,0i}^2\,{}^2P_{1j} \end{bmatrix}$$

From dynamic equilibrium:

$${}^BM_A\,{}^A\vec{F}_A - {}^B\vec{F}_B - {}^B\vec{f}_A + \varepsilon\left({}^BM_A\,{}^A\vec{T}_A + {}^BD_A\,{}^A\vec{F}_A - {}^B\vec{T}_B - {}^B\vec{t}_A\right) = 0$$

$${}^BM_A\,{}^A\vec{F}_A - {}^B\vec{F}_B - {}^B\vec{f}_A = 0$$

$${}^BM_A\,{}^A\vec{T}_A + {}^BD_A\,{}^A\vec{F}_A - {}^B\vec{T}_B - {}^B\vec{t}_A = 0$$

$${}^2\vec{F}_2 = \left[{}^2M_1\right]^{-1}\left\{{}^2\vec{f}_1\right\}$$

$${}^1\vec{F}_1 = \left[{}^1M_0\right]^{-1}\left\{{}^2\vec{F}_2 + {}^1\vec{f}_0\right\}$$

$${}^2\vec{T}_2 = \left[{}^2M_1\right]^{-1}\left\{{}^2\vec{t}_1 - {}^2D_1\,{}^2\vec{F}_2\right\}$$

$${}^1\vec{T}_1 = \left[{}^1M_0\right]^{-1}\left\{{}^2\vec{T}_2 + {}^1\vec{t}_0 - {}^1D_0\,{}^1\vec{F}_1\right\}$$

7. Conclusions

The presented method, based on dual-number representation, has demonstrated be a powerful tool for solving a great variety of problems, that imply motions simultaneity off rotation and translation of rigid bodies in the space; the aforementioned, allows establishing dual rotation matrices. Robotics is a field wherein dual numbers have been employed to describe the motion of a rigid body, in particular of serial robotic arms. The methodology proposed is useful for robotic arms with cylindrical, prismatic and rotational joints. Once established the dual angles $\hat{\theta}$ and $\hat{\alpha}$, if the dual part of $\hat{\theta}$ is zero, the mechanism has only revolute joints, otherwise if the primary part of $\hat{\theta}$ is zero, only exist prismatic joints. So the developed methodology can be generalized to different topologies, which is a great advantage that allows that only one program solves a great variety of topologies.

The dynamic model is treated by using the dual momentum, wherein the inertial forces are computed by means of a set of linear equations, thus a $6 \times n$ vector of forces is calculated, and in consequence one obtains a complete description of the robotic manipulator. An appropriate way of dual numbers programming will yield a suitable software alternative to simulate and analyze different serial robotic manipulators topologies.

8. Acknowledgments

The authors gratefully acknowledge the support of CONACYT, IPN and ICyTDF for research projects and scholarships. They also would like to thank the anonymous reviewers for their valuable comments and suggestions.

9. References

Al-Widyan, K.; Qing Ma, Xiao & Angeles, J. (2011). The robust design of parallel spherical robots.

Bandyopadhyay, S. (2004). Analytical determination of principal twists in serial, parallel and hybrid manipulators using dual vectors and matrices. *Journal of Mechanism and Machine Theory*, Vol. 39, (2004), pp. (1289-1305), ISSN: 0094-114X.

Bandyopadhyay, S. (2006). *Analysis and Design of Spatial Manipulators: An Exact Algebraic Approach using Dual Numbers and Symbolic Computation*. Ph. D. Thesis, Department of Mechanica Engineeriring, Indian Institute of Science.

Bayro-Corrochano, E. & Kähler, D. (2000). Motor algebra approach for computing the kinematics of robot manipulators. *Journal of Robotic Systems*, Vol. 17, 9, (September 2000), pp. (495-516), DOI: 10.1002/1097-4563

Bayro-Corrochano, E. & Kähler, D. (2001), Kinematics of Robot Manipulators in the Motor Algebra, In: *Geometric computing with Clifford algebras*, Springer, ISBN: 3-540-41198-4 Institute of Computer Science and Applied Mathematics, University of Kiel.

Brodsky, V. &Shoham, M. (1998). *Derivation of dual forces in robot manipulators*. Journal of Mechanism and Machine Theory, 33: 1241-1248.

Cecchini, E., Pennestri, E., & Vergata, T. (2004). A dual number approach to the Kinematic analysis of spatial linkages with dimensional and geometric tolerances. *Proceedings of Design Engineering Technical Conferences and Computers and Information in Engineering Conference*, Salt Lake City, Utah, USA.

Cheng, H.H. (1994). Programming with dual numbers and its applications in mechanisms design. *Engineering with Computers*, Springer.

Fischer, I. (2003). Velocity analysis of mechanisms with ball joints. *Journal of Mechanics Research Communications*, Vol. 30, 1, January-February 2003, pp. (69-78), doi: 10.1016/S0093-6413(02)00350-6

Fisher, I. S. (1998). *Dual Number Methods in Kinematics, Statics and dynamics*. (1st Edition), CRC Press, ISBN: 9780849391156, U. S. A.

Fisher, I. S. (1998). The dual angle and axis of a screw motion. *Journal of Mechanisms and Machine Theory*, Vol. 33, 3, (April 1998), pp. (331 - 340), DOI: 10.1016/S0094-114X(97)00039-6

Fisher, I. S. (2000). Numerical analysis of displacements in spatial mechanisms with ball joints. *Journal of Mechanisms and Machine Theory*, Vol. 35, 11, (November 2000), pp. (1623 - 1640), doi:10.1016/S0094-114X(99)00058-0

Funda, J. (1988). *A Computational Analysis of Line-Oriented Screw Transformations in Robotics.* Technical Report, University of Pennsylvania, U. S. A.

Gu, Y.L. & Luh, J. Y. S. (1987). Dual-Number Transformation and Its Applications To Robotics.

Keler, M. L. (2000). *On the theory of screws and the dual method.* Proceedings of A Symposium Commemorating the Legacy, Words and Life of Sir Robert Stawell Ball Upon the 100th Anniversary of a Treatise on the Theory of Screws, University of Cambridge, Trinnity College.

Kiat Teu, K.; Kim, W.; Fuss, F. K. & Tan, J.(2006). The analysis of golf swing as a kinematic chain using dual Euler angle algorithm.

Moon, Y-M. & Kota, S.(2002). Automated synthesis of mechanisms using dual-vector algebra. *Mechanisms and Machine Theory,* (February 2002), pp. (143-166), doi: 10.1016/S0094-114X(01)00073-8

Page, A.; Mata, V.; Hoyos, J. V. & Porcar, Rosa. (2007) Experimental depermination of instantaneous screw axis in human motions. Error analysis

Pennestrí, E. &Stefanelli, R. (2007). Linear Algebra and numerical algorithms using dual numbers. *Journal of Multi-body System Dynamics,* Vol. 18, 3, pp. (323-344)

Pennock, G. R. & Mattson, K. G. (1996). Forward position problem of two PUMA-type robots manipulating a planar four-bar linkage payload.

Pennock, G. R. & Meehan P. J. (2000). Geometric insight into dynamics of a rigid body using the theory of screws.

Sai-Kai, C. (2000). Simbolic computation of Jacobian of manipulators using dual number transformations.

Seilig, J.M. (1999). *Geometrical Methods in Robotics.* (1st Edition), Springer, ISBN: 0387947280, New Jersey, U. S. A.

Wang, J.; Liang, H-Z. & Sun Z. (2010). Relative Coupled Dynamics and Control using Dual Number. *Systems and Control in Aeronautics and Astronautics (ISSCAA), 2010 3rd International Symposium on*

Yang, J. & Wang, X. (2010). The application of the dual number methods to scara kinematics.

Ying, N.; Kim, W.; Wong, Y. & Kan, H. K. (2004). Analysis of passive motion characteristics of the ankle joint complex using dual Euler angle parameters. Clinical Biomechanics, Vol. 19, 2, (February 2004), pp. (153-160), doi: 10.1016/j.clinbiomech. 2003.10.005

6

On the Stiffness Analysis and Elastodynamics of Parallel Kinematic Machines

Alessandro Cammarata
University of Catania, Department of Mechanical and Industrial Engineering
Italy

1. Introduction

Accurate models to describe the elasticity of robots are becoming essential for those applications involving high accelerations or high precision to improve quality in positioning and tracking of trajectories. Stiffness analysis not only involves the mechanical structure of a robot but even the control system necessary to drive actuators. Strategies aimed to reduce noise and dangerous bouncing effects could be implemented to make control systems more robust to flexibility disturbances, foreseing mechanical interaction with the control system because of regenerative and modal chattering (1). The most used approaches to study elasticity in the literature encompass: the *Finite Element Analysis* (FEA), the *Matrix Structural Analysis* (MSA), the *Virtual Joint Method* (VJM), the *Floating Frame of Reference Formulation* (FFRF) and the *Absolute Nodal Coordinates Formulation* (ANCF).

FEA is largely used to analyze the structural behavior of a mechanical system. The reliability and precision of the method allow to describe each part of a mechanical system with great detail (2). Applying FEA to a robotic system implies a time-consuming process of re-meshing in the pre-processing phase every time that the robot posture has changed. Optimization all over the workspace of a robot would require very long computational time, thus FEA models are often employed to verify components or subparts of a complex robotic system.

The MSA includes some simplifications to FEA using complex elements like beams, arcs, cables or superelements (3–5). This choice reduces the computational time and makes this method more efficient for optimization tasks. Some authors recurred to the superposition principle along with the virtual work principle to achieve the global stiffness model (6–8). Others considered the minimization of the potential energy of a PKM to find the global stiffness matrix (9), while some approaches used the total potential energy augmented adding the kinematic constraints by means of the Lagrange multipliers (10; 11).

The first papers on VJM are based on pseudo-rigid body models with "virtual joints" (12–14). More recent papers include link flexibility and linear/torsional springs to take into account bending contributes (15–19). These approaches recur to the Jacobian matrix to map the stiffness of the actuators of a PKM inside its workspace; especially for PKMs with reduced mobility, it implies that the stiffness is limited to a subspace defined by the dofs of its end-effector. Pashkevich *et al.* tried to overcome this issue by introducing a multidimensional lumped-parameter model with localized 6-dof virtual springs (20).

Finally, the FFRF and ANCF are powerful and accurate formulations, based on FEM and continuum mechanics, to study any flexible mechanical system (21). The FFRF is suitable

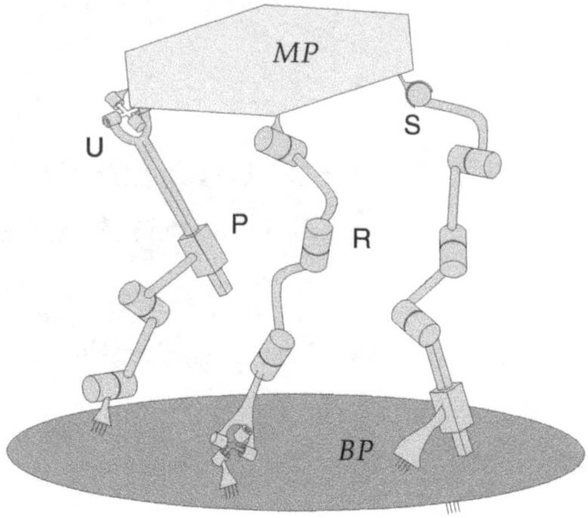

Fig. 1. Schematic drawing of a PKM: P prismatic joint, R revolute joint, S spherical joint, U universal joint

for large rotations and small deformations, while the ANCF is preferred for large rotations and large deformations. Notwithstanding, both the two formulations usually require great computational efforts to be extended to complex robots or to optimization techniques involving the whole robot's workspace.

In this chapter we propose a formulation to study the elastostatics and elastodynamics of PKMs. The method is linear and tries to combine some feature existing in the literature to build a solid framework, the outlines of which are described in Section 2. We start from a MSA approach based on the minimization of the strain energy in which all joints are introduced by means of constraints between nodal displacements. In this way we avoid Lagrange multipliers and the introduction of joints becomes straightforward. Unlike VJM methods based on lumped stiffness, we use 3D Euler beams to simulate links and to distribute stiffness to the flexible structure of a robot, as described in Section 3. The same set of nodal displacement arrays is used to obtain the generalized stiffness and inertia matrix, the latter being lumped or distributed, as discussed in Sections 4 and 5. Besides, the flexibility of the proposed method allows us to provide useful extension to compliant PKMs with joint flexures. The ease in setting up, the direct control of joints and the speed of execution make the procedure adapt for optimization routines, as described in Section 6. Finally, some feasible applications are described in Section 7 studying an articulated PKMs with four dofs.

2. Outlines of the algorithm

Before introducing the reader to the proposed methodology, we have to clarify the reasons of this work. The first question that might arise is: *Why use a new method without recurring to well confirmed formulations existing in the literature?* The right answer is essentially tied to the simplicity of the proposed formulation. The method is addressed to designers that want to implement software to study the elastodynamics of PKMs, as well as to students and researchers that wish to create their own customized algorithm. The author experienced that the elastodynamics study of a complex robotic system is not a trivial issue, mainly when dealing with some features concerning the optimization of performances in terms of elasticity or admissible range of eigenfrequencies. Performing these analyzes often needs for a global optimization all over the workspace of a robot too cumbersome to be faced with conventional formulations. Further, considering that the elastodynamics behavior of a robot changes according to its pose, an accurate analysis should be carried out several times to capture the right response inside the robot's workspace, drastically affecting the computational cost. The focused issue becomes even more complicated when constrained optimization routines, based on indices of elastostatics/dynamics performances, are implemented to work in all the workspace. In the latter case the search for local minimum configurations needs for a simple and robust mathematical framework adapt to iterative procedures. Working with ordinary resources, in terms of CPU speed and memory, can make optimization prohibitive unless the complex flexible multibody formulations would be simplified to meet requirements. Therefore, the second question might be: *Why simplify a complex formulation and do not create a simpler one instead?* The sought formulation is what the author is going to explain in this chapter.

Let us start considering a generic PKM. It is essentially a complex robotic system with parallel kinematics in which one or more limbs connect a base platform to a moving platform, as shown in Fig. 1. The latter contains a tool, often referred to as *end-effector*, necessary to perform a certain task or, sometimes, as in the case of flight simulators or assembly stations, the end effector is the moving platform itself. The limbs connecting the two platforms are composed of links constrained by joints. A limb can be a serial kinematic chain or an articulated linkage with one or more closed kinematic chains or loops.

In performing our analysis we have to choose what is flexible and what is rigid as well. Generally, each body is flexible and the notion of rigid body is an abstraction that becomes a good approximation if strains of the structure are small when compared to displacements. Thus, considering a link flexible or rigid depends on many aspects as: material, geometry, wrenches involved in the process. Besides, some tasks might need for high precision to be accomplished, then an accurate analysis should take care of deformations to fulfill the requirements without gross errors. Here, we model the MP and BP as rigid bodies because, for the most part of industrial PKMs, these are usually one order of magnitude stiffer than the remaining links. The latter will be either flexible or rigid depending on the assumptions made by the designer. As already pointed out in the Introduction, we use 3D Euler beams to represent links, even if the treatment can be extended to superelements, as reported in (22). The formulation is linear and only small deformations are considered. Given a starting pose of the MP, the IKP allows us to find the robot's configuration; then, the elastodynamics—statics—is performed around this undeformed configuration. It might be useful to lock the actuated joints in order to avoid rigid movements and to isolate only the flexible modes.

Below, we summarize all steps necessary to find the elastodynamics equations of a PKM:

1. Solve the inverse kinematic problem (IKP) in order to know the starting pose, i.e. the undeformed configuration, of the PKM. Here, we make the assumption that the configuration is *frozen*, meaning that no coupling of rigid body and elastic motions is considered.
2. Distinguish rigid and flexible links, and discretize the latter into the desired number of flexible elements. The base and moving platforms are modeled as rigid.
3. Introduce nodes and then nodal-arrays for each node. Each flexible part has two end-nodes at its end-sections; the rigid links have a single node at their center of mass.
4. Introduce joint-matrices and arrays for each joint.

5. Individuate the couples of bodies linked by a joint and distinguish among three cases: rigid-flexible, flexible-flexible and flexible-rigid.
6. Find the equations expressing dependent nodal-array in terms of independent nodal-arrays, then find the generalized stiffness and inertia matrices. The latter change whether lumped or distributed method is used.
7. Introduce the global array $\mathbf{q}$ of independent nodal coordinates. This array contains all the independent nodal arrays in the order defined by the reader.
8. Expand all matrices expressing each generalized stiffness and inertia matrix in terms of $\mathbf{q}$ by means of the Boolean matrices $\mathbf{B}_1$ and $\mathbf{B}_2$. Then, sum all contributes to find the generalized stiffness matrix $\mathbf{K}_{PKM}$ and inertia matrix $\mathbf{M}_{PKM}$ of the PKM.
9. Introduce the array $\mathbf{f}$ of generalized nodal wrenches and, finally, write the elastodynamics equations.

3. Mathematical background and key concepts

In a mechanical system flexible bodies storage and exchange energy like a tank is able to storage and supply water. The energy associated to deformation is termed *strain energy* while the aptitude of a body for deformation is tied to the property of *stiffness*. For a continuum body the stiffness is distributed and the strain energy changes according to the variation of the displacement field of its points. In a discrete flexible element inner points' displacements depend on displacements of some points or *nodes*.

3.1 Nodal and joint-array

Robotic links can be well described by means of 3D-beams, that is, flexible elements with two end-nodes, as shown in Fig. 2. The expression of the strain energy depends on the kind of displacements chosen for rotations: slopes, Euler angles, quaternions, and so on, and on the entity of displacement: large or small. Here, we choose a linear formulation based on small displacements and Euler angles to describe rotations. Based on these assumptions, the strain energy is a positive-definite quadratic form in the nodal displacement coordinates of the end-nodes arrays of the beam, where a nodal displacement array $\mathbf{u} = \begin{bmatrix} u_x & u_y & u_z & u_\varphi & u_\theta & u_\psi \end{bmatrix}^T$ has six scalar *displacements*, three translational and three rotational. Hereafter, subscripts and superscripts will be, respectively, referred to the beam and to one of the two end-nodes of a beam.

Beams belonging to the same link are contiguous, while joints couple beams belonging to two adjoining links. To express the kinematic bond existing between the two nodes, or sections, coupled by the joint, a constraint equation is introduced in which the nodal displacements of the two nodes are tied through the means of an array of joint displacements. Figure 2 describes two beams linked by a prismatic joint of axis parallel to the unit vector $\mathbf{e}$. The two nodal arrays $\mathbf{u}_1^2$ and $\mathbf{u}_2^1$ are tied together by means of the translational displacement s of the prismatic joint P and by the 6-dimensional joint-array $\mathbf{h}^P = \begin{bmatrix} \mathbf{e}^T & \mathbf{0}^T \end{bmatrix}^T$, i.e.,

$$\mathbf{u}_1^2 = \mathbf{u}_2^1 + \mathbf{h}^P s \tag{1}$$

We stress that eq.(1) describes a constraint among displacements, then deformations, and not nodal coordinates. In general, $\mathbf{H}^j$ is a $6 \times m(j)$ joint-matrix where $m(j)$ is the dimension of the joint-array $\boldsymbol{\theta}^j$. The joint-matrix $\mathbf{H}^j$ and the joint-array $\boldsymbol{\theta}^j$ depend on the type of joint: the former

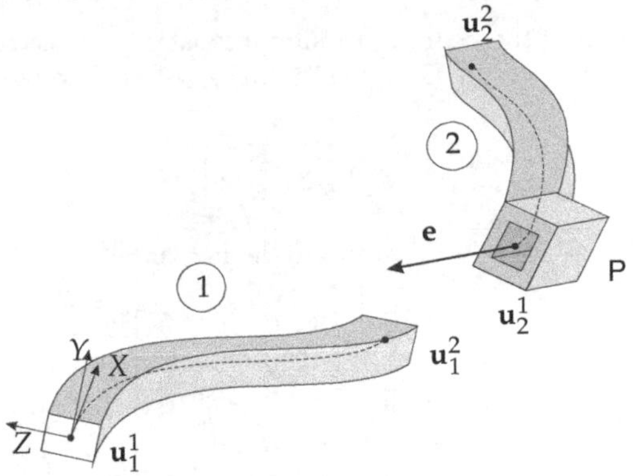

Fig. 2. Notation of two 3D Euler beams coupled by a joint.

containing unit vectors indicating geometric axes, the latter containing joint displacements, either linear s^j, for translations, or angular θ^j, for rotations. Below, two more examples of joints are provided:

Revolute joint:

$$\mathbf{h}^R = \left[\mathbf{0}^T \ \mathbf{e}^T \right]^T, \quad \theta^R = \theta \tag{2}$$

where $\mathbf{e}$ is the unit vector along the axis of the revolute joint R and θ is the angular displacement about the said axis.

Universal joint:

$$\mathbf{H}^U = \begin{bmatrix} \mathbf{0} & \mathbf{0} \\ \mathbf{e}^1 & \mathbf{e}^2 \end{bmatrix}, \quad \boldsymbol{\theta}^U = \left[\theta^1 \ \theta^2 \right]^T \tag{3}$$

where $\mathbf{e}^1$ and $\mathbf{e}^2$ are the unit vectors along the axes of the universal joint U and θ^1 and θ^2 are the angular displacements about the axes of U.

Other joints, i.e. cylindrical, spherical and so on, can be created combining together elementary prismatic and/or revolute joints. The described constraint equations are used to consider joints contribute to elastodynamics in a direct way without recurring to Lagrangian multipliers to introduce joint constraints, (10). In the next section the above equation will be used to obtain joint displacements in terms of nodal displacements.

3.2 Strain energy and stiffness matrix

The strain energy $V_i(\mathbf{u}_i^1, \mathbf{u}_i^2)$ of a flexible body is a nonlinear scalar function of nodal deformations, here expressed via nodal displacements. For the case of a 3D Euler beam, considered in this analysis, the strain energy V_i of the ith-beam turns into a positive-definite quadratic form of the stiffness matrix $\mathbf{K}_i$ in twelve variables, i.e. the nodal displacements of the end-nodes arrays $\mathbf{u}_i^1$ and $\mathbf{u}_i^2$. By introducing the 12-dimensional array $\tilde{\mathbf{u}}_i = \left[\mathbf{u}_i^{1^T} \ \mathbf{u}_i^{2^T} \right]^T$, the strain energy assumes the following expression

$$V_i = \frac{1}{2} \tilde{\mathbf{u}}_i^T \mathbf{K}_i \tilde{\mathbf{u}}_i \tag{4}$$

The 12×12 stiffness matrix $\mathbf{K}_i$ of a 3D Euler beam with circular cross-section, and for the case of homogeneous and isotropic material, depends on geometrical and stiffness parameters as:

cross section area A, length of the beam L, torsional constant J, mass moment of inertia I, the Young module E and shear module G. For our purposes, it is convenient to divide $\mathbf{K}_i$ into blocks, i.e.

$$\mathbf{K}_i = \begin{bmatrix} \mathbf{K}_i^{1,1} & \mathbf{K}_i^{1,2} \\ \mathbf{K}_i^{2,1} & \mathbf{K}_i^{2,2} \end{bmatrix} \tag{5}$$

in which $\mathbf{K}_i^{2,1} = (\mathbf{K}_i^{1,2})^T$ and the other blocks are defined as

$$\mathbf{K}_i^{1,1} = \frac{E}{L} \begin{bmatrix} A & 0 & 0 & 0 & 0 & 0 \\ 0 & \frac{12I}{L^2} & 0 & 0 & 0 & \frac{6I}{L} \\ 0 & 0 & \frac{12I}{L^2} & 0 & -\frac{6I}{L} & 0 \\ 0 & 0 & 0 & \frac{GJ}{E} & 0 & 0 \\ 0 & 0 & -\frac{6I}{L} & 0 & 4I & 0 \\ 0 & \frac{6I}{L} & 0 & 0 & 0 & 4I \end{bmatrix} \tag{6a}$$

$$\mathbf{K}_i^{1,2} = \frac{E}{L} \begin{bmatrix} -A & 0 & 0 & 0 & 0 & 0 \\ 0 & -\frac{12I}{L^2} & 0 & 0 & 0 & \frac{6I}{L} \\ 0 & 0 & -\frac{12I}{L^2} & 0 & -\frac{6I}{L} & 0 \\ 0 & 0 & 0 & -\frac{GJ}{E} & 0 & 0 \\ 0 & 0 & \frac{6I}{L} & 0 & 4I & 0 \\ 0 & -\frac{6I}{L} & 0 & 0 & 0 & 4I \end{bmatrix} \tag{6b}$$

$$\mathbf{K}_i^{2,2} = \frac{E}{L} \begin{bmatrix} A & 0 & 0 & 0 & 0 & 0 \\ 0 & \frac{12I}{L^2} & 0 & 0 & 0 & -\frac{6I}{L} \\ 0 & 0 & \frac{12I}{L^2} & 0 & \frac{6I}{L} & 0 \\ 0 & 0 & 0 & \frac{GJ}{E} & 0 & 0 \\ 0 & 0 & \frac{6I}{L} & 0 & 4I & 0 \\ 0 & -\frac{6I}{L} & 0 & 0 & 0 & 4I \end{bmatrix} \tag{6c}$$

where the two diagonal block matrices, respectively, refer to the nodal displacements of the two end-nodes while the extra-diagonal blocks refer to the coupling among nodal displacements of different nodes: in fact, each entry of the generic 6×6 block-matrix $\mathbf{K}_i^{l,m}$ can be thought as a force— torque—at the lth-node of the ith-beam when a unit displacement—rotation—is applied to the mth-node.

3.3 Rigid body displacement

Let us consider a rigid body with center of mass at the point G and a generic point P inside its volume, besides let $\mathbf{d}_P$ be the vector pointing from G to the point P. If the body can accomplish only small displacements/rotations, let $\mathbf{p}$ be the displacement of point G and $\mathbf{r}$, be the axial vector of the small rotation matrix $\mathbf{R}$, (9). Upon these assumptions, the displacement array $\mathbf{u}_G$ of G is defined as $\mathbf{u}_G = \begin{bmatrix} \mathbf{p}^T & \mathbf{r}^T \end{bmatrix}^T$.
the displacement array $\mathbf{u}_P$ can be expressed in terms of $\mathbf{u}_G$ by means of the following equation:

$$\mathbf{u}_P = \mathbf{G}_P \mathbf{u}_G, \quad \mathbf{G}_P = \begin{bmatrix} \mathbf{1} & -\mathbf{D}_P \\ \mathbf{O} & \mathbf{1} \end{bmatrix} \tag{7}$$

where $\mathbf{1}$ and $\mathbf{O}$, respectively, are the 3×3 identity- and zero-matrices and $\mathbf{D}_P$ is the Cross-Product Matrix (C.P.M.) of $\mathbf{d}_P$, (23).

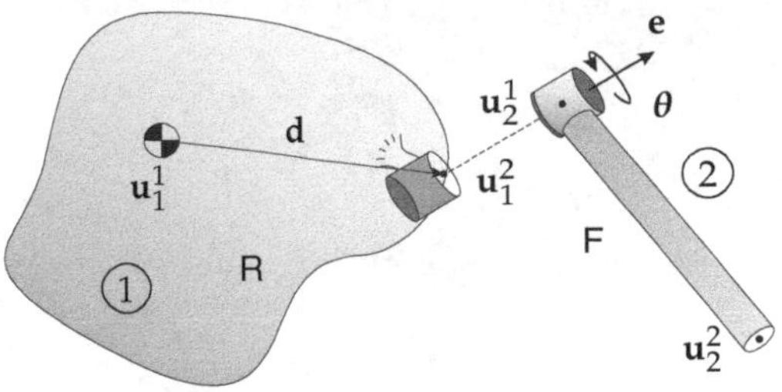

Fig. 3. Case a): Rigid body-flexible beam.

4. Stiffness matrix determination

Let us consider a linkage of flexible beams and rigid bodies connected by means of joints. The strain energy $V(\mathbf{u}_1^1, \mathbf{u}_1^2, \mathbf{u}_2^1, \ldots, \mathbf{u}_{G1}, \mathbf{u}_{G2}, \ldots, \boldsymbol{\theta}^1, \boldsymbol{\theta}^2, \ldots)$ of this system depends on nodal displacement arrays of both rigid and flexible parts and on joint displacement arrays. In the following, we will show how to express V in terms of a reduced set of independent nodal coordinates. In this process, we will start from three elementary blocks composed of rigid and flexible bodies that will be combined to build a generic linkage, for instance, the limb of a PKM. The first and the third case pertain the coupling between a rigid body and a flexible beam by means of a joint; the second case describes the coupling of two beams belonging to two different links. The choice to use two cases to describe the rigid body-flexible body connection is only due to convenience to follow the order of bodies from the base to the moving platform of a PKM. The reader might recur to a unique case to simplify the treatment. The last part of the section is devoted to some insight on joints' stiffness and feasible application to flexures.

4.1 Case a) Rigid body-flexible beam

Figure 3 describes a rigid body R coupled to a flexible beam F by means of a joint. The strain energy V_a of the beam is function of the nodal displacement arrays $\mathbf{u}_2^1$ and $\mathbf{u}_2^2$, i.e.

$$V_a = \frac{1}{2}\begin{bmatrix}\mathbf{u}_2^1\\ \mathbf{u}_2^2\end{bmatrix}^T \begin{bmatrix}\mathbf{K}_2^{1,1} & \mathbf{K}_2^{1,2}\\ \mathbf{K}_2^{2,1} & \mathbf{K}_2^{2,2}\end{bmatrix}\begin{bmatrix}\mathbf{u}_2^1\\ \mathbf{u}_2^2\end{bmatrix} \tag{8}$$

The array $\mathbf{u}_2^1$ can be expressed in terms of $\mathbf{u}_1^2$ and $\boldsymbol{\theta}$ recalling eq.(1), while $\mathbf{u}_1^2$, in turn, is tied to $\mathbf{u}_1^1$ from eq.(7): upon combining both the equations, we obtain

$$\mathbf{u}_2^1 = \mathbf{G}\mathbf{u}_1^1 + \mathbf{H}\boldsymbol{\theta} \tag{9}$$

where $\mathbf{G}$ depends on $\mathbf{d}$. By substituting the previous equation into eq.(8) we find that $V_a = V(\mathbf{u}_1^1, \mathbf{u}_2^2, \boldsymbol{\theta})$. If the joint is passive, its displacement $\boldsymbol{\theta}$ depends on the elastic properties of the system and, therefore, on the two displacements $\mathbf{u}_1^1$ and $\mathbf{u}_2^2$: thus, it implies that V_a is only function of the two mentioned array, i.e. $V_a = V(\mathbf{u}_1^1, \mathbf{u}_2^2)$. In order to obtain the law for $\boldsymbol{\theta}$ we minimize the strain energy V_a w.r.t. $\boldsymbol{\theta}$:

$$dV_a / d\boldsymbol{\theta} = \mathbf{0}_6^T \tag{10}$$

where $\mathbf{0}_6$ is the six-dimensional zero array. After rearrangements and simplifications, we find

$$\boldsymbol{\theta} = \mathbf{Y}^1\mathbf{u}_1^1 + \mathbf{Y}^2\mathbf{u}_2^2 \tag{11}$$

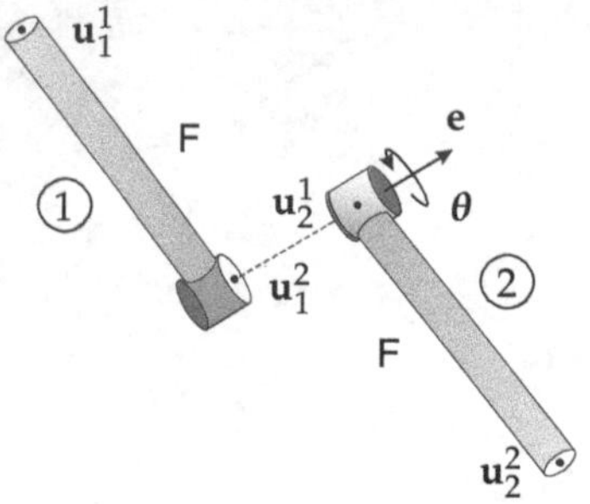

Fig. 4. Case b): Flexible beam-flexible beam.

where

$$\mathbf{Y}^1 = -(\mathbf{H}^T\mathbf{K}_2^{1,1}\mathbf{H})^{-1}\mathbf{H}^T\mathbf{K}_2^{1,1}\mathbf{G} \tag{12a}$$
$$\mathbf{Y}^2 = -(\mathbf{H}^T\mathbf{K}_2^{1,1}\mathbf{H})^{-1}\mathbf{H}^T\mathbf{K}_2^{1,2} \tag{12b}$$

Then, by substituting eq.(11) into eq.(9), we obtain

$$\mathbf{u}_2^1 = \mathbf{X}^1\mathbf{u}_1^1 + \mathbf{X}^2\mathbf{u}_2^2 \tag{13}$$
$$\mathbf{X}^1 = \mathbf{G} + \mathbf{H}\mathbf{Y}^1, \quad \mathbf{X}^2 = \mathbf{H}\mathbf{Y}^2 \tag{14}$$

Let us define the 12-dimensional array $\tilde{\mathbf{u}}_a = \begin{bmatrix}\mathbf{u}_1^{1^T} & \mathbf{u}_2^{2^T}\end{bmatrix}^T$ and let us substitute the above expression into V_a, thereby obtaining

$$V_a = \frac{1}{2}\tilde{\mathbf{u}}_a^T\mathbf{K}_a\tilde{\mathbf{u}}_a,\ \mathbf{K}_a = \begin{bmatrix}\mathbf{X}^1 & \mathbf{X}^2\\ \mathbf{O} & \mathbf{1}\end{bmatrix}^T \begin{bmatrix}\mathbf{K}_2^{1,1} & \mathbf{K}_2^{1,2}\\ \mathbf{K}_2^{2,1} & \mathbf{K}_2^{2,2}\end{bmatrix} \begin{bmatrix}\mathbf{X}^1 & \mathbf{X}^2\\ \mathbf{O} & \mathbf{1}\end{bmatrix} \tag{15}$$

where $\mathbf{K}_a$ is the 12×12 stiffness matrix sought.

4.2 Case b) Flexible body-flexible body

For the case b) of Fig.4 two beams are coupled by a joint. The strain energy V_b is function of the nodal displacement arrays of the two flexible bodies, i.e.

$$V_b = \frac{1}{2}\begin{bmatrix}\mathbf{u}_1^1\\ \mathbf{u}_1^2\end{bmatrix}^T \begin{bmatrix}\mathbf{K}_1^{1,1} & \mathbf{K}_1^{1,2}\\ \mathbf{K}_1^{2,1} & \mathbf{K}_1^{2,2}\end{bmatrix} \begin{bmatrix}\mathbf{u}_1^1\\ \mathbf{u}_1^2\end{bmatrix} + \frac{1}{2}\begin{bmatrix}\mathbf{u}_2^1\\ \mathbf{u}_2^2\end{bmatrix}^T \begin{bmatrix}\mathbf{K}_2^{1,1} & \mathbf{K}_2^{1,2}\\ \mathbf{K}_2^{2,1} & \mathbf{K}_2^{2,2}\end{bmatrix} \begin{bmatrix}\mathbf{u}_2^1\\ \mathbf{u}_2^2\end{bmatrix} \tag{16}$$

The four arrays are not all independent as the following equation stands:

$$\mathbf{u}_1^2 = \mathbf{u}_2^1 + \mathbf{H}\boldsymbol{\theta} \tag{17}$$

The strain energy is, thus, dependent on $\mathbf{u}_1^1, \mathbf{u}_2^1, \mathbf{u}_2^2$ and $\boldsymbol{\theta}$: $V_b = V(\mathbf{u}_1^1, \mathbf{u}_1^2, \mathbf{u}_2^1, \mathbf{u}_2^2) \equiv V(\mathbf{u}_1^1, \mathbf{u}_2^1, \mathbf{u}_2^2, \boldsymbol{\theta})$. Here, we choose to minimize V_b w.r.t. $\boldsymbol{\theta}$, as for the case a), and $\mathbf{u}_2^1$. The reader should notice that our choice is not unique, it is only a particular reduction process necessary for our purposes, it means that the reader might develop a treatment in which $\mathbf{u}_2^1$ is not dependent anymore.

Now, by imposing that the derivative of V_b w.r.t. $\boldsymbol{\theta}$ vanishes, we obtain

$$\boldsymbol{\theta} = \mathbf{F}^1\mathbf{u}_1^1 + \mathbf{F}^2\mathbf{u}_2^1 \tag{18}$$

where

$$\mathbf{F}^1 = -(\mathbf{H}^T\mathbf{K}_1^{2,2}\mathbf{H})^{-1}\mathbf{H}^T\mathbf{K}_1^{2,1} \tag{19a}$$

$$\mathbf{F}^2 = -(\mathbf{H}^T\mathbf{K}_1^{2,2}\mathbf{H})^{-1}\mathbf{H}^T\mathbf{K}_1^{2,2} \tag{19b}$$

Following the previous condition, even the dependent nodal-array $\mathbf{u}_2^1$ must minimize $V_b = V(\mathbf{u}_1^1, \mathbf{u}_1^2(\mathbf{u}_2^1, \boldsymbol{\theta}(\mathbf{u}_1^1, \mathbf{u}_2^1)), \mathbf{u}_2^1, \mathbf{u}_2^2)$; therefore, by applying the chain-rule, we obtain

$$\frac{dV_b}{d\mathbf{u}_2^1} \equiv \frac{\partial V_b}{\partial \mathbf{u}_1^2}\frac{\partial \mathbf{u}_1^2}{\partial \mathbf{u}_2^1} + \frac{\partial V_b}{\partial \mathbf{u}_2^1} \equiv \frac{\partial V_b}{\partial \mathbf{u}_1^2}(\mathbf{1}_6 + \mathbf{H}\mathbf{F}^2) + \frac{\partial V_b}{\partial \mathbf{u}_2^1} = \mathbf{0}_6^T \tag{20}$$

The array $\mathbf{u}_2^1$ is then written in terms of $\mathbf{u}_1^1, \mathbf{u}_2^2$, i.e.

$$\mathbf{u}_2^1 = \mathbf{G}^1\mathbf{u}_1^1 + \mathbf{G}^2\mathbf{u}_2^2 \tag{21}$$

where the 6×6 matrices $\mathbf{G}^1$ and $\mathbf{G}^2$ have the following expressions:

$$\begin{aligned}\mathbf{G}^1 = -&(\mathbf{K}_1^{2,2} + \mathbf{K}_2^{1,1} + \mathbf{F}^{2^T}\mathbf{H}^T\mathbf{K}_1^{2,2} + \mathbf{F}^{2^T}\mathbf{H}^T\mathbf{K}_1^{2,2}\mathbf{H}\mathbf{F}^2)^{-1}\\ &(\mathbf{K}_1^{2,1} + \mathbf{K}_1^{2,2}\mathbf{H}\mathbf{F}^1 + \mathbf{F}^{2^T}\mathbf{H}^T\mathbf{K}_1^{2,1} + \mathbf{F}^{2^T}\mathbf{H}^T\mathbf{K}_1^{2,2}\mathbf{H}\mathbf{F}^1)\end{aligned} \tag{22a}$$

$$\mathbf{G}^2 = -(\mathbf{K}_1^{2,2} + \mathbf{K}_2^{1,1} + \mathbf{F}^{2^T}\mathbf{H}^T\mathbf{K}_1^{2,2} + \mathbf{F}^{2^T}\mathbf{H}^T\mathbf{K}_1^{2,2}\mathbf{H}\mathbf{F}^2)^{-1}\mathbf{K}_1^{1,2} \tag{22b}$$

The joint-arrays $\boldsymbol{\theta}^j$ becomes,

$$\boldsymbol{\theta} = \mathbf{Y}^1\mathbf{u}_1^1 + \mathbf{Y}^2\mathbf{u}_2^2 \tag{23}$$

$$\mathbf{Y}^1 = \mathbf{F}^1 + \mathbf{F}^2\mathbf{G}^1,\ \mathbf{Y}^2 = \mathbf{F}^2\mathbf{G}^2 \tag{24}$$

The dependent nodal-array $\mathbf{u}_1^2$ is obtained by substituting eq.(21) and eq.(23) into eq.(17):

$$\mathbf{u}_1^2 = \mathbf{X}^1\mathbf{u}_1^1 + \mathbf{X}^2\mathbf{u}_2^2 \tag{25}$$

$$\mathbf{X}^1 = \mathbf{G}^1 + \mathbf{H}\mathbf{Y}^1,\ \mathbf{X}^2 = \mathbf{G}^2 + \mathbf{H}\mathbf{Y}^2 \tag{26}$$

Let $\tilde{\mathbf{u}}_b = \begin{bmatrix}\mathbf{u}_1^{1^T} & \mathbf{u}_2^{2^T}\end{bmatrix}^T$ be the 12-dimensional array of independent nodal displacements, then the final expression of V_b in terms $\tilde{\mathbf{u}}_b$ is

$$V_b = \frac{1}{2}\tilde{\mathbf{u}}_b^T\mathbf{K}_b\tilde{\mathbf{u}}_b \tag{27}$$

$$\mathbf{K}_b = \begin{bmatrix}\mathbf{1} & \mathbf{O}\\ \mathbf{X}^1 & \mathbf{X}^2\end{bmatrix}^T \begin{bmatrix}\mathbf{K}_1^{1,1} & \mathbf{K}_1^{1,2}\\ \mathbf{K}_1^{2,1} & \mathbf{K}_1^{2,2}\end{bmatrix} \begin{bmatrix}\mathbf{1} & \mathbf{O}\\ \mathbf{X}^1 & \mathbf{X}^2\end{bmatrix} + \begin{bmatrix}\mathbf{G}^1 & \mathbf{G}^2\\ \mathbf{O} & \mathbf{1}\end{bmatrix}^T \begin{bmatrix}\mathbf{K}_2^{1,1} & \mathbf{K}_2^{1,2}\\ \mathbf{K}_2^{2,1} & \mathbf{K}_2^{2,2}\end{bmatrix} \begin{bmatrix}\mathbf{G}^1 & \mathbf{G}^2\\ \mathbf{O} & \mathbf{1}\end{bmatrix} \tag{28}$$

where $\mathbf{K}_b$ is the 12×12 stiffness matrix.

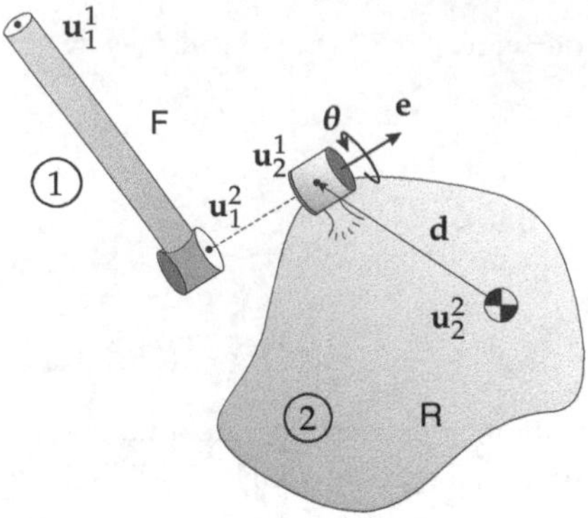

Fig. 5. Case c): Flexible beam-rigid body.

4.3 Case c) Flexible body-rigid body

The case c) describes a beam coupled to a rigid body by means of a joint, as shown in Figure 5. The expressions involving the case c) can be easily obtained by extension of those of the case a), hence, we write only the final results for brevity:

$$\mathbf{Y}^1 = -(\mathbf{H}^T\mathbf{K}_1^{2,2}\mathbf{H})^{-1}\mathbf{H}^T\mathbf{K}_1^{2,1} \tag{29a}$$
$$\mathbf{Y}^2 = -(\mathbf{H}^T\mathbf{K}_1^{2,2}\mathbf{H})^{-1}\mathbf{H}^T\mathbf{K}_1^{2,2}\mathbf{G} \tag{29b}$$

$$\mathbf{X}^1 = \mathbf{H}\mathbf{Y}^1 \tag{30a}$$
$$\mathbf{X}^2 = \mathbf{G} + \mathbf{H}\mathbf{Y}^2 \tag{30b}$$

The strain energy V_c associated to the beam becomes

$$V_c = \frac{1}{2}\tilde{\mathbf{u}}_c^T\mathbf{K}_c\tilde{\mathbf{u}}_c, \quad \mathbf{K}_c = \begin{bmatrix} \mathbf{1} & \mathbf{O} \\ \mathbf{X}^1 & \mathbf{X}^2 \end{bmatrix}^T \begin{bmatrix} \mathbf{K}_1^{1,1} & \mathbf{K}_1^{1,2} \\ \mathbf{K}_1^{2,1} & \mathbf{K}_1^{2,2} \end{bmatrix} \begin{bmatrix} \mathbf{1} & \mathbf{O} \\ \mathbf{X}^1 & \mathbf{X}^2 \end{bmatrix} \tag{31}$$

where $\tilde{\mathbf{u}}_c = \begin{bmatrix} \mathbf{u}_1^{1\,T} & \mathbf{u}_2^{2\,T} \end{bmatrix}^T$ is the 12-dimensional array of independent nodal displacements and $\mathbf{K}_c$ is the 12×12 generalized stiffness matrix of the case c).

4.4 Joint's stiffness and flexure joints

The final part of the present section is devoted to show some feasible extension of the formulation to flexure mechanisms. Similar concepts can be applied even to ordinary joints to take into account joint stiffness. In order to reproduce the counterparts of mechanical joints, in a continuous structure it is a common strategy to recur to flexure joints, in fact, zones where the geometry and shape are designed to increase the compliance along specified degrees of freedom (*dofs*). An ideal flexure joint should allow for only motions along its *dofs*, while withstanding to remaining motions along its degrees of constraint (*docs*).

Figure 6 shows the cases of ideal and real flexure revolute joints. While for the ideal case the nodal displacements $\mathbf{u}_1^2$ and $\mathbf{u}_2^1$ of the two flexible beams (1) and (2), coupled by the joint, are tied by the usual constraint equation

$$\mathbf{u}_1^2 = \mathbf{u}_2^1 + \mathbf{H}\delta\boldsymbol{\theta} \tag{32}$$

for the real case $\mathbf{H} \equiv \mathbf{1}_6$. The joint displacement array is now denoted with $\delta\boldsymbol{\theta}$.
Referring to Figure 7, let us consider three different configurations of the flexure joints: an undeformed and unloaded configuration in which the joint-array is $\boldsymbol{\theta}_o$; a preloaded initial configuration with the joint-array $\boldsymbol{\theta}_i$ and a final configuration in which $\boldsymbol{\theta}_f = \boldsymbol{\theta}_i + \delta\boldsymbol{\theta}$, being $\delta\boldsymbol{\theta}$ an array of small displacements around the initial joint-array $\boldsymbol{\theta}_i$. For the following explanation, we refer only to the ideal case of Figure 6. Let $\mathbf{K}_\theta$ be the $m(j) \times m(j)$ joint stiffness matrix and let $\mathbf{w}$ be the generic wrench acting on the flexure joint; then, for the three configurations, we can write

$$\mathbf{w}_o = \mathbf{0}_6 \tag{33a}$$
$$\mathbf{w}_i = \mathbf{K}_\theta(\boldsymbol{\theta}_i - \boldsymbol{\theta}_o) \equiv \mathbf{K}_\theta \Delta\boldsymbol{\theta}_i \tag{33b}$$
$$\mathbf{w}_f = \mathbf{K}_\theta(\boldsymbol{\theta}_f - \boldsymbol{\theta}_o) \equiv \mathbf{K}_\theta \Delta\boldsymbol{\theta}_f \tag{33c}$$

where $\Delta\boldsymbol{\theta}_f = \Delta\boldsymbol{\theta}_i + \delta\boldsymbol{\theta}$. The strain energy V_{θ_f} of the flexure joint in its final configuration simply reduces to

$$V_{\theta_f} = \frac{1}{2}\Delta\boldsymbol{\theta}_f^T \mathbf{K}_\theta \Delta\boldsymbol{\theta}_f \tag{34}$$

The above expressions can be simplified considering $\boldsymbol{\theta}_o = \mathbf{0}_6$ for the unloaded configuration. We do not further discuss on flexure joints, leaving the reader to derive three new cases, similar to those discussed above, taking into account joint stiffness.

5. Mass matrix determination

In this section two ways to include masses/inertias are discussed: the *lumped approach* and the *distributed approach*. The former concentrates masses on nodes of both rigid and flexible parts; the latter considers the real distribution of masses inside beams. As will be explained in the text, the two methods produce good results, particularly when an accurate degree of partitioning is chosen for flexible bodies. Finally, we focus our attention on a way to consider joints with mass and inertia.

5.1 Lumped approach

Reducing mass and inertia of a rigid body to a particular point, the center of mass, without changing dynamic properties of the system, is a common procedure. On the contrary, for flexible bodies every reduction process is an approximation generating mistakes. Let us refer to Fig. 8 in which a link is divided into four beams. In the lumped approach the mass

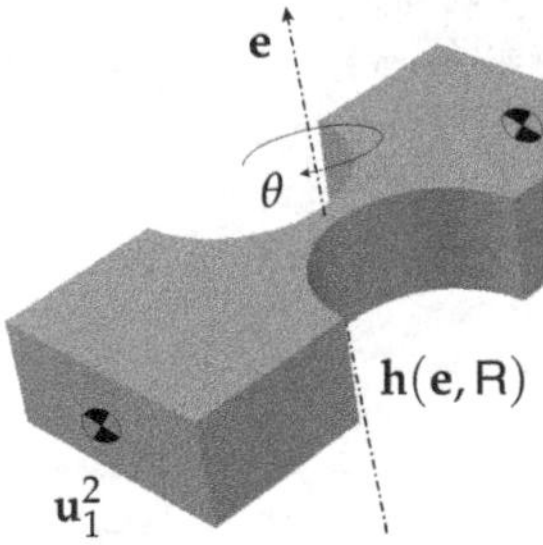

(a) Ideal flexure joint

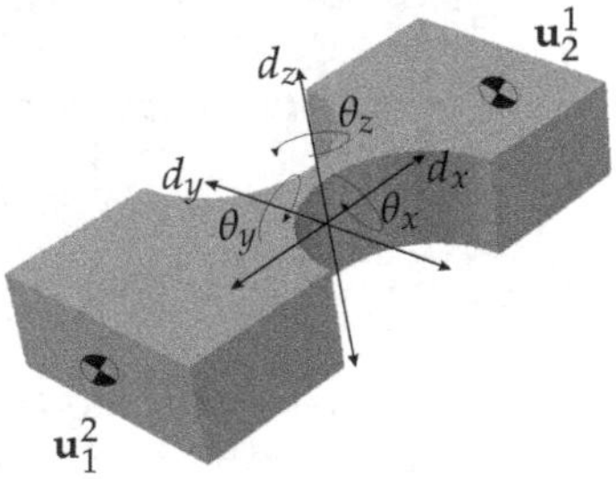

(b) Real flexure joint

Fig. 6. Flexure revolute joint.

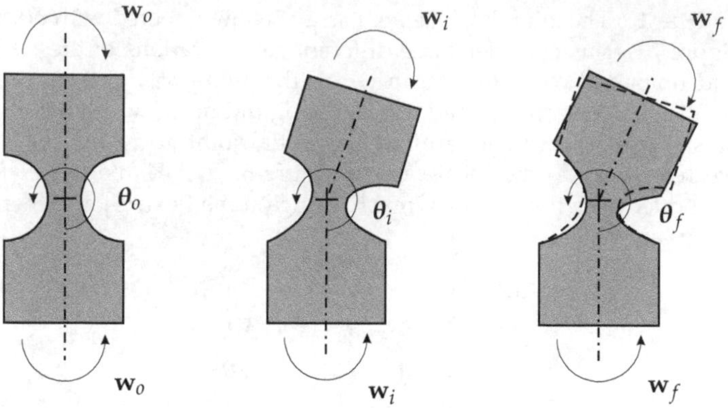

Fig. 7. Flexure joint: a) Undeformed configuration; b) starting preloaded configuration; c) small rotated configuration about the starting preloaded configuration.

and inertia of each beam is concentrated on its end-nodes. Here, we choose a symmetric distribution in which the mass m is divided by two, but the reader can use another distribution according to the case to be examined, as instance beams with varying cross section. The first node has a mass of $m/2$ while any other node, but the fourth, bears a mass m because it receives half a mass from each of the two contiguous beams coupled at its section. The fourth beam of the link is attached to a joint. According to what explained in the previous section, even the mass is concentrated only on the independent joints: it means that the mass of the last beam has to be concentrated on the fourth node, thereby the latter carrying a mass equal to $1/2m + m = 3/2m$. Analogous arguments, not reported here for conciseness, may be used to describe inertias.

The lumped approach concentrates masses and inertias on all the independent nodes, belonging to both rigid and flexible bodies. Mathematically, a 6×6 mass dyad $\widetilde{\mathbf{M}}_i$ is associated at the ith-independent node, defined as

$$\widetilde{\mathbf{M}}_i = \begin{bmatrix} m_i\mathbf{1} & \mathbf{O} \\ \mathbf{O} & \mathbf{J}_i \end{bmatrix} \tag{35}$$

where m_i and $\mathbf{J}_i$ are the mass and the 3×3 inertia matrix, respectively, of either a rigid body, if the independent node is at the center of mass of the said body, or of a beam's end-node. The generalized inertia matrix $\mathbf{M}_{PKM}$ of a PKM is readily derived upon assembling in diagonal blocks the previous inertia dyads, following the order chosen to enumerate all independent nodes inside the global array of independent nodal coordinates, hence

$$\mathbf{M}_{PKM} = diag(\widetilde{\mathbf{M}}_1, \widetilde{\mathbf{M}}_2, \ldots, \widetilde{\mathbf{M}}_n) \tag{36}$$

5.2 Distributed approach

The distributed approach considers the true distribution of mass inside a flexible beam. Particularly, it is possible to find a 12×12 matrix $\mathbf{M}_i$ referred to the twelve nodal coordinates of a beam's end-nodes. This matrix can be divided into blocks, whose entries are reported below, as already done for the stiffness matrix $\mathbf{K}_i$, i.e.

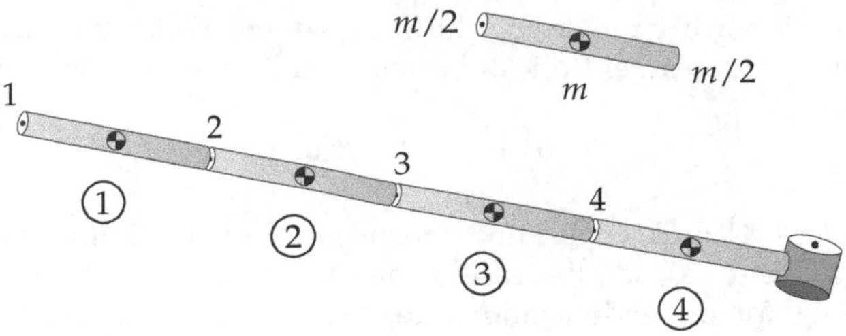

Fig. 8. Lumped distribution of mass.

$$\mathbf{M}_i^{1,1} = \rho A_i L_i \begin{bmatrix} \frac{1}{3} & 0 & 0 & 0 & 0 & 0 \\ 0 & \frac{13}{35} + \frac{6I_z}{5\rho A_i L_i^2} & 0 & 0 & 0 & \frac{11L_i}{210} + \frac{I_z}{10\rho A_i L_i} \\ 0 & 0 & \frac{13}{35} + \frac{6I_z}{5\rho A_i L_i^2} & 0 & \frac{-11L_i}{210} - \frac{I_y}{10\rho A_i L_i} & 0 \\ 0 & 0 & 0 & \frac{I_x}{3\rho A_i} & 0 & 0 \\ 0 & 0 & \frac{-11L_i}{210} - \frac{I_y}{10\rho A_i L_i} & 0 & \frac{L_i^2}{105} + \frac{2I_y}{15\rho A_i} & 0 \\ 0 & \frac{11L_i}{210} + \frac{I_z}{10\rho A_i L_i} & 0 & 0 & 0 & \frac{L_i^2}{105} + \frac{2I_z}{15\rho A_i} \end{bmatrix} \tag{37}$$

$$\mathbf{M}_i^{1,2} = \rho A_i L_i \begin{bmatrix} \frac{1}{6} & 0 & 0 & 0 & 0 & 0 \\ 0 & \frac{9}{70} - \frac{6I_z}{5\rho A_i L_i^2} & 0 & 0 & 0 & -\frac{13L_i}{420} + \frac{I_z}{10\rho A_i L_i} \\ 0 & 0 & \frac{9}{70} - \frac{6I_y}{5\rho A_i L_i^2} & 0 & \frac{13L}{420} - \frac{I_y}{10\rho A_i L_i} & 0 \\ 0 & 0 & 0 & \frac{I_x}{3\rho A_i} & 0 & 0 \\ 0 & 0 & -\frac{13L_i}{420} + \frac{I_z}{10\rho A_i L_i} & 0 & \frac{-L_i^2}{140} - \frac{I_y}{30\rho A_i} & 0 \\ 0 & \frac{13L_i}{420} - \frac{I_z}{10\rho A_i L_i} & 0 & 0 & 0 & \frac{-L_i^2}{140} - \frac{I_z}{30\rho A_i} \end{bmatrix} \tag{38}$$

$$\mathbf{M}_i^{2,2} = \rho A_i L_i \begin{bmatrix} \frac{1}{3} & 0 & 0 & 0 & 0 & 0 \\ 0 & \frac{13}{35} + \frac{6I_z}{5\rho A_i L_i^2} & 0 & 0 & 0 & -\frac{11L_i}{210} - \frac{I_z}{10\rho A_i L_i} \\ 0 & 0 & \frac{13}{35} + \frac{6I_y}{5\rho A_i L_i^2} & 0 & \frac{11L_i}{210} + \frac{I_y}{10\rho A_i L_i} & 0 \\ 0 & 0 & 0 & \frac{I_x}{3\rho A_i} & 0 & 0 \\ 0 & 0 & \frac{11L_i}{210} + \frac{I_y}{10\rho A_i L_i} & 0 & \frac{L_i^2}{105} + \frac{2I_y}{15\rho A_i} & 0 \\ 0 & -\frac{11L_i}{210} - \frac{I_z}{10\rho A_i L_i} & 0 & 0 & 0 & \frac{L_i^2}{105} + \frac{2I_z}{15\rho A_i} \end{bmatrix} \tag{39}$$

where ρ, L_i, A_i, I_x, I_y and I_z, respectively, are the density, the length, the cross section area and the mass moments of inertia for unit of length of the ith-beam. The matrix $\mathbf{M}_i$ is associated to the kinetic energy T_i of a beam, the latter being defined as a quadratic forms into the time-derivatives $\dot{\mathbf{u}}_i^1$, $\dot{\mathbf{u}}_i^2$ of the nodal displacement arrays $\mathbf{u}_i^1$ and $\mathbf{u}_i^2$:

$$T_i = \frac{1}{2} \begin{bmatrix} \dot{\mathbf{u}}_i^1 \\ \dot{\mathbf{u}}_i^2 \end{bmatrix}^T \begin{bmatrix} \mathbf{M}_i^{1,1} & \mathbf{M}_i^{1,2} \\ \mathbf{M}_i^{2,1} & \mathbf{M}_i^{2,2} \end{bmatrix} \begin{bmatrix} \dot{\mathbf{u}}_i^1 \\ \dot{\mathbf{u}}_i^2 \end{bmatrix} \tag{40}$$

In the previous section we have found how dependent nodal displacement arrays are expressed in terms of independent ones. Let us consider the time-derivative of both sides of eq.(13), we write

$$\dot{\mathbf{u}}_2^1 = \mathbf{X}^1\dot{\mathbf{u}}_1^1 + \mathbf{X}^2\dot{\mathbf{u}}_2^2 \tag{41}$$

in which the matrices $\mathbf{X}^1$ and $\mathbf{X}^2$ are not dependent on time. Similar expressions can be obtained for eqs.(21) and (26) for the case b) and for the case c). It means that the same expressions standing for dependent and independent displacements can be extended at velocity and acceleration level. Therefore, the following matrices $\mathbf{M}_a$, $\mathbf{M}_b$ and $\mathbf{M}_c$ can be written with perfect analogy to their counterparts $\mathbf{K}_a$, $\mathbf{K}_b$ and $\mathbf{K}_c$:

$$\mathbf{M}_a = \begin{bmatrix} \mathbf{X}^1 & \mathbf{X}^2 \\ \mathbf{O} & \mathbf{1} \end{bmatrix}^T \begin{bmatrix} \mathbf{M}_2^{1,1} & \mathbf{M}_2^{1,2} \\ \mathbf{M}_2^{2,1} & \mathbf{M}_2^{2,2} \end{bmatrix} \begin{bmatrix} \mathbf{X}^1 & \mathbf{X}^2 \\ \mathbf{O} & \mathbf{1} \end{bmatrix} \tag{42a}$$

$$\mathbf{M}_b = \begin{bmatrix} \mathbf{1} & \mathbf{O} \\ \mathbf{X}^1 & \mathbf{X}^2 \end{bmatrix}^T \begin{bmatrix} \mathbf{M}_1^{1,1} & \mathbf{M}_1^{1,2} \\ \mathbf{M}_1^{2,1} & \mathbf{M}_1^{2,2} \end{bmatrix} \begin{bmatrix} \mathbf{1} & \mathbf{O} \\ \mathbf{X}^1 & \mathbf{X}^2 \end{bmatrix} + \begin{bmatrix} \mathbf{G}^1 & \mathbf{G}^2 \\ \mathbf{O} & \mathbf{1} \end{bmatrix}^T \begin{bmatrix} \mathbf{M}_2^{1,1} & \mathbf{M}_2^{1,2} \\ \mathbf{M}_2^{2,1} & \mathbf{M}_2^{2,2} \end{bmatrix} \begin{bmatrix} \mathbf{G}^1 & \mathbf{G}^2 \\ \mathbf{O} & \mathbf{1} \end{bmatrix} \tag{42b}$$

$$\mathbf{M}_c = \begin{bmatrix} \mathbf{1} & \mathbf{O} \\ \mathbf{X}^1 & \mathbf{X}^2 \end{bmatrix}^T \begin{bmatrix} \mathbf{M}_1^{1,1} & \mathbf{M}_1^{1,2} \\ \mathbf{M}_1^{2,1} & \mathbf{M}_1^{2,2} \end{bmatrix} \begin{bmatrix} \mathbf{1} & \mathbf{O} \\ \mathbf{X}^1 & \mathbf{X}^2 \end{bmatrix} \tag{42c}$$

with obvious meaning of all terms. The three mass matrices, above defined, are referred to the time-derivatives $\dot{\tilde{\mathbf{u}}}_a$, $\dot{\tilde{\mathbf{u}}}_b$ and $\dot{\tilde{\mathbf{u}}}_c$ of the independent nodal displacement arrays $\tilde{\mathbf{u}}_a$, $\tilde{\mathbf{u}}_b$ and $\tilde{\mathbf{u}}_c$. To clarify some doubt that might arise let us refer to Fig. 3. In this case the mass matrix of the rigid body, see eq.(35), is referred to its center of mass with displacement array $\mathbf{u}_1^1$. The distributed approach first allows us to find the 12×12 mass matrix $\mathbf{M}_2$ of the beam, then the latter is expressed in terms of only independent displacements by means of $\mathbf{M}_a$. Obviously, $\mathbf{M}_2$ and $\mathbf{M}_a$ refer to the same object, the beam; but $\mathbf{M}_a$ *distributes* $\mathbf{M}_2$ into the independent nodes with displacement arrays $\mathbf{u}_1^1$ and $\mathbf{u}_2^2$. This implies that the center of mass of the rigid body carries its own mass/inertia and part of the mass/inertia of the beam. Similar conclusions may be sought for the cases b) and c).

5.3 Joint's mass and inertia

In this final subsection we describe a method to consider mass and inertia of joints. For convenience, let us refer to Fig. 4. The joint is split into two parts, one belonging to the first beam, the remaining to the second beam. Let $\widetilde{\mathbf{M}}_L$ and $\widetilde{\mathbf{M}}_R$, where capital letters stand for left and right, be the mass dyads of the two half-parts of the joint, defined as

$$\widetilde{\mathbf{M}}_L = \begin{bmatrix} m_L\mathbf{1} & \mathbf{O} \\ \mathbf{O} & \mathbf{J}_L \end{bmatrix} \quad \widetilde{\mathbf{M}}_R = \begin{bmatrix} m_R\mathbf{1} & \mathbf{O} \\ \mathbf{O} & \mathbf{J}_R \end{bmatrix} \tag{43}$$

The kinetic energy T_J of the joint can be written in terms of the above matrices $\widetilde{\mathbf{M}}_L$ and $\widetilde{\mathbf{M}}_R$ and of the time-derivatives of the nodal arrays $\dot{\mathbf{u}}_1^2$, $\dot{\mathbf{u}}_2^1$, thus

$$T_J = \frac{1}{2}(\dot{\mathbf{u}}_1^2)^T\widetilde{\mathbf{M}}_L\dot{\mathbf{u}}_1^2 + \frac{1}{2}(\dot{\mathbf{u}}_2^1)^T\widetilde{\mathbf{M}}_R\dot{\mathbf{u}}_2^1 \tag{44}$$

Then, upon recalling eqs.(21) and (26), T_J can be expressed in terms of $\tilde{\mathbf{u}} = \begin{bmatrix} \mathbf{u}_1^{1^T} & \mathbf{u}_2^{2^T} \end{bmatrix}^T$, i.e.

$$T_J = \frac{1}{2}(\tilde{\mathbf{u}})^T \mathbf{M}_J \tilde{\mathbf{u}} \tag{45}$$

where $\mathbf{M}_J$ is the 12×12 generalized inertia matrix of the joint expressed in terms of independent nodal displacements, respectively, defined as

$$\mathbf{M}_a = \begin{bmatrix} \mathbf{X}^{1^T}\widetilde{\mathbf{M}}_R\mathbf{X}^1 + \mathbf{G}^T\widetilde{\mathbf{M}}_L\mathbf{G} & \mathbf{X}^{1^T}\widetilde{\mathbf{M}}_R\mathbf{X}^2 \\ \mathbf{X}^{2^T}\widetilde{\mathbf{M}}_R\mathbf{X}^1 & \mathbf{X}^{2^T}\widetilde{\mathbf{M}}_R\mathbf{X}^2 \end{bmatrix} \tag{46a}$$

$$\mathbf{M}_b = \begin{bmatrix} \mathbf{X}^{1^T}\widetilde{\mathbf{M}}_L\mathbf{X}^1 + \mathbf{G}^{1^T}\widetilde{\mathbf{M}}_R\mathbf{G}^1 & \mathbf{X}^{1^T}\widetilde{\mathbf{M}}_L\mathbf{X}^2 + \mathbf{G}^{1^T}\widetilde{\mathbf{M}}_R\mathbf{G}^2 \\ \mathbf{X}^{2^T}\widetilde{\mathbf{M}}_L\mathbf{X}^1 + \mathbf{G}^{2^T}\widetilde{\mathbf{M}}_R\mathbf{G}^1 & \mathbf{X}^{2^T}\widetilde{\mathbf{M}}_L\mathbf{X}^2 + \mathbf{G}^{2^T}\widetilde{\mathbf{M}}_R\mathbf{G}^2 \end{bmatrix} \tag{46b}$$

$$\mathbf{M}_c = \begin{bmatrix} \mathbf{X}^{1^T}\widetilde{\mathbf{M}}_L\mathbf{X}^1 & \mathbf{X}^{1^T}\widetilde{\mathbf{M}}_L\mathbf{X}^2 \\ \mathbf{X}^{2^T}\widetilde{\mathbf{M}}_L\mathbf{X}^1 & \mathbf{X}^{2^T}\widetilde{\mathbf{M}}_L\mathbf{X}^2 + \mathbf{G}^T\widetilde{\mathbf{M}}_R\mathbf{G} \end{bmatrix} \tag{46c}$$

for the three cases in exam.

6. Linearized elastodynamics equations

In this section we will derive the generalized inertia and stiffness matrices necessary to write the linearized elastodynamics equations. As described in the two previous sections stiffness and inertia matrices of rigid bodies and flexible beams are referred to the independent nodal displacements of the system. Now, one might ask how to combine different matrices to build the global stiffness and inertia matrices. In order to solve this issue let us introduce two Boolean matrices $\mathbf{B}_1$ and $\mathbf{B}_2$ able to map a 6-dimensional displacement array $\mathbf{u}_i$ or a 12-dimensional displacement array $\tilde{\mathbf{u}}$ in terms of a $6n$-dimensional global array $\mathbf{q} = \begin{bmatrix} \mathbf{u}_1 \; \mathbf{u}_2 \ldots \mathbf{u}_n \end{bmatrix}$ containing all the independent displacement arrays of the robot. It is important that the reader would define the order in which every single array $\mathbf{u}_i$ appears inside $\mathbf{q}$. The $6 \times 6n$ matrix $\mathbf{B}_1$ and the $12 \times 6n$ matrix $\mathbf{B}_2$ are defined as:

$$\mathbf{u}_i = \mathbf{B}_1(i)\mathbf{q}, \quad \tilde{\mathbf{u}} \equiv \begin{bmatrix} \mathbf{u}_i \\ \mathbf{u}_j \end{bmatrix} = \mathbf{B}_2(i,j)\mathbf{q} \tag{47}$$

$$B_1(i) = \begin{bmatrix} \mathbf{O}_6 \; \mathbf{O}_6 \ldots \mathbf{1}_6(i) \ldots \mathbf{O}_6 \end{bmatrix}, \quad B_2(i,j) = \begin{bmatrix} \mathbf{O}_6 & \mathbf{O}_6 & \ldots \mathbf{1}_6(i) \ldots & \mathbf{O}_6 \\ \mathbf{O}_6 & \mathbf{1}_6(j) \ldots & \mathbf{O}_6 & \ldots \mathbf{O}_6 \end{bmatrix} \tag{48}$$

The above expressions allow us to convert each nodal displacement array in term of $\mathbf{q}$. As instance, let us take into exam the strain energy V_b of eq.(27), it simply turns into $V_b = \frac{1}{2}\mathbf{q}^T\mathbf{B}_2(i,j)^T\mathbf{K}_b\mathbf{B}_2(i,j)\mathbf{q}$, where i and j indicate the position indices of $\mathbf{u}_1^1$ and $\mathbf{u}_2^2$ inside $\mathbf{q}$. From V_b we can extract a new $6n \times 6n$ matrix $\overline{\mathbf{K}}_b$, that is a stiffness matrix expanded to the final dimension of the problem, defined as

$$\overline{\mathbf{K}}_b = \mathbf{B}_2(i,j)^T\mathbf{K}_b\mathbf{B}_2(i,j) \tag{49}$$

Likewise, the generic expanded $6n \times 6n$ mass matrix $\overline{\mathbf{M}}_i$ of $\mathbf{M}_i$, see eq.(35), can be written as $\overline{\mathbf{M}}_i = \mathbf{B}_1(i)^T\mathbf{M}_i\mathbf{B}_1(i)$, where i is the position index of $\mathbf{u}_i$, i.e. the displacement array of the ith-rigid body's center of mass, inside $\mathbf{q}$. By the same strategy, it is possible to expand every stiffness and inertia matrix. This operation is essential as, *only when referred to the same set*

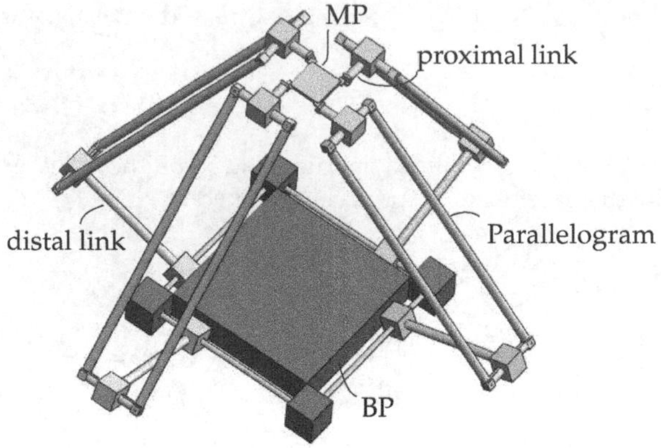

Fig. 9. CAD model of the robot.

of nodal displacement coordinates, these matrices can be summed and combined to obtain the global generalized matrices $\mathbf{K}_{PKM}$ and $\mathbf{M}_{PKM}$ of the robot. The way to use $\mathbf{B}_1$ and $\mathbf{B}_2$ will be shown in detail in the case study of the next section.
Let us introduce the $6n$-dimensional array $\mathbf{f}$ of generalized nodal wrenches, i.e. nodal forces and torques, then, the system of linearized elastodynamics equations becomes

$$\mathbf{M}_{PKM}\ddot{\mathbf{q}} + \mathbf{K}_{PKM}\mathbf{q} = \mathbf{f} \tag{50}$$

The previous system may be used to solve statics around a starting posture of the robot. The homogeneous part of eq.(50), i.e. the left side, may be used to find eigenfrequencies and eigenmodes of the robot, or the zero-input response. As already pointed out, the natural frequencies of the robot change with regard to the pose that the robot is attaining. To determine how stiffness or natural frequencies vary inside the workspace local and global performance indices may be introduced to investigate elastodynamic behavior of PKMs. As instance, let us consider the first natural frequency f_1 mapped inside a robot's workspace. The latter can be analyzed to differentiate areas with high range of frequency from areas in which the elastodynamic performances worsen. The multidimensional integral Ω_1 of f_1, extended to the whole workspace W or part of it, can be a good global index to show how global performances increase or decrease changing some geometrical, structural or inertial parameter, i.e.

$$\Omega_1 = \frac{\int_W f_1 dW}{W} \approx \frac{\sum_{i=1}^{n_w} W_i f_{1i}}{\sum_{i=1}^{n_w} W_i} \tag{51}$$

where, in the approximated discrete formula of the right side, W is divided into n_w hypercubes W_i, while the frequency f_{1i} is calculated at the center point of W_i.
We conclude this section considering an useful application of the method to find singularities. It is well known that as a robot approaches to a singularity configuration it gains mobility. As a consequence, its generalized stiffness matrix becomes singular and its first eigenfrequency goes to zero. Analyzing the variation of f_1 inside the workspace may be useful to avoid zones close to singularity or with low values of frequency. We stress that singularities directly come from kinematics: stiffness and inertial parameters, respectively influencing the elasticity

and the dynamics of a mechanical system, can only amplify or reduce, without creating or nullifying, the effects of the kinematics.
Different considerations can be made for the boundary surface of the workspace. For the latter case, a robot undergoes poses in which one or more legs are completely extended, thus the robot loses dofs associated to those directions normal to the tangent plane at the boundary surface of the workspace. If the displacement of the MP for the corresponding eigenmode, when evaluated at a pose lying on the boundary surface, is normal to the said tangent plane the robot is virtually locked, it means that its stiffness along that direction is high, thus reflecting into an increment of frequency. On the contrary, if the displacement of the MP lies onto the tangent plane its in-plane stiffness, and its natural frequency as well, in theory goes to zero. All remaining cases are included between zero and a maximum value depending on the projection of displacements on the tangent plane. This important conclusion can be summarized by the following statement: ***the value of a given natural frequency of a PKM on the boundary surface of its workspace can be interpreted through the projection of the moving platform's displacements, for the corresponding eigenmode, on the tangent plane to the said surface***. The previous statement is strict only when a three dimensional workspace, e.g. the constant orientation workspace, can be considered. In other cases, one should recur to concepts of projection based on twist theory, beyond the scopes of this chapter.

7. Case study

In this section we apply the method to a complex PKM with 4-dofs. The robot, shown in Fig. 9, is composed of four legs connecting the base frame to a moving platform. Due to the particular kind of constraints generated by the limbs, the moving platform can accomplish three translations and a rotation around the vertical axis. This motion, even referred to as Schönflies motion, is typical of SCARA robots and it is largely used in industry for pick and place applications. Each leg is composed of a distal link connected to the base frame by means of an actuated revolute joint and to a parallelogram linkage, i.e. a Π-joint, by means of a revolute joint. The parallelogram, in turn, is linked to a proximal link by another revolute. Finally, the proximal link is coupled to the MP by a vertical axis revolute joint. All revolute joints, apart from the one fixed to the inertial frame, are passive. We do not further go inside the analysis of the kinematics, citing (24) for more explanations and for any detail pertaining the robot's geometry.

7.1 Application of the method to a generic leg

In order to apply the proposed method each leg is decomposed into three modules. The first module includes the BP and the proximal link, the second the parallelogram linkage, while the last includes the distal link and the MP. We recur to some simplifications that do not change the meaning of the treatment. In general, the two bases of a PKM are one order stiffer than the links composing the structure and can be modeled as rigid without loss of accuracy. Even the short input and output links of a Π-joint can be considered rigid when compared to the slender coupler links. As verified by FEM, the said assumptions are good approximations that simplify the final model introducing rigid-flexible and flexible-rigid connections inside and between each module.
Let us analyze the first module shown in Fig. 10. The distal link is split into three flexible bodies, the choice of discretization being absolutely arbitrary. In this simple case the end-bodies F_1 and F_3 of the link are coupled to two rigid bodies, i.e., the base BP and the input link I of the Π-joint, by means of two revolute joints of axes parallel to the unit vector

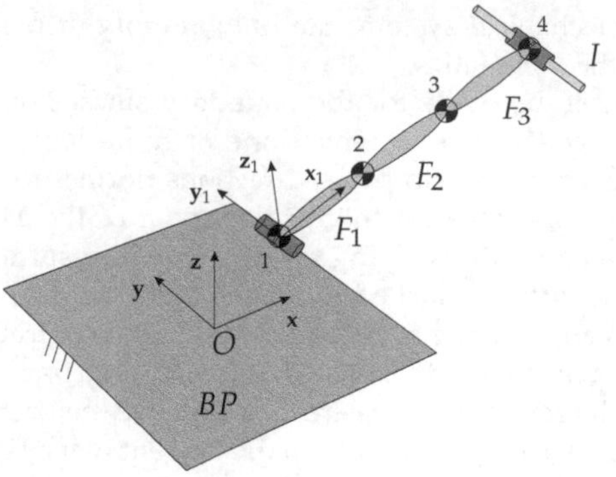

Fig. 10. Drawing of a the first module: Base platform-distal link

e. The remaining flexible body F_2 is internal to the proximal link and coupled to the others bodies by means of fixed connections. Following what explained in the previous sections, the strain energy V_{m1} of the first module is the sum of three components, i.e. the strain energies V_{F_i} of the three beams in which is decomposed, namely

$$V_{m1} = \sum_{i=1}^{3} V_{F_i} = \sum_{i=1}^{3} \frac{1}{2} \tilde{\mathbf{u}}_i^T \mathbf{K}_i \tilde{\mathbf{u}}_i \tag{52}$$

with $\tilde{\mathbf{u}}_1 = \left[\mathbf{u}_1^{1^T} \ \mathbf{u}_1^{2^T} \right]^T$, $\tilde{\mathbf{u}}_2 = \left[\mathbf{u}_2^{2^T} \ \mathbf{u}_2^{3^T} \right]^T$ and $\tilde{\mathbf{u}}_3 = \left[\mathbf{u}_3^{3^T} \ \mathbf{u}_3^{4^T} \right]^T$. The first actuated revolute joint is considered locked to perform the elastodynamics analysis, thereby, $\mathbf{u}_1^1 \equiv \mathbf{u}_{BP}^1 \equiv \overline{\mathbf{u}}^1$. In this way the generalized stiffness matrix of the robot is not singular as rigid body motions of the robot are deleted and the analysis is performed at a frozen configuration. Even for the other three fixed connections at points 2 and 3, similar equations stand: $\mathbf{u}_1^2 \equiv \mathbf{u}_2^2 \equiv \overline{\mathbf{u}}^2$ and $\mathbf{u}_2^3 \equiv \mathbf{u}_3^3 \equiv \overline{\mathbf{u}}^3$. Finally, for the displacement array of the input link R, we can write: $\mathbf{u}_R \equiv \overline{\mathbf{u}}^4$. By substituting into the previous expressions, we derive $\tilde{\mathbf{u}}_1 = \left[\overline{\mathbf{u}}^{1^T} \ \overline{\mathbf{u}}^{2^T} \right]^T$, $\tilde{\mathbf{u}}_2 = \left[\overline{\mathbf{u}}^{2^T} \ \overline{\mathbf{u}}^{3^T} \right]^T$ and $\tilde{\mathbf{u}}_3 = \left[\overline{\mathbf{u}}^{3^T} \ \mathbf{u}_3^{4^T} \right]^T$. The local stiffness matrix $\mathbf{K}_i$ of the ith-body F_i is expressed into the global reference frame. In Figure 10 the first local frame $\{\mathbf{x}_1, \mathbf{y}_1, \mathbf{z}_1\}$, with origin at point 1, and the base global reference frame $\{\mathbf{x}, \mathbf{y}, \mathbf{z}\}$, with origin at point O, are displayed. The set of independent displacements are defined inside the independent displacement array $\mathbf{q}_{M1}$ of the first module: $\mathbf{q}_{M1} = \left[\overline{\mathbf{u}}^1 \ \overline{\mathbf{u}}^2 \ \overline{\mathbf{u}}^3 \ \overline{\mathbf{u}}^4 \right]$, where $\overline{\mathbf{u}}^4$ is the nodal displacement array of the center of mass of the rigid body I. While the first and the second term, i.e. V_{F_1} and V_{F_2}, of the strain energy V_{m1} contain only independent displacements, the last term V_{F_3} is function of the dependent nodal array $\mathbf{u}_3^4$. The flexible body F_3, indeed, is coupled by a revolute joint to the rigid body I, thus, the case c), flexible-rigid, can be applied to express its strain energy only in terms of independent displacements. Following the notation above introduced, we write:

$$V_{F_3} = \frac{1}{2} \tilde{\mathbf{u}}_3^T \mathbf{K}_3 \tilde{\mathbf{u}}_3 \equiv \frac{1}{2} \tilde{\mathbf{u}}_{c3}^T \mathbf{K}_{c3} \tilde{\mathbf{u}}_{c3} \tag{53}$$

where $\tilde{\mathbf{u}}_{c3} \equiv \left[\overline{\mathbf{u}}^{3^T} \ \overline{\mathbf{u}}^{4^T} \right]^T$ and $\mathbf{K}_{c3}$ has been defined in eq.(31). Notice that in order to find $\mathbf{K}_{c3}$ we have to use the 6-dimensional joint array $\mathbf{h}^R(\mathbf{e})$; besides, the position vector $\mathbf{d}$ going

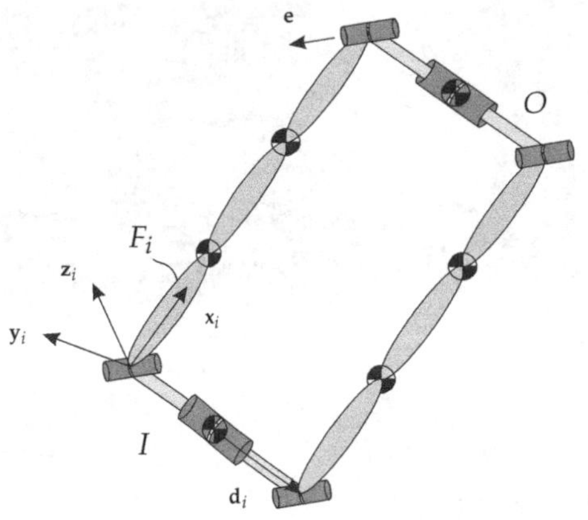

(a) Drawing of a the second module: parallelogram linkage

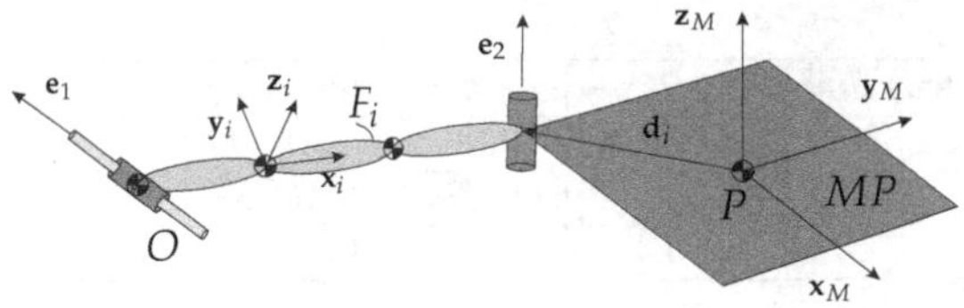

(b) Drawing of a the third module: proximal link-moving platform

from the center of mass of I to the center of the revolute joint, in this case, is the zero vector. Then, by introducing the $12 \times n_{qm1}$ binary-entry matrix $\mathbf{B}_2(\bullet, \bullet)$ mapping the independent nodal-arrays in terms of the array $\mathbf{q}_{M1}$, the expression of the generalized stiffness matrix $\mathbf{K}_{M1}$ can be calculated as

$$\begin{aligned} V_{m1} &= \frac{1}{2}\mathbf{q}_{M1}{}^T\mathbf{K}_{M1}\mathbf{q}_{M1} \\ \mathbf{K}_{M1} &= \mathbf{B}_2(1,2)^T\mathbf{K}_1\mathbf{B}_2(1,2) + \mathbf{B}_2(2,3)^T\mathbf{K}_2\mathbf{B}_2(2,3) + \mathbf{B}_2(3,4)^T\mathbf{K}_{c3}\mathbf{B}_2(3,4) \end{aligned} \tag{54}$$

Here, the matrix $\mathbf{K}_{M1}$ is expressed in terms of $\mathbf{q}_{M1}$. The reader must notice that to obtain $\mathbf{K}_{PKM}$ each matrix has to be expressed in terms of a final n_q-dimensional array $\mathbf{q}$ including *all* the independent nodal displacements. In order to define $\mathbf{q}$, as the modules of a leg are in series, the last node of a module coincides with the first one of the next module and it must be denoted by a unique nodal displacement array. Extension to the final dimension of $\mathbf{q}$ is readily obtained by substituting into eq.(54) a new $12 \times n_q$ binary-entry matrix $\mathbf{B}_2(\bullet, \bullet)$ mapping the independent nodal-arrays in terms of the global array $\mathbf{q}$.

The second module is shown in Fig. 11(a). In this case two rigid-flexible and two flexible-rigid connections must be used to describe the couplings between flexible couplers and input-output rigid bodies I and O. The joint array $\mathbf{h}^R(\mathbf{e})$ remains the same for all the four revolute joints as a matter of fact of the parallelogram architecture. The third module, shown in Fig. 11(b), includes one rigid-flexible and one flexible-rigid connection, respectively, between the distal link and the output link O of the Π-joint and the distal link and the MP. As displayed in the same figure, the two revolute joints to be used to find the joint arrays $\mathbf{h}^R$ have axes of unit vectors $\mathbf{e}_1$ and $\mathbf{e}_2$. We do not describe the calculations to find the generalized

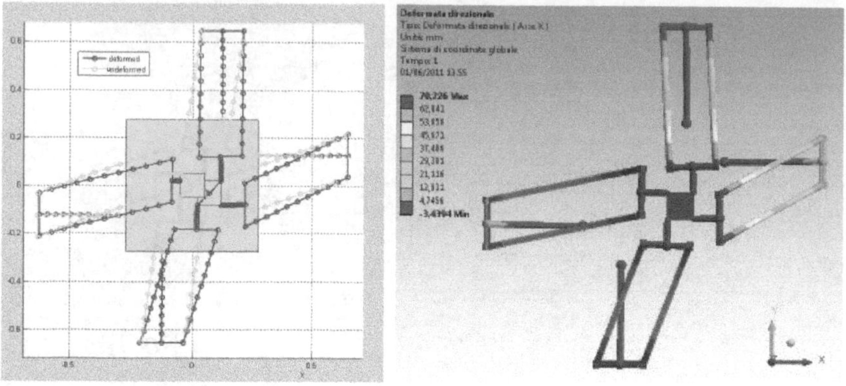

(c) Algorithm model (d) Ansys© model

Fig. 11. Static deformation of the robot.

Disp.	*Algorithm*	*Marc©*	*Ansys©*	*err% Marc©*	*err% Ansys©*
u_x	0.068066	0.068065	0.067436	−0.0008815	−0.933626
u_y	−0.030545	−0.030545	−0.029890	0.003274	−2.191368
u_z	0	0	0	0	0

Table 1. Translational displacements of the end-effector's center node.

stiffness matrices of the two modules as can be readily obtained following procedures similar to the case of the first module.

7.2 Comparison to FEA software and validation

In this subsection the proposed elastodynamics model is compared to FEA results for validation. Two FEA models, with increasing complexity, are implemented in order to establish the nearness of the examined case to the real case. The first model, developed by the commercial software Marc©, considers only 3D-beams and rigid bodies, while joints are modeled by means of relative degrees of freedom between common nodes belonging to two coupled bodies. This model perfectly fits the simplifications of the method in exam. The second model, developed by the commercial software Ansys©, describes a robot with a complex structure closer to reality. Links are solids while joints have finite burdens and provide their function by means of the coupling of surfaces and screws.

7.2.1 Statics

We first compared the static deformation of the robot when an external force of 1000 N, along the x-direction, is applied on the end-effector, when the latter is positioned at its home pose with angle of rotation of MP equal to zero. Figure 11 shows the displacements of our model when compared to Ansys© model, while in Tab. 1 the translational displacements of the end-effector center node are reported.
It can be observed how the first MARC© model perfectly fit to the results of our method. The relative error grows, but still remaining limited, when displacements are compared to Ansys© model. The reason is well understood as it comes from the use of solid bodies and joints to simulate the robot's structure.

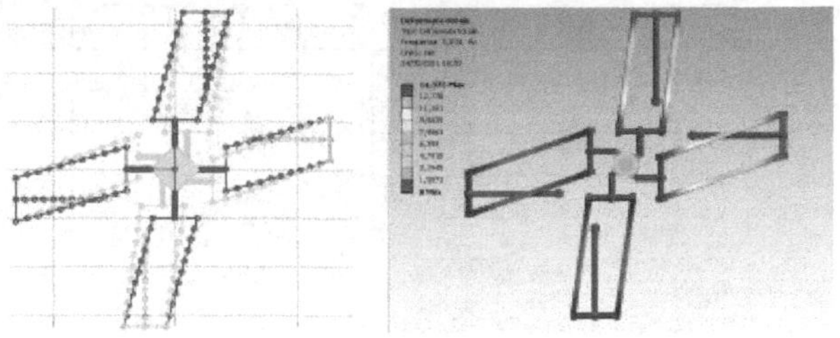

(a) First mode

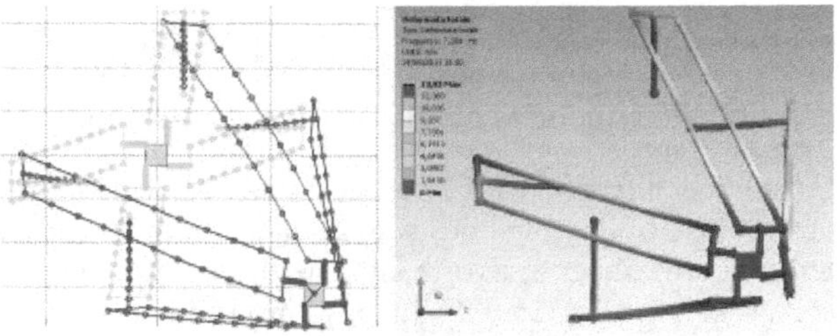

(b) Second mode

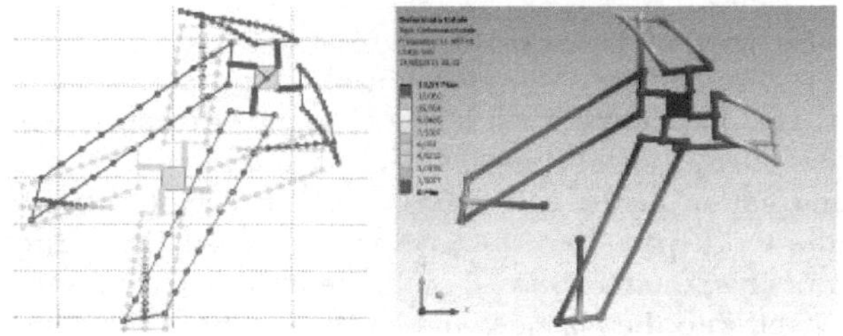

(c) Third mode

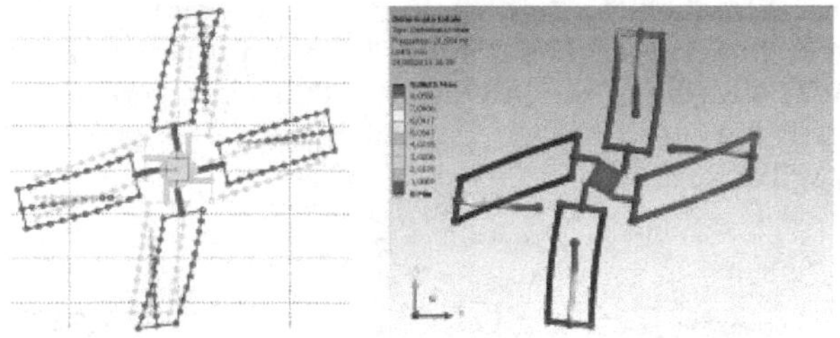

(d) Fourth mode

Fig. 12. Modes comparison: Algorithm *vs.* Ansys©.

7.2.2 Natural modes and frequencies

As in the case of statics, the elastodynamics model of the 3T1R robot is used to compare natural modes and frequencies to FEA software. We report only the comparison to Ansys©, as more indicative of a feasible application of the method to a real case. Figure 12 shows the first four natural modes obtained when the robot is attaining its home pose. In Table 2 the first ten natural frequency of the robot, at the same pose, are reported. Results show good agreement

Freq.	Algorithm [Hz]	Ansys© [Hz]	err%
1	2.920	3.019	3.28%
2	7.062	7.204	1.97%
3	11.275	11.487	1.85%
4	20.834	21.554	3.34%
5	25.480	25.722	0.94%
6	25.545	25.804	1.00%
7	26.028	26.181	0.58%
8	28.306	28.830	1.82%
9	30.368	31.295	2.96%
10	31.060	32.286	3.80%

Table 2. The first ten natural frequencies of the robot at the home pose.

with Ansys© model. Some discrepancies still occur due to the same reasons explained for the static case and to the use of flexible bodies to simulate all links and platforms in Ansys©. This choice has been taken in order to avoid asymmetric contacts between rigid and flexible solids leading to convergence mistakes. In turn, we used a stiffer material to approximate rigid behavior of the MP and rigid links.

7.2.3 Distribution of frequencies inside the workspace

In order to provide a further useful extension of the proposed method, the first natural frequency is calculated and plotted inside the constant orientation workspace of the robot (25). We have chosen elementary cubes of side 5 cm to discretize the workspace of the robot while the angle of rotation of the MP is $\theta_z = 0$. It can be observed that two privileged diagonals divide the workspace into four symmetric areas. These directions coincide with the two diagonals of the squared BP of Fig. 9, meaning that, at the home pose, geometric symmetry reflects itself into the elastic behavior of the robot. The boundary of the constant orientation workspace shows areas with high range of frequency along with other areas in which the first natural frequency reaches values close to zero (Hz). As already outlined in the text, this behavior is due to the MP's displacement (deformation) in correspondence of the first eigenmode: if at a certain pose, in which the analysis is performed, the displacement is along a *doc* of the MP, the ensuing frequency will be high; conversely, if it is along a *dof* of the MP, the frequency will come near zero.

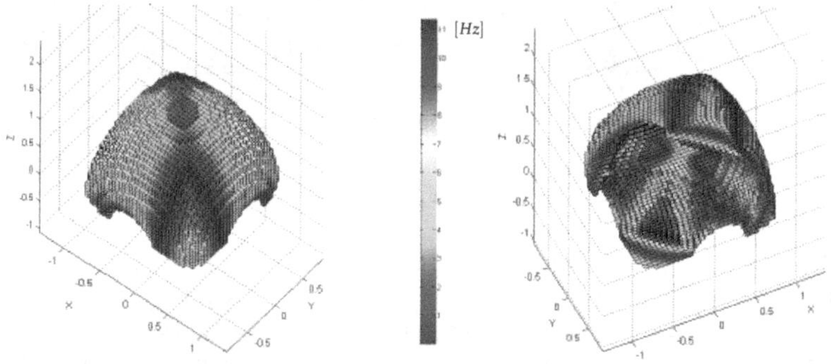

Fig. 13. Distribution of the first natural frequency inside the robot's constant orientation workspace.

8. Conclusions

This chapter has discussed a method, based on the Matrix Structural Analysis, to study the linearized elastodynamics of PKMs. Base and moving platforms are considered rigid, while links can be modeled as rigid or flexible parts, the latter being decomposed into two or more flexible bodies. Here, we used 3D Euler beams but the method can be extended to superelements with two end-nodes. Joints are directly included, without recurring to Lagrange multipliers, by means of kinematic constraints between nodal displacement arrays. Three cases have been taken into account to model the rigid-flexible, flexible-flexible and flexible-rigid coupling of bodies by means joints. Each case yields equations, linking dependent, independent nodal displacement arrays and joint displacements as well, to be used to find generalized stiffness and inertia matrices. The latter are then combined as elementary blocks to find the global matrices of the whole system. Some useful extension to compliant mechanism has been introduced, while two strategy, the lumped and the distributed one, have been explained to include mass/inertia into the model. Feasible applications of the method pertain: the study of natural frequencies inside the robot's workspace by means of local and global indices, the singularity finding, the optimization of elastodynamic performances varying geometric, structural or inertial parameters.
Finally, the method has been applied to an articulated four-dofs PKMs with Schönflies motions. A modular approach is used to split each of the four legs into three modules. Results, compared to commercial software, revealed good accuracy in determining natural frequency range, while drastically reducing the time of computation avoiding the annoying and time-consuming FEM meshing routines.

9. References

[1] Tyapin, I., Hovland, G. & Brogårdh, T. (2008). Kinematic and Elastodynamic Design Optimisation of the 3-DOF Gantry-Tau Parallel Kinematic Manipulator, In: *Proceedings of the Second International Workshop on Fundamental Issues and Future Research Directions for Parallel Mechanisms and Manipulators*, September 21-22, Montpellier, France.

[2] Zienkiewicz, O. C. & Taylor, R. L. (2000). *Solid mechanics - Volume 2*, Butterworth Heinemann, London.

[3] Martin, H. C. (1966). *Introduction to matrix methods of structural analysis*, McGraw-Hill Book Company.

[4] Wang, C.K. (1966). *Matrix methods of structural analysis*, International Textbook Company.

[5] Przemieniecki, J. S. (1985). *Theory of Matrix Structural Analysis*, Dover Publications, Inc, New York.

[6] Huang, T., Zhao, X. & Withehouse, D.J. (2002). Stiffness estimation of a tripod-based parallel kinematic machine, *IEEE Trans. on Robotics and Automation*, Vol. 18(1).

[7] Li, Y.W., Wang, J.S. & Wang, L.P. (2002). Stiffness analysis of a Stewart platform-based parallel kinematic machine, In: *Proceedings of IEEE ICRA: Int. Conf. On Robotics and Automation*, May 11-15, Washington, US.

[8] Clinton, C.M., Zhang, G. & Wavering, A.J. (1997). Stiffness modeling of a Stewart-platform-based milling machine, In: *Trans. of the North America Manufacturing Research Institution of SME*, May 20-23, Vol. XXV, pp 335-340, Lincoln, NB, US.

[9] Al Bassit, L., Angeles, J., Al-Wydyan, K. & Morozov., A. (2002). The elastodynamics of a Schönflies -Motion generator, *Technical Report TR-CIM-02-06*, Centre for Intelligent Machines, McGill University, Montreal, Canada.

[10] Deblaise, D., Hernot, X. & Maurine, P. (2006). A Systematic Analytical Method for PKM Stiffness Matrix Calculation, In: *Proceedings of the 2006 IEEE International Conference on Robotics and Automation*, May, Orlando, Florida.

[11] Gonçalves, R. S. & Carvalho, J. C. M. (2008). Stiffness analysis of parallel manipulator using matrix structural analysis, In: *EUCOMES 2008, 2-nd European Conference on Mechanism Science*, Cassino, Italy.

[12] Wittbrodt, E., Adamiec-Wójcik, I. & Wojciech, S. (2006). *Dynamics of Flexible Multibody Systems*, Springer.

[13] Gosselin, C.M. (1990). Stiffness mapping for parallel manipulator, *IEEE Trans. on Robotics and Automation*, vol. 6, pp 377-382.

[14] Quennouelle, C. & Gosselin, C.M. (2008). Kinemato-Static Modelling of Compliant Parallel Mechanisms: Application to a 3-PRRR Mechanism, the Tripteron, In: *Proceedings of the Second International Workshop on Fundamental Issues and Future Research Directions for Parallel Mechanisms and Manipulators*, September 21-22, Montpellier, France.

[15] Briot, S., Pashkevich, A. & Chablat, D. (2009). On the optimal design of parallel robots taking into account their deformations and natural frequencies, In: *Proceedings of the ASME 2009 International Design Engineering Technical Conferences & Computers and Information in Engineering Conference IDETC/CIE*, August 30 - September 2, San Diego, California, USA.

[16] Yoon, W.K., Suehiro, T., Tsumaki, Y. & Uchiyama, M. (2004). Stiffness analysis and design of a compact modified Delta parallel mechanism, *Robotica*, 22(5), pp. 463-475.

[17] Majou, F., Gosselin, C.M., Wenger, P. & Chablat, D. (2004). Parametric stiffness analysis of the orthoglide, In: *Proceedings of the 35th International Symposium on Robotics*, March, Paris, France.

[18] El-Khasawneh, B.S. & Ferreira, P.M. (1999). Computation of stiffness and stiffness bounds for parallel link manipulator, *Int. J. Machine Tools and manufacture*, vol. 39(2), pp 321-342.

[19] Gosselin, C.M. & Zhang, D. (2002). Stiffness analysis of parallel mechanisms using a lumped model, *Int. J. of Robotics and Automation*, vol. 17, pp 17-27.

[20] Pashkevich, A., Chablat, D. & Wenger, P. (2009). Stiffness analysis of overconstrained parallel manipulators, *Mechanism and Machine Theory*, Vol. 44, pp. 966-982.

[21] Shabana, A.A. (2008). *Computational continuum mechanics*, Cambridge University Press, ISBN 978-0-521-88569-0, USA.

[22] Cammarata, A. & Angeles, J. (2011). The elastodynamics model of the McGill Schönflies Motion Generator, In: *Multibody Dynamics 2011*, 4-7 July, Bruxelles, Belgium.

[23] Angeles, J. (2007). *Fundamentals of Robotic Mechanical Systems: Theory, Methods and Algorithms, Third edition*, Springer, ISBN 0-387-29412-0, New York.

[24] Salgado, O., Altuzarra, O., Petuya, V. & Hernández, A. (2008). Synthesis and design of a novel 3T1R fullyparallel manipulator, *ASME Journal of Mechanical Design*, Vol. 130, Issue 4, pp. 1-8.

[25] Merlet, J.-P. (2006). *Parallel robots, Second Edition*, Kluwer Academic Publisher, ISBN-10 1-4020-4132-2(HB), the Netherlands.

7

Design and Postures of a Serial Robot Composed by Closed-Loop Kinematics Chains

David Úbeda, José María Marín, Arturo Gil and Óscar Reinoso
Universidad Miguel Hernández de Elche
Spain

1. Introduction

The robot presented in this chapter is a new binary hybrid (parallel-serial) type climbing system composed by several closed-chains arranged in an open-chain. The originality of the robot resides in the possibility of combining several parallel modules to build a new configuration of the robot according to the intended application. In this chapter, one possible morphology will be covered, destinated to climb metallic cross-linked structures, but we will also study the kinematics and the structure of the simplest modules apart from the final robot.

Hybrid climbing robots are hard to find, but hybrid climbing robots with binary actuators are still less common. In our opinion, binary actuators are interesting, since they allow for an easier control of the robot: only two different positions need to be controlled. This advantage comes at a price, since less points in the workspace can be reached, but hereby, large motion workspace, small volume and multiple degrees of freedom will be some of the new challenges of this kind of robots (Lichter et al., 2002).

In the robot proposed in this chapter, the activation of every lineal actuator of the closed-loop generates a planar and rotational movement of the output link respect to the input one. Several parallel modules can be connected in a serial mode. In this sense, the robot is freely reconfigurable, thus, new modules (translational or rotational) can be added.

A great variety of applications can be reached with the final 2+2 closed-chain disposed in a serial mode proposed system.

As it was mentioned above, it is possible to build complex structures combining parallel modules. Attending some researches, it is not difficult to find many papers of climbing or walking robots (Qi et al., 2009; Nagakubo & Hirose, 1994; Aracil et al., 2006b), but mostly adopt serial mechanism, what makes them less robust and with compromised stability. However, few authors have researched about the use of the hybrid mechanisms to add the desired characteristics with a simple and sensor-less control (Lichter et al., 2002; Chen & Yeo, 2002; Lees & Chirikjian, 1996; Erdmann & Mason, 1988; Goldberg, 1989; Craig, 1989).

The mechanism proposed combines parallel modules to set up a climbing robot, in addition with improve the mentioned characteristics, to outfit the robot with best stability and strength.

Another main feature that contributes originality compared with the related works of similar robots, it is aimed to a discrete workspace. Also we will demonstrate that it is possible to reach enough points to accomplish the required application (climb metallic

structures for inspection and maintenance tasks, construction, petroleum, bridges, etc.), and to achieve similar characteristics according to usual robots destinated to these tasks.
In order to demonstrate that this robot could be used in a continuous workspace, and could be positioned in different planes and surfaces of the metallic structure, we will study the closed chain forward and inverse kinematics, and direct kinematics of the final system consisting of hybrid leg mechanism.
This chapter reviews related works, main important features and applications of some climbing and walking robots. A description of the geometry of a new developed climbing robot with a discrete hybrid (parallel-serial) moving mechanism composed by a closed kinematic chain and binary actuators will be explained. The schematic design of the robot and main postures will be also provided. Moreover, an analysis of the robot's workspace, forward and inverse kinematics of the parallel module, and forward kinematics of the complete serial robot will be also discussed.

2. Related work

Climbing and walking robots are common in industry applications, like as, maintenance activities of nuclear plants, oil refineries, bridges, high voltage towers, medical fields, endoscope devices, and surgical instruments in general. The applications mentioned above are practiced with some problems to access and with hazardous environmental conditions.
To accomplish these tasks, robots need some important features, as reliable or robust, to be able to move over 3D structures, including walls, ceilings, pipes or cylindrical structures (Reinoso et. al, 2001), and they also need to be adaptable into different terrains.
Parallel robots have good performance, and they are perfect to manage manipulation tasks with short manipulation cycles, high speeds and accelerations. These characteristics are very difficult to obtain in serial robots. However, the proposal system combines advantages of serial and parallel robots. For example, thanks to the linear actuators, a parallel robot has a high ratio of payload and deadweight, so using linear actuators is totally justified.
Another desirable characteristics of these robots are that they need powerful torque in the actuators, mainly if they use serial legs for climbing. Although our system uses serial legs for climbing, the combination with parallel modules does that the torque of the actuators is not needed to be very high.
Higher speed is a desirable characteristic, but it is reduced when using legs for climbing, however, in our system, velocity is not a problem because of the parallel modules that are used, similar to Stewart-Gough platforms.
Most common problem in the walking and climbing robots is how to negotiate the boundary of two plain surfaces such as convex or concave corners in 3D. In this paper, we propose a robot model that solves this problem in a right way, using hybrid legs.
The main purpose of this project was to design a very simple robot similar to the Stewart-Gough platform, but combining serial robot abilities that allowed this to do climbing tasks. Probably the most important task to solve in this robot is the control system. To solve this, it is proposed a binary actuation. The binary degrees of freedom in this system are quasi-exponential compared with serial robots, but the approach with a serial robot is increased too. As opposed, our system does not require feedback of any sensor, among other features. Some researchers have studied theses kind of binary actuation robots, but none of them has used hybrid serial-parallel technology.

Similar to parallel robots (Aracil et al., 2006a) this robot can climb the exterior and interior of tubes or metallic structures. According to the kind of structure, the end-effector and the base could carry magnetic foots or suction pads (Kim et al., 2005).
In this chapter, we do not dedicate attention to the fixation system of the end-effector and base of the robot, although a suction pad or magnetic feet are recommended. We are going to concentrate on the kinematics analysis and the postures it can achieve.

3. Schematic design of the robot

As it was introduced above, the main goal of this project was to assemble different parallel modules to set up some new more complex serial ones. All of the linear actuators will be binary, and they will work into ON (stretched)/OFF (shrunken) position, to accomplish a wide workspace.

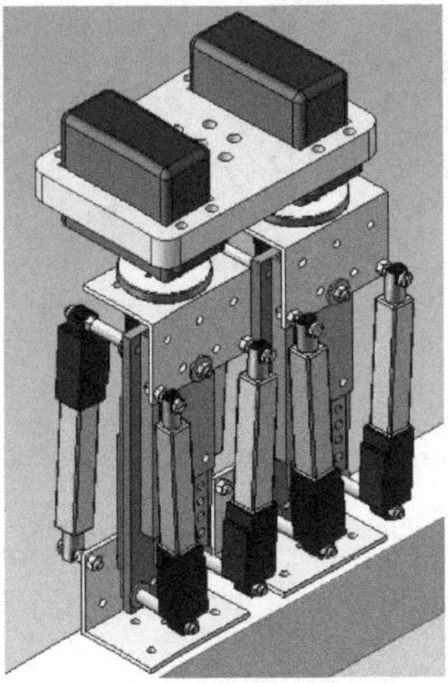

Fig. 1. Hybrid climbing robot

The purpose of this section will be the kinematics study of the closed-chain module, two closed-chain modules to set up a planar serial module, and the 2+2 closed-chain modules to set up a serial open-chain for three-dimensional movements. Another types of structures are possible to run up with the disposal and the number of the closed-chain modules.

3.1 Parallel module

If we focus in the study of a simple closed-chain module, the Figure 2(a) shows such structure, that is composed by two prismatic actuated joints, with two central sliders to limit the lateral movement, and in turn a free rotational joint in the upper side to provide a rotational movement of the end-effector of the module.
As a consequence of the use of a discrete workspace, the actuated joints will be binary, and these modules will have a d_1 translation and a 45 degrees rotation (Figure 2(b)), when their configuration will be P_1 ON, P_2 OFF or vice versa.

The Figure 2(c) refers to both actuators P_1 and P_2 set to ON.

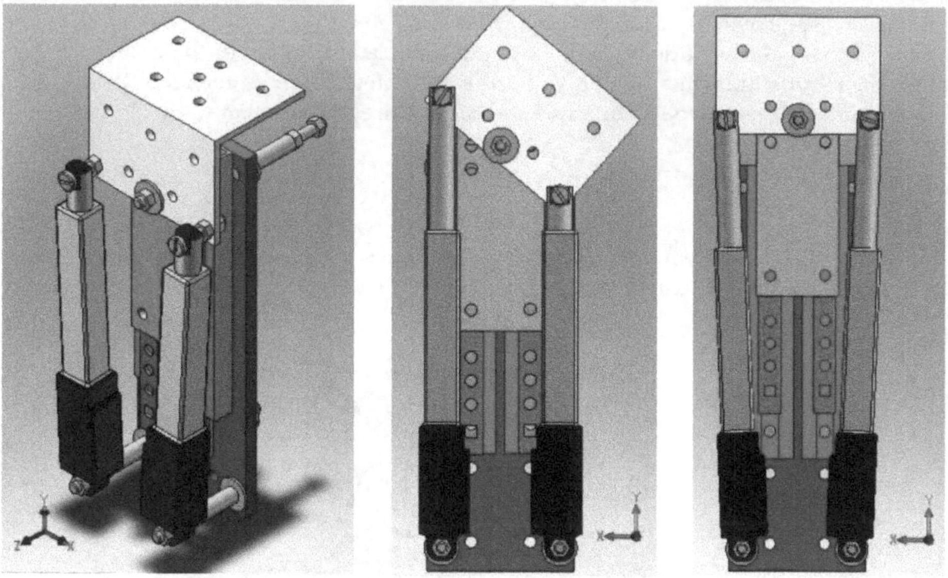

a) Isometric view of the parallel module

b) п/4 rotation: Actuator 2 ON

c) Actuator 1 and 2 ON

Fig. 2. Some views of the closed-chain module

Obviously, the different positions that this module is able to reach are given in Table 1:

Actuator 1	*Actuator 2*	*Translation*	*Rotation*
0	0	0	0°
0	1	d_1	45°
1	0	d_1	-45°
1	1	d_2	0°

Table 1. Positions from one parallel module

3.2 Two parallel modules disposed in a serial mode

In the Figure 3, two parallel modules can be observed. We have disposed them in a serial configuration, leaving one parallel module rotated 180 degrees and disposed in a mirror mode behind the other, to achieve robust and stable postures, and at the same time, it allows to set up 90 degrees rotations (figure 5(b)) of the end-effector around of the free joint axis.

All different positions that these two attached modules are able to reach are summarized in Table 2.

In Figure 4(a), we can observe that the four linear actuators are set to off, and therefore the minimum elongation of the module is obtained.

However, if we need to perform a translation around the axis of the actuators without performing a rotation of the end-effector, we could use the image configuration 4(b) with the 4 actuators set to ON. In that way, there will not be any movement induced by rotational

Actuator P1	*Actuator P2*	*Actuator P3*	*Actuator P4*	*Translation*	*Rotation*
0	0	0	0	0	0°
0	0	0	1	d_1	-45°
0	0	1	0	d_1	45°
0	0	1	1	d_2	0°
0	1	0	0	d_1	-45°
0	1	0	1	$d_1+ d_1$	-90°
0	1	1	0	$d_1+ d_1$	0°
0	1	1	1	$d_1+ d_2$	-45°
1	0	0	0	d_1	45°
1	0	0	1	$d_1+ d_1$	0°
1	0	1	0	$d_1+ d_1$	90°
1	0	1	1	$d_1+ d_2$	45°
1	1	0	0	d_2	0°
1	1	0	1	$d_2+ d_1$	-45°
1	1	1	0	$d_2+ d_1$	45°
1	1	1	1	$d_2+ d_2$	0°

Table 2. Positions from two parallel modules in a serial mode

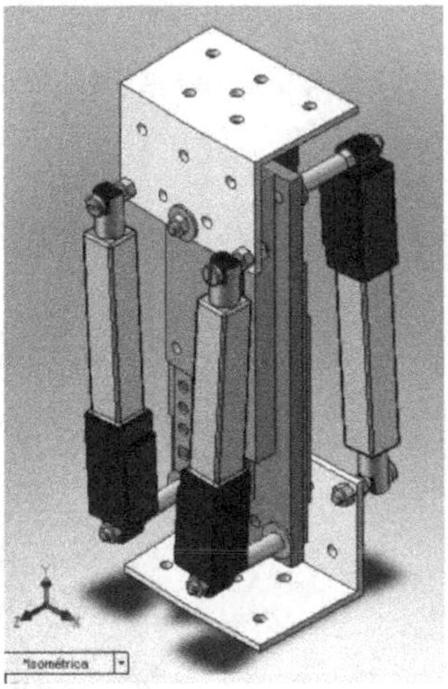

Fig. 3. Two parallel modules to draw up a serial structure

joints. In this pose, it can be observed that the central sliders indicate the distance d_2 between the base and the end-effector, obviously, they will be at its highest point of stretching.

However, it could be needed to reach some points in the workspace that are inclined at 90 degrees according to the base. In order to achieve this, and according to the Table 2, it could be observed in the Figure 5(b), that it can be obtained an inclination of 90 degrees, if the two mirror drives of each parallel module are switched alternately to ON and OFF.

In a similar way, if a 45 degrees inclination according to the base link was needed to reach, we could perform an actuation to ON of one of the linear actuators (figure 5(a)). This could be observed in the Table 2.

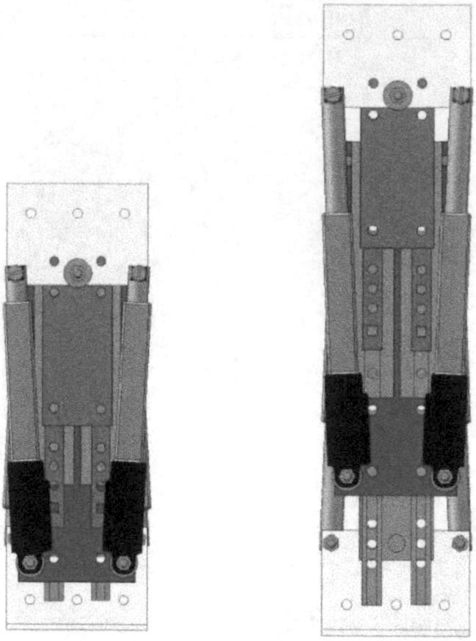

a) Minimum length. Frontal view.

b) Maximum length. Frontal view.

Fig. 4. Minimum and maximum prolongation of the open-chain module

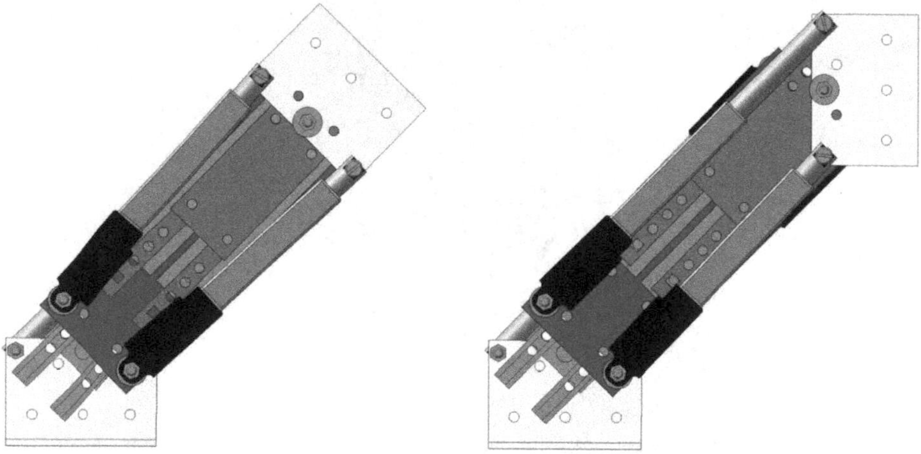

a) 45 degrees inclination

b) 90 degrees inclination

Fig. 5. Different postures of the open-chain module

3.3 2+2 Parallel modules arranged in a serial mode

As was mentioned before, the robot is composed of several parallel modules, and they are able to attach themselves in series to perform more complex structures in order to carry out a particular operation.

The robot in the Figure 1 consists of 2 + 2 parallel modules arranged to each other in a serial configuration and connected by a top link between two actuated rotational joints. According to this configuration, a three-dimensional movement of 90 and 180 degrees is possible to achieve around the rotational axis of the joint, as shown in the Table 3:

Actuator R1	*Actuator R2*	*Translation*	*Rotation*
0	0	0	0°
0	1	d_1	90°
1	0	d_1	-90°
1	1	d_2	180°

Table 3. Positions of the rotational actuator

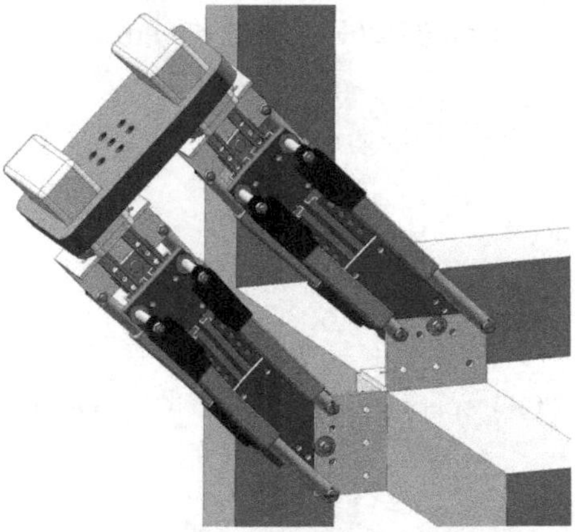

Fig. 6. Posture to a surface change of the structural frame

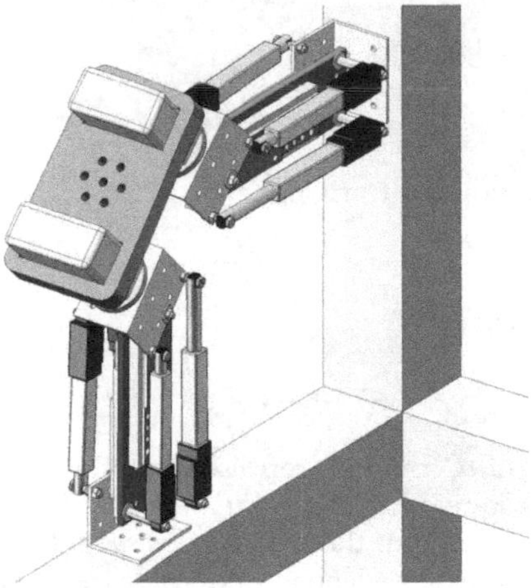

Fig. 7. Posture to evade a structural node: changing the plane

A basic problem that must be solved in the development of this robot, is that it should be able to get around structural nodes. According to the Figure 7, a posture to evade a structural node is shown. Also, the Figure 6 shows a posture to achieve a surface change. Later in this chapter, we will discuss the kinematics of the modules, so we will not focus on all possible combinations of the actuators of the robot with 2+2 parallel modules arranged in a serial mode.

4. Closed-chain module

4.1 Degrees of freedom of the module

Figure 8(a) shows a CAD model of the closed-chain, and it is composed by two active links with variable length l_1 (link e_2 with e_3) and l_2 (link e_4 with e_5) , and two passive links, e_1 and e_6, with fixed length $2a_1$ and $2a_2$, respectively. A slider keeps the midpoint of the upper passive link e_6 on the line $x=a_1$. There is a fixed link e_1, four revolute joints (1, 3, 5 and 7) and two prismatic joints (2 and 6).

According to the Grübler criterion (Grübler, 1883), the number of active degrees of freedom is given by:

$$F = \lambda(n - j - 1) + \sum_i f_i \tag{1}$$

where:

- n, the number of links in the mechanism, including the fixed link (from e_1 to e_7):

$$n = 7 \tag{2}$$

- λ, degrees of freedom of the space (planar) in which the mechanism is intended to work:

$$\lambda = 3 \tag{3}$$

- j_1, number of prismatic joints

$$j_1 = 3 \tag{4}$$

- j_3, number of revolute joints

$$j_3 = 5 \tag{5}$$

- f_i, DoF permitted by revolute or prismatic joint i

$$f_i = 1 \tag{6}$$

replacing (2), (3), (4), (5), (6 in (1)

$$F = 3(7 - 8 - 1) + 5*1 + 3*1 = 2 \tag{7}$$

These Grübler criterion results could be represented with an equivalent model (figure 8(b)) composed by a prismatic joint plus a solidary revolute joint, that represents reliably the model of figure 8(a). In this model, the P part represents the 3D prismatic joint and the R part represents the 3D revolute joint. The total distance traveled by the slider is equivalent to the prismatic joint distance *d* of figure 8(a).

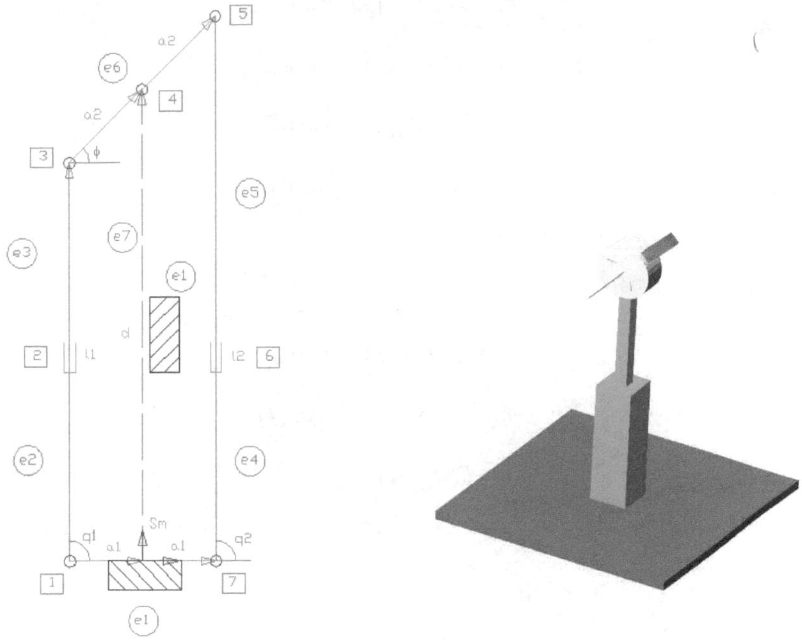

a) DoF: Grübler Criterion b) Equivalent Model

Fig. 8. Closed-chain module

4.2 Inverse kinematics of the parallel module

To study the inverse kinematics of the parallel module of the robot, a series of vectorial equations have been defined to estimate the position and orientation of the base according to the length of the linear actuators and the rotational angles of the rotational joints.

Inverse kinematics pretends to solve the position of the link e_6 of figure 8(a) according to a reference system S_m that is solidary to the ground link. It could be defined as a translation d and a rotation ϕ. In this way, inverse kinematics proposes to solve (l1, l2) as a function of (d, ϕ) parameters.

We could define a movement for the parallel module in different alternative ways. In the next one it is indicated that the mid point of the e_6 link should be at the position (a_1, d) according to the reference system S_m solidary to the ground link.

According to the figure 8(a), two vectorial equations are obtained:

$$\vec{a_1} + \vec{d} = \vec{a_2} + \vec{l_1} \tag{8}$$

$$\vec{a_1} + \vec{l_2} = \vec{d} + \vec{a_2} \tag{9}$$

where, a restriction on e_6 or d link can be observed, because this vector angle will be always 90° according to the ground link a_1. Through a vectorial decomposition of the equations (8), (9), (10) and (11) these new equations are obtained:

$$a_1 = a_2 \cos(\phi) + l_1 \cos(q_1) \tag{10}$$

$$d = l_1 \sin(q_1) + a_2 \sin(\phi) \tag{11}$$

$$a_1 + l_2 \cos(q_2) = a_2 \cos(\phi) \tag{12}$$

$$l_2 \sin(q_2) = d + a_2 \sin(\phi) \tag{13}$$

and with a new disposition of the equations:

$$a_1 = l_1 \cos(q_1) + a_2 \cos(\phi) \tag{14}$$

$$d = l_1 \sin(q_1) + a_2 \sin(\phi) \tag{15}$$

$$-a_1 = l_2 \cos(q_2) - a_2 \cos(\phi) \tag{16}$$

$$d = l_2 \sin(q_2) - a_2 \sin(\phi) \tag{17}$$

it could be obtained l_1 link length based on d and ϕ from (14) and (15) equations:

$$(a_1 - a_2 \cos(\phi))^2 = (l_1 \cos(q_1))^2 \tag{18}$$

$$(d - a_2 \sin(\phi))^2 = (l_1 \sin(q_1))^2 \tag{19}$$

where the addition of both of them:

$$(a_1 - a_2 \cos(\phi))^2 + (d - a_2 \sin(\phi))^2 = l_1{}^2 \cos^2(q_1) + l_1{}^2 \sin^2(q_1) \tag{20}$$

and in the next step we will find the value of l_1, in order that q_1 disappears from the equation:

$$l_1 = +\sqrt{(a_1 - a_2 \cos(\phi))^2 + (d - a_2 \sin(\phi))^2} \tag{21}$$

Likewise, for l_2:

$$l_2 = +\sqrt{(-a_1 + a_2 \cos(\phi))^2 + (d + a_2 \sin(\phi))^2} \tag{22}$$

Only positive solutions are of interest for (l_1, l_2). As well, looking at the equations, with a pair (d, ϕ), a unique pair (l_1, l_2) is obtained and, backwards, with (l_1, l_2) a unique pair (d, ϕ) is obtained too.

4.3 Forward kinematics of the parallel module

Forward kinematics problem has been solved with a numerical method based on least squares method.

The equations (14), (15), (16) and (17) don't allow to find the value of ϕ and d as a length l_1 and l_2 function, because q_1 and q_2 are themselves function of d and ϕ.

We are going to set out the Levenberg-Marquardt numerical method (Levenberg, 1944), (Marquardt, 1963) so the equations (14), (15), (16) and (17) should be redefined as:

$$a_1 = l_1 \cos(q_1) + a_2 \cos(\phi) \tag{23}$$

$$0 = l_1 \sin(q_1) + a_2 \sin(\phi) - d \tag{24}$$

$$-a_1 = l_2 \cos(q_2) - a_2 \cos(\phi) \tag{25}$$

$$0 = l_2 \sin(q_2) - a_2 \sin(\phi) - d \tag{26}$$

An optimization process could be defined with the next functions:

$$f_1(\vec{s}) = l_1 \cos(q_1) + a_2 \cos(\phi) \tag{27}$$

$$f_2(\vec{s}) = l_1 \sin(q_1) + a_2 \sin(\phi) - d \tag{28}$$

$$f_3(\vec{s}) = l_2 \cos(q_2) - a_2 \cos(\phi) \tag{29}$$

$$f_4(\vec{s}) = l_2 \sin(q_2) - a_2 \sin(\phi) - d \tag{30}$$

defining:

$$\vec{f} = \begin{bmatrix} f_1(\vec{s}) \\ f_2(\vec{s}) \\ f_3(\vec{s}) \\ f_4(\vec{s}) \end{bmatrix} \tag{31}$$

where the state vector $\vec{s}$ is defined as:

$$\vec{s} = \begin{bmatrix} d \\ \phi \\ q_1 \\ q_2 \end{bmatrix} \tag{32}$$

and the constraints vector:

$$\vec{y} = \begin{bmatrix} a_1 \\ 0 \\ -a_1 \\ 0 \end{bmatrix} \tag{33}$$

Wherewith, the problem is raised as a minimization of the function:

$$F(\vec{s}) = \sum_{i=1}^{4} \left(f_i(\vec{s}) - y_i \right)^2 \tag{34}$$

For the minimization, we could raise an initial solution $\vec{s_0}$ and update in the direction of the gradient $\frac{\partial J}{\partial s}$

$$\vec{s}_{k+1} = \vec{s}_k + \delta \tag{35}$$

with:

$$\delta = (JJ^T)^{-1} J^T (\vec{y} - \vec{f}) \tag{36}$$

where J is the Jacobian matrix of the robot.

For the algorithm converging, it must indicate an initial solution of the $\vec{s}$. An approximation in order to achieve good results is:

$$d_0 = \frac{l_1 + l_2}{2} \tag{37}$$

$$\phi_0 = \frac{l_2 + l_1}{2a_2} \tag{38}$$

$$q_{10} = \arctan \frac{d_0 - a_2 \sin \phi_0}{a_1 - a_2 \cos \phi_0} \tag{39}$$

$$q_{20} = \arctan \frac{d_0 + a_2 \sin \phi_0}{-a_1 + a_2 \cos \phi_0} \tag{40}$$

The Jacobian matrix is calculated as follows:

$$J = \begin{bmatrix} \frac{\partial f_1}{\partial d} & \frac{\partial f_1}{\partial \phi} & \frac{\partial f_1}{\partial q_1} & \frac{\partial f_1}{\partial q_2} \\ \frac{\partial f_2}{\partial d} & \frac{\partial f_2}{\partial \phi} & \frac{\partial f_2}{\partial q_1} & \frac{\partial f_2}{\partial q_2} \\ \frac{\partial f_3}{\partial d} & \frac{\partial f_3}{\partial \phi} & \frac{\partial f_3}{\partial q_1} & \frac{\partial f_3}{\partial q_2} \\ \frac{\partial f_4}{\partial d} & \frac{\partial f_4}{\partial \phi} & \frac{\partial f_4}{\partial q_1} & \frac{\partial f_4}{\partial q_2} \end{bmatrix} = \begin{bmatrix} 0 & -a_2 \sin \phi & -l_1 \sin q_1 & 0 \\ -1 & a_2 \cos \phi & l_1 \cos q_1 & 0 \\ 0 & a_2 \sin \phi & 0 & -l_2 \sin q_2 \\ -1 & -a_2 \cos \phi & 0 & l_2 \sin q_2 \end{bmatrix} \tag{41}$$

In the performed tests in order to solve forward kinematics, the algorithm converges to the correct solution in only two iterations, with an $F<0.0001$ error.

Others ways to find the value of δ could be probed, like:

$$\delta = (JJ^T + \lambda I)^{-1} J^T (\vec{y} - \vec{f}) \tag{42}$$

where λ permits to regulate the speed as the function converges. Another way is:

$$\delta = (JJ^T + \lambda diag(J^T J))^{-1} J^T (\vec{y} - \vec{f}) \tag{43}$$

this is known as the Levenberg-Marquardt algorithm.

5. Forward kinematics of the open-chain module

According to the Subsection 3.3 a 2+2 parallel modules disposed in a serial structure could be set up. We are going to start from a complete serial model of the robot. The Figure 1 shows this set up of the robot. An equivalent model is shown in the Figure 9.

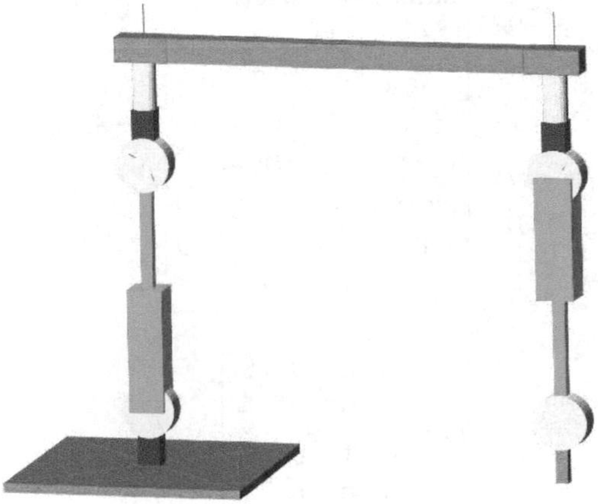

Fig. 9. 3D model of the serial robot

To solve forward kinematics of the open-chain, Denavit-Hartenberg algorithm (Denavit & Hartenberg, 1955; Uicker et al., 1964; Hartenberg & Denavit, 1964) has been used. In order to solve transformation matrices, *The Robotics Toolbox for Matlab* (Corke, 1996) has been used. Firstly, and following D-H convention, we could establish the coordinate axes as follows in the Figure 10. Secondly, the transformation matrix will be solved:

$$ {}^{0}T_1 = \begin{bmatrix} -\sin\phi_1 & 0 & \cos\phi_1 & 0 \\ \cos\phi_1 & 0 & \sin\phi_1 & 0 \\ 0 & 1 & 0 & H_1 \\ 0 & 0 & 0 & 1 \end{bmatrix} \tag{44} $$

$$ {}^{1}T_2 = \begin{bmatrix} 1 & 0 & 0 & 0 \\ 0 & 0 & 1 & 0 \\ 0 & -1 & 0 & L_2 \\ 0 & 0 & 0 & 1 \end{bmatrix} \tag{45} $$

$$ {}^{2}T_3 = \begin{bmatrix} \cos\phi_3 & 0 & \sin\phi_3 & 0 \\ \sin\phi_3 & 0 & -\cos\phi_3 & 0 \\ 0 & 1 & 0 & 0 \\ 0 & 0 & 0 & 1 \end{bmatrix} \tag{46} $$

$$^{3}T_4 = \begin{bmatrix} \cos\phi_4 & -\sin\phi_4 & 0 & \cos\phi_4 * H_4 \\ \sin\phi_4 & \cos\phi_4 & 0 & \sin\phi_4 * H_4 \\ 0 & 0 & 1 & 0 \\ 0 & 0 & 0 & 1 \end{bmatrix} \quad (47)$$

$$^{4}T_5 = \begin{bmatrix} \cos\phi_5 & 0 & -\sin\phi_5 & 0 \\ \sin\phi_5 & 0 & \cos\phi_5 & 0 \\ 0 & -1 & 0 & 0 \\ 0 & 0 & 0 & 1 \end{bmatrix} \quad (48)$$

$$^{5}T_6 = \begin{bmatrix} \cos\phi_6 & 0 & -\sin\phi_6 & 0 \\ \sin\phi_6 & 0 & \cos\phi_6 & 0 \\ 0 & -1 & 0 & 0 \\ 0 & 0 & 0 & 1 \end{bmatrix} \quad (49)$$

$$^{6}T_7 = \begin{bmatrix} 1 & 0 & 0 & 0 \\ 0 & 0 & -1 & 0 \\ 0 & 1 & 0 & L_6 \\ 0 & 0 & 0 & 1 \end{bmatrix} \quad (50)$$

$$^{7}T_8 = \begin{bmatrix} \cos\phi_8 & 0 & -\sin\phi_8 & 0 \\ \sin\phi_8 & 0 & \cos\phi_8 & 0 \\ 0 & -1 & 0 & 0 \\ 0 & 0 & 0 & 1 \end{bmatrix} \quad (51)$$

Below, the Table 4 indicates the D-H Parameters of the 8 DoF of the model:

Joint	ϕ	d	a	α
1R	ϕ_1+90°	H_1	0	90°
2R	0	L_2	0	-90°
3R	ϕ_3	0	0	90
4R	ϕ_4	0	H_4	0
5R	ϕ_5	0	0	-90
6R	ϕ_6	0	0	-90
7R	0	L_6	0	90
8R	ϕ_8	0	0	-90

Table 4. D-H Parameters table of the serial model

6. Robot workspace

A preliminary study of the workspace that could reach a set up of 2+2 parallel modules arranged in a serial mode, with actuated rotational joints, have been performed. The goal was to check if it was able to climb a three-dimensional cross-linked structure.

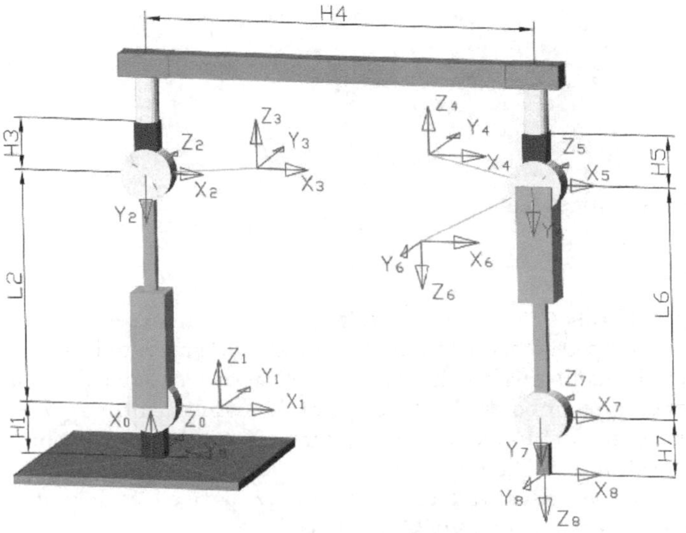

Fig. 10. 3D model of the serial robot with D-H convention axis

For one of the possible workspaces, some of the mathematical combinations of the robot actuators have been obtained, and a vector with all of them has been generated. This vector has been used to obtain different final points of the end-effector. Therefore, every element of the vector will be every final point of the previously described.

We have obtained a 2^{10} elements vector (Figure 11) as a result of the ten joints of the real model. In this vector, the interferences between links were not taken into consideration.

On the other hand, a cross-linked structure and a robot model have been simulated through SolidWorks. The goal was to check if the robot was able to reach enough workspace points and, at the same time, to perform a plane change in the cross-linked structure, and all of this has been shown in the Figures 6 and 7 with the simulated model.

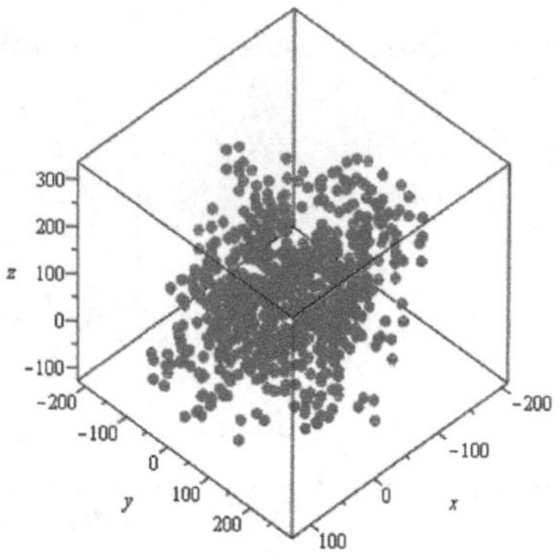

Fig. 11. 1024 points work space

7. Conclusions

Some years ago, new robotic devices with a large number of degrees of freedom and binary actuators were developed to achieve a large motion workspace, to be capable of large fine motion, and have a small-stowed volume. Applications as inspection and maintenance tasks require them to adapt the robot to this kind of hostile environment.

In this way, a new reconfigurable binary climbing robot with closed-chains disposed in an open-chain architecture, has been presented.

Related works has been reviewed at the Section 2, as well as important features and applications of some climbing and walking robots.

Sometimes, the kinematics solution of parallel mechanisms requires using redundant sensors to establish a control loop because it becomes quite complicated. In this chapter, a binary actuators solution is presented, so a sensor-less feature is included.

Linear actuators are directly connected to the base and to the end-effector of the parallel modules, so these actuators are at the same time structural elements of the complete serial robot, and they work in a simultaneous way, which gives them the ability to handle loads much greater than its own weight.

The schematic design of the robot, description of the geometry and main postures have been also provided in the Section 3. In this Section, also has been studied in an independent way the parallel module, the two parallel modules disposed in a serial mode, and the complete robot composed by 2+2 Parallel modules arranged in a serial mode.

Moreover, an analysis of the forward and inverse kinematics of the parallel module, and forward kinematics of the complete serial robot are discussed in the Sections 4 and 5.

Finally the discrete workspace of the robot has been represented in the Section 6.

Future works will consist on determine the inverse kinematics solution of the serial robot, to implement the control system and applying path-planning algorithms to move the robot around of the cross-linked structure.

8. References

Aracil, R.; Saltaren, R. & Reinoso, O. (2006a). A climbing parallel robot. *IEEE Robotics & Automation Magazine,* (March 2006), pp. 16–22, ISSN 1070-9932

Aracil, R.; Saltaren, R.; Sabater, J. & Reinoso, O. (2006b). Robots paralelos: Máquinas con un pasado para una robótica del futuro. *Revista Iberoamericana de automática e informática industrial,* Vol.3, No.1, (January 2006), pp. 16–28, ISSN 1697-7912

Chen, I.M. & Yeo, S. H. (2002). Locomotion and navigation of a planar walker based on binary actuation, *Proceedings of the IEEE International Conference on Robotics and Automation*, pp. 329-334, ISBN 0-7803-7272-7, Washington DC, USA, May, 2002

Corke, P. (1996). A robotics toolbox for MATLAB. *IEEE Robotics and Automation Magazine* Vol.3, No.1, (March 1996), pp: 24-32, ISSN 1070-9932

Craig, J. J. (1989). *Introduction to Robotics : Mechanics and Control* (2nd edition), Prentice-Hall, ISBN 9780131236295

Denavit, J. & Hartenberg, R. S. (1955). A kinematic notation for lower-pair mechanisms based on matrices. *Journal of Applied Mechanics,* Vol.86, pp. 215–221

Erdmann, M. A. & Mason, M. T. (1988). An exploration of sensor-less manipulation. *IEEE Journal of Robotics and Automation,* Vol.4, (August 1988), pp. 369–379

Goldberg, D. (1989). *Genetic algorithms in search, optimization, and Machine learning,* Addison-Wesley, ISBN 0201157675

Grübler, M. (1883). Allgemeine Eigenschaften der ZwanglŁufigen ebenen kinematischen Ketten. *Part I: Zivilingenieur,* Vol.29, pp. 167-200

Hartenberg, R.S. & Denavit, J. (1964). *Kinematic Synthesis of Linkages,* McGraw-Hill, ISBN 978-0070269101, New York

Kim, H.; Kang, T.; Loc, V. G. & Choi, H. R. (2005). Gait planning of quadruped walking and climbing robot for locomotion in 3d environment, *Proceedings of the IEEE International Conference on Robotics and Automation,* pp. 2733–2738, ISBN 0-7803-8914-X, Barcelona, Spain, April, 2005

Lees, D. S. & Chirikjian, G. S. (1996). A combinatorial approach to trajectory planning for binary manipulators, *Proceedings of the IEEE International Conference on Robotics and Automation,* pp. 2749-2754, ISBN: 0-7803-2988-0, Washington DC, USA, April 22-28, 1996

Levenberg, K. (1944). A method for the solution of certain non-linear problems in least squares. *The Quarterly of Applied Mathematics,* Vol.2, (July 1944), pp. 164-168

Lichter, M. D.; Sujan, V. A. & Dubowsky, S. (2002). Computational issues in the planning and kinematics of binary robots, *Proceedings of the IEEE International Conference on Robotics and Automation,* pp. 341-346, ISBN 0-7803-7272-7, Washington DC, USA, May, 2002

Marquardt, D. (1963). An algorithm for least-squares estimation of nonlinear parameters. *SIAM Journal on Applied Mathematics,* Vol.11, No.2, pp: 431-441, ISSN 431-41

Nagakubo, A. & Hirose, S. (1994). Walking and running of the quadruped wall-climbing robot, *Proceedings of the IEEE International Conference on Robotics and Automation,* pp. 1005–1012, ISBN 0-8186-5330-2, Washington DC, USA, May 8-13, 1994

Qi, Z.; Wang, H.; Huang, Z. & Zhang, L. (2009). Kinematics of a quadruped/biped reconfigurable walking robot with parallel leg mechanisms, *ASME/IFToMM International Conference on Reconfigurable Mechanisms and Robots,* pp. 558–564, ISBN 978-88-89007-37-2, London, UK, June 22-24, 2009

Reinoso, O.; Saltaren, R.; Aracil, R.; Almonacid, M. & Perez, C. (2001). Avances en el desarrollo de un robot trepador de estructuras cilíndricas, *XXII Jornadas de Automática*, ISBN 84-699-4593-9, Barcelona, Spain, September, 2001

Uicker, J. J.; Denavit, J. & Hartenberg, R. S. (1964). An interactive method for the displacement analysis of spatial mechanisms. *Journal of Applied Mechanics ASME*, Vol.86, pp: 215–221

8

A Reactive Anticipation for Autonomous Robot Navigation

Emna Ayari, Sameh El Hadouaj and Khaled Ghedira
High Institute of Management, Tunis
Tunisia

1. Introduction

Nowadays, mobile robots are expected to carry out various tasks in all kinds of application fields ranging from manufacturing plants, transportation, nursing service, resource or underwater exploration. In all these applications, robots should navigate autonomously in uncertain and dynamic environments in order to achieve their goals. So, the most current challenge in the development of autonomous robot control systems is making them respond intelligently to changing environments. Navigation in such environments involves many mechanisms such as: object detection, perception, internal building model, decision making, prediction of the future state of the environment and on-line navigation. To attend its goal safely, the robot should minimise interaction with other actors in order to avoid conflict situations. Generally, this problem comes up when many robots and/or actors would have access to the same space at the same time. In this case, the control of autonomous robotic navigation for conflict resolution has been widely studied. Some researchers have been focused on navigation in dynamic environments, where either reactive systems (producing real time behaviour), deliberative systems (introduce reasoning and need much more time to calculate a suitable decision) or hybrid systems (combine deliberative and reactive approaches) have been used in order to attend a known goal.

Typically, reactive systems are used to deal with simple problems (detect an obstacle, go away from an obstacle, follow a wall, etc.). Nevertheless, reactive systems are typically less affected by errors and do not require an explicit model of the environment in order to navigate inside an unknown space. Furthermore, they usually deal only with local information that may be captured at real time. However, in reactive systems, the robot can be derived to a conflict situation with other actors because they ignore prediction and reasoning in the decision process. To face this problem, global planning approaches are used. They consist of elaborating a global plan from beginning state to goal state. These approaches need prior complete information about the state of the environment, so they do not take into account the environment uncertainty. So, in more complex situations, hybrid approaches are used. They combine reactive approaches according to a higher level in order to include anticipation of the state of the environment in the decision process. In these cases, low level control operates in a reactive way (local navigation) whereas high level systems tend to be deliberative. It provides, at each step, a partial moving plan to the robot. But these systems are complex because they need much more time to calculate or to update a suitable trajectory toward a predefined goal. In this situation, it is interesting to introduce prediction

in reactive schemas, without using planning techniques, in order to consider the environment's evolution in the future.
To deal with uncertainty, autonomous navigation involves using systems for navigation control that must be not too computationally expensive, as this would result in a sluggish response. In this case, we choose the fuzzy logic technique because it is faster when the frequency of the environment's changes is high.
In the field of multi-agent systems, several works have encouraged researchers to develop models simulating robot's behaviours in order to achieve a known target (Posadas et al., 2008). They are more flexible and fault tolerant as several simple agents are easier to handle and cheaper to build. Indeed, the term "agent" has been defined as hardware or a software system with certain properties such as autonomy, social ability, reactivity and pro-activity (Ferber, 1995). In fact, there are similarities between robot and agent because they share the same characteristics. So, the mapping from a robot to an agent seems straightforward because each robot represents a physical entity, independent from other robots, with a specific task.
In this paper, we present a reactive anticipation model based on fuzzy logic that takes into account: (1) the reactive navigation in an environment composed of local minimum and moving obstacles, and (2) the anticipation of blocking situations and the prediction of the nature, the velocity and the position of obstacles in the future. This information is used to predict conflict situations without using a motion planning method. In order to validate our work, we evaluate our approach by simulating various scenarios. We also give a comparative study between results obtained by our model and those of some other approaches.
In the remainder of the article, we give, in section 2, an overview of the existing approaches for conflict resolution applied to autonomous robot navigation. In section 3, we present the most used techniques to deal with the uncertainty of perception and we justify our choice of the fuzzy logic method. In section 4, we describe our model that combines reactivity with anticipation in order to deal with the problem of conflict resolution without using a motion planning. In Section 5, we present our experimental results. Finally, we conclude in section 6 and we give perspectives for this work.

2. Existing approaches for conflict resolution

When navigation occurs in an environment that is totally or partially unknown or even dynamically changing, higher degree of autonomy for a mobile robot is required. Thus, a mobile robot should be able to take decisions on-line and to minimise conflict situations in such uncertain and dynamic environments using only sensors' limited information.
In this context, a wide range of approaches have been adopted to suggest solutions to the problem of space conflict when a robot shares the same environment with other actors.

2.1 Reactive approach

Many reactive approaches have been proposed to resolve the problem of autonomous robot navigation. In the following, we mention the most important among them.

- Potential Fields (Huang et al., 2006; Khatib, 1986): this approach relies on creating an artificial repulsion field around obstacles and an attraction field around the goal.
- Vector Field Histogram (Ulrich & Borenstein, 2000): it uses a heading dependent histogram to represent the obstacle's density. So, the robot can move in the direction where there are less obstacles in order to minimise its interaction with them.

- Dynamic Window (Fox et al., 1997) and Curvature Velocity (Simmons, 1996): they search a space of the robot's translational and rotational velocities. Obstacles near the robot are transformed into this space by eliminating all commanded velocities that would cause a collision within a certain time period. Both these methods take into account the kinematics and the dynamics of the robot. Commanded velocities are chosen based on an objective function that considers both progress towards the goal and robot safety.
- Nearness Diagram (Minguez & Montano, 2000): it consists in analysing a situation from two polar diagrams. One diagram is used to extract information of environmental characteristics and identify the immediate goal valley. The second diagram is used to define the safety level between the robot and the obstacles by identifying the closest one.
- Elastic Bands (Quinlan & Khatib, 1993): this method modifies the trajectory of the robot, originally provided by a planner, by using artificial forces which depend on the layout of the obstacles in the way.
- Behaviour-based methods (Jing et al., 2003; Langer et al., 1994): they define fundamental behaviour sets and establish mappings between sensors' information under different situations and reactive behaviours of the mobile robot. The reactive behaviour of the robot can be regarded as a weighted sum of these fundamental behaviours or can be chosen from the behaviour set according to an evaluation function.
- The Velocity Obstacles method (Fiorini & Shiller, 1998): the moving obstacle is transformed into a velocity obstacle. This approach uses the distance between robot and obstacle to minimise conflict at real time. It is considered as a mapping between sensory data and control commands without introducing reasoning in the decision process.

The main drawback of these strategies is that the robot gets into an infinite loop or local minimum when it is moving among multiple obstacles or in conflict situations when it shares the same resource with moving obstacles. To overcome this problem, many methods are proposed. We mention:

- Memory state method that uses a state memory strategy (Zhu & Yang, 2004). The variables from which this method makes ultimate decisions are: "the distance between the robot and the target" and "the distance between the robot and the nearest obstacle".
- Minimum risk approach is based on trial-return behaviour phenomenon (Omid et al., 2009; Wang & Liu, 2005).
- Virtual wall method directs the robot away from the dead-end by placing a virtual wall that bars the limit cycle path (Ordonez et al., 2008). However, these strategies do not take into account the dynamic nature of the environment and the uncertainty of perception. So the robot can be driven into dangerous or conflict situations.

2.2 Global planning approach

Hence, other approaches applied in dynamic environments are proposed with the idea of combining the reactive techniques with global planning methods (Stachniss & Burgard, 2002). At the beginning, a complete plan from present state to goal state is computed. These approaches need prior complete information about the state of the environment, so they do not take into account the uncertainty.

To face this limit, deliberative approaches are appeared. They use a global world model provided by user input or sensory information to generate appropriate actions for the mobile robot to reach the target (Pruski & Rohmer, 1997). This kind of approach enables prediction and reasoning in the decision process by considering the current information and the past information (Nilsson, 1980; Sahota, 1994). The deliberative control architecture comprises three modules: sensing, planning and action modules. First, robot sense it's surrounding and creates a world model of static environment by combining sensory information. Then, it employs planning module to search an optimal path toward the goal and generate a plan for robot to follow. Finally, robot executes the desired actions to reach the target. After a successful action, robot stops and updates information to perform the next motion. Then, it repeats the process until it reaches the goal. It can coordinate multiple goals and constraints within a complex environment (Huq et al., 2008; Yang et al., 2006).

However, in deliberative navigation, accurate model of environment is needed to plan a globally feasible path. The computational complexity of such systems is generally too great to attain the cycle rates needed for the resulting action to keep pace with the changing environments (it is difficult to obtain a completely known map). To perform necessary calculations, enormous processing capabilities and memory is needed.

Therefore, these approaches are not proper in the presence of uncertainty in dynamic or real world.

2.3 Hybrid approach

In more complex situations, other control architectures appeared (Oreback & Christensen, 2003), including the anticipation in the decision process. This combination is known as hybrid architecture (Arkin, 1990) which has two levels: reactive and deliberative. The deliberative level has to determine and offer the reactive level those behavioural patterns that are necessary for the robot to achieve its objectives. The reactive level has to execute these behavioural patterns by guaranteeing real-time restrictions (Fulgenzi et al., 2008; Fulgenzi, 2009).

Practically, these methods are more complex and require more time calculations. They usually use local methods of obstacle avoidance in order to expand the tree and cover the search space. However, local methods are less suited to dynamic environments (Minguez et al., 2002).

In (Schwartz & Sharir, 1983), a general algorithm was developed. It is doubly exponential time, this limit has been lowered by the Canny algorithm described in (Canny et al., 1990) whose running time is exponential in dimension.

The algorithm D* proposed by Stentz (stentz, 1995) is a generalisation of the A* search algorithm in the case of partially known environments. In a finite discretized configuration space, the initial cost of shipping is to move from one configuration to another. A path is first found, and then the robot begins to execute it. If a change in the cost of a path is detected, only the relevant configurations are considered and the optimal path is updated in less time.

Partial motion planning (PMP) (Petti & Fraichard, 2005), the algorithm explicitly takes into account the computation time, constraints and problems safety of navigation in dynamic environment. A tree is grown using a conventional algorithm based on sampling. A node is added to the tree if there is not an inevitable consequence collision. This enables PMP to provide a safe partial path at any time. During the execution of this partial path chosen, another partial path is developed from the end of the previous path developed. This method is applied at real time in dynamic environments, but its major drawback is that it does not consider the uncertainty of the information collected. So it can produce non-executable partial paths.

2.4 Multi-agent systems and robotic simulation

In the field of multi-agent systems, several works have encouraged researchers to develop models simulating robot's behaviours in order to achieve a known target. In fact, there are similarities between robot and agent. So, the mapping from a robot to an agent seems straightforward because each robot represents a physical entity, independent from other robots, with a specific task. Hence, multi-agent simulations models are widely used to solve problems in mobile robotics under dynamic environments. In this way, many approaches are using multi-agent systems to simulate and control autonomous mobile robot in order to resolve the problem of space's conflict by minimising the interaction with agents. For example, (Ros, 2005) proposes a multi-agent system for autonomous robot navigation in unknown environments. In this work, the authors use the case-based reasoning (CBR) techniques in order to solve the problem of local minimum and avoid only static obstacles. In (Ono, 2004), a multi-agent architecture is also proposed to control an intelligent wheelchair. It takes into account only three behaviours: obstacle avoidance (using distance), door passage and wall following. However, it cannot combine behaviours together and it is applied only in static environment. In (Innocenti, 2007), a multi-agent system is proposed as the robot control architecture divided into four sub-systems of agents: perception, behaviour, deliberative and actuator. This architecture is applied in static and dynamic environments and uses the distance between the robot and the moving obstacle in order to predict conflict situations. But this information gives a limited knowledge about the state of the environment and does not allow the robot to take an intelligent decision while navigating in a dynamic and unknown environment. In (Posadas et al., 2008) a multi-agent system is proposed to control the navigation of robot by using a hybrid approach (it uses a motion planning). The architecture is composed of two levels; reactive and deliberative. The communications between these levels are assured by the agents.

2.5 Discussion

Reactive systems are applied in local navigation and used to deal with simple problems. Nevertheless, reactive systems are typically less affected by errors and do not require an explicit models of the environment in order to navigate inside an unknown space. Furthermore, they usually deal only with local information that may be captured at real time. However, reactive systems may fall into local traps or blocking situations because they ignore prediction and reasoning in the decision process. For this reason, the robot can be derived to a conflict situation with other actors sharing the same space. In more complex situations, they are combined according to a higher level in order to include anticipation in the decision process. In these cases, low level control operates in a reactive way (local navigation) whereas high level systems tend to be deliberative and use motion planning method to update the trajectory when there are modifications in the state of the environment. These approaches are called hybrid approaches and are well applied in uncertain and dynamic environments by producing, at each step, a motion planning method that minimise the conflict (Petti & Fraichard, 2005). Nevertheless, building a movement planning requires an important computation time in order to find the most appropriate way. However, if the frequency of environment's changes is high, the use of a planning method becomes not only inefficient, but also useless (if the robot generates plans without using them).

So, in this situation it is interesting to introduce prediction in reactive schemas in order to consider the environment's evolution in the future. Hence, the main concern of this paper is to propose a model that combines reactivity and anticipation in order to resolve the problem of conflict without using a motion planning. Thus, the robot should anticipate the nature and velocity of the obstacles in order to minimise interactions. We use the multi-agent system to simulate the robot's behaviour because there is a mapping between robots, while navigating in dynamic environments, and multi-agent systems.

3. Autonomous navigation in uncertain environment

Autonomous navigation in an uncertain environment involves using systems for navigation control that must not be too computationally expensive, as this would result in a sluggish response. In this case, soft computing methods have played important roles in the design of the robot controllers. The commonly used soft computing methods are: Bayesian Networks and Fuzzy Logic.

3.1 Bayesian networks

Bayesian Networks are models which capture uncertainties in terms of probabilities that can be used to perform reasoning under uncertainty. It epitomises probabilistic dependency models that represent random stochastic uncertainty via its nodes (Darwiche, 2009). The Bayesian inference has been widely used especially in the localisation problem (Thrun, 1998), in the mapping and in the learning mechanisms in mobile robotics. But this technique is very slow and complex (NP-Complex) because it uses a probabilistic reasoning that require much more time to choose a suitable decision. Also, this technique requires a causal model of reality that is not always given. So, its application in autonomous navigation in the case of dynamic environments can be unprofitable.

3.2 Fuzzy logic technique for autonomous navigation

Fuzzy Logic (FL) has been investigated by several researchers to treat the problem of uncertainty in perception. This technique represents uncertainty via fuzzy sets and membership functions. It gives robustness to the system with respect to inaccuracy in data acquisition and uncertainty, and makes definition behaviour and their interactions quite easy by using simple linguistic rules (Klir et al., 1997; Sajotti et al., 1995). It also allows implementing of human knowledge and experience without requiring a precise analytical model of the environment. Probably, the greatest strength of behaviour-based fuzzy approaches is that they operate on/and reason with uncertain perception-based information, which makes them suitable even for difficult environments.

We present in table 1 comparison results between the characteristics of the Bayesian Networks and the Fuzzy Logic. According to this comparison study, we can conclude that the fuzzy logic technique is faster, easier to implement and well applied to autonomous robot navigation.

	Fuzzy Logic	Bayesian Network
Complexity	Simple rules	NP-complex
Reasoning	Fuzzy reasoning	Probabilistic reasoning
Application	Avoidance collision	Localization and mapping
Controller	Fuzzy inference	Bayesian inference
Rapidity	Fast (simple rules)	Slow

Table 1. Comparison between fuzzy logic and Bayesian network.

However, the majority of existing fuzzy logic methods deal only with static environments, and only use the distance between the robot and the obstacle to avoid collision and to minimise conflict with other agents sharing the same space (Bonarini et al, 2003; Selekwa et al., 2008). This kind of information gives a limited knowledge about the state of the environment, so the robot cannot take intelligent decisions. For this reason, in our work, we adopt a fuzzy logic technique to deal with uncertainty. We propose a fuzzy model for autonomous navigation in a dynamic and uncertain environment based on the nature and the velocity of obstacles.

4. Proposed approach

According to Ferber (Ferber, 1995), a multi-agent system is composed of:

1. An environment 'E',
2. A set of objects 'O' in 'E',
3. A set of agents 'A' in 'E',
4. A set of relationships 'R' between agents,
5. A set of operation 'op' enabling agents 'A' to collect, produce, consume and manipulate objects of 'O'.

Basing on this definition, we define our multi-agent system as a set of objects which represent the static objects (have a fixed position) and a set of agents (have a moving position) that represent dynamic objects and the robots. In our model, the principal agent is the "robot". It has its own physical parts including sensors, processors, actuators and communication devices. So, in order to guarantee the safety of the mobile robot, it should be able to navigate by minimising interactions with the other agents navigating in the same space in order to minimise conflict and to achieve a known goal.

This section is organised as follows. We present in subsection 4.1 the perception model of the robot. It allows the robot to perceive its environment in order to take the suitable decision autonomously. In subsection 4.2, we present the kinematic model of the robot. It allows the robot to localise in its environment. In subsection 4.3, we describe the principle of the fuzzy controller technique. In subsection 4.4, we present our fuzzy controllers model. The first controller is described in subsection 4.4.1. It allows the robot to determine its angular velocity. The second controller is presented in subsection 4.4.2. It allows the robot to calculate its linear velocity. We present in subsection 4.4.3 the system rules applied in our fuzzy controllers used to avoid conflict situations.

4.1 The perception model

One of the ultimate goals of robotics is to realise autonomous robots that are able to organise their own internal structure and to achieve their goals through interactions with dynamically changing environments. The robot should take decisions on-line using only sensors providing limited information (without prior information about the position of obstacles and their trajectories) when it navigates autonomously in dynamic spaces.

Our main objective is to make the robot able to predict the nature and the velocity of obstacles (static or moving) in order to minimise conflict situations, to avoid collision and to achieve a specific goal. At this stage, we use a perception model composed of eight ultrasonic sensors. We choose this type of sensors because it has some attractive properties, e.g. cheapness, reliability and soon, which makes it widely used in mobile robots. This type of sensor calculates the distance between the robots.

According to this model of perception, we divide the robot's space into three subspaces controlled by three groups of sensors (see figure 1). The subspaces are: (1) the front area "F" controlled by the first group composed of sensors number 1,2,7,8, (2) the left area "L" controlled by the second group composed of sensors number 3, 4, and (3) the right area "R" controlled by the third group composed of sensors number 5, 6.

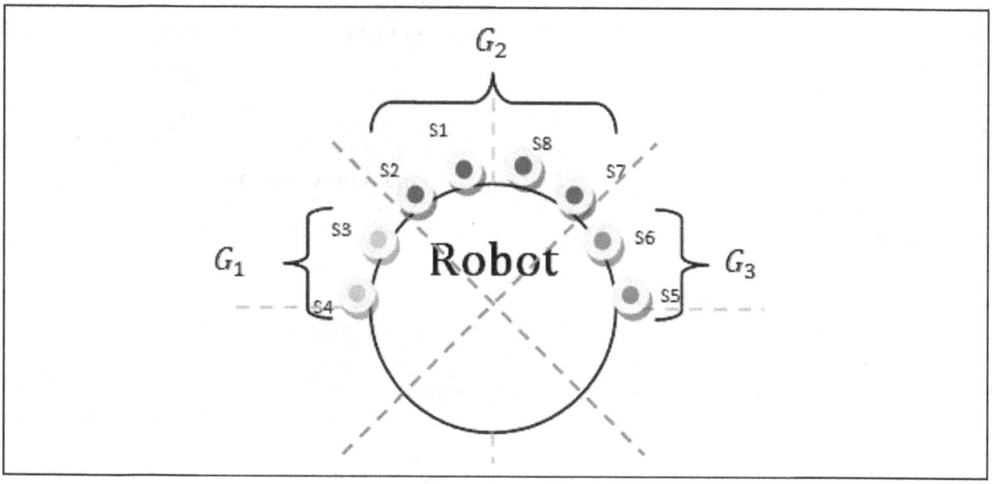

Fig. 1. The Robot's perception areas.

4.2 The kinematic model

In this study, the mobile robot is a system, which is subject to non holonomic constraints. Its position in the environment is represented by P=(x,y,θ), where (x, y) is the position of the robot in the reference coordinate system XOY, and the heading direction θ is taken counter clockwise from the positive direction of X-axis (see figure 2). X'O'Y' is the reference coordinate according to the robot system. It allows the robot to locate the objects in the environment.

The robot is composed of two wheels (left and right). The velocity, acceleration and orientation angle of the robot is assured by the fuzzy controllers.

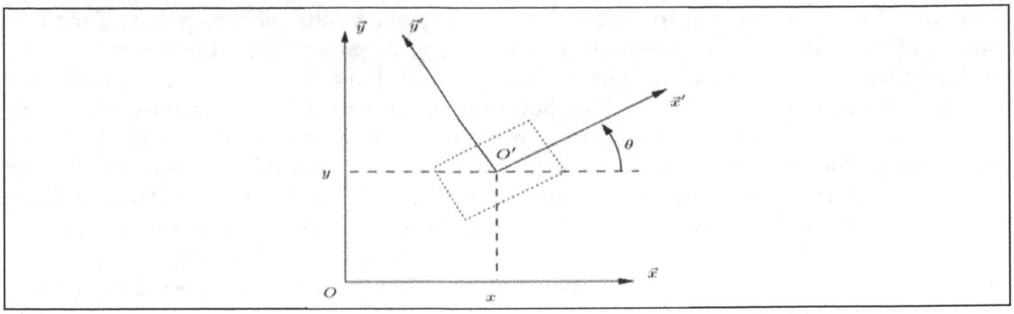

Fig. 2. Robot's kinematic model.

4.3 The fuzzy controller

A Fuzzy Controller (FC) is a control system, whose input is the output of the process to be controlled (sensory data, internal state). Its outputs are commands for the actuators of the process. The ranges of the input and output parameters/variables are represented by membership functions and fuzzy sets. In addition, the interactions between input and output variables/parameters are represented by fuzzy rules. In a nutshell, system input parameters and variables are encoded into fuzzy representations using well defined "If/Then" rules which are converted into their mathematical equivalents. These rules would then determine actions to be taken based on Implication Operators such as Zadeh Min/Max (Zadeh, 1973), or Mamdani Min (Mamdani, 1974). The fuzzified data is then put through a defuzzification process via Center of Gravity, Center of Sum or Mean of Maxima methods to obtain actual commands for the actuators of the process.

In our work, we use the Mamdani fuzzy inference methodology in order to provide the adequate way of modelling the relevance of the controller. For the defuzzification process we choose the Center of Gravity method by applying the formula 1. The crisp value U* is the geometrical center of the output fuzzy value μ(y), where μ(y) is the union of all the contributions of rules whose Degree Of Freedom (DOF) is more than zero, y is the universe of discourse and b is the number of samples.

$$U^* = \frac{\sum_{i=a}^{b} \mu(y) \times y}{\sum_{i=a}^{b} \mu(y)} \quad (1)$$

4.4 The principle of the behaviour controller based on Fuzzy Logic

The use of fuzzy logic in the design of autonomous navigation behaviours is nowadays quite popular in robotic. The set of behaviours that are being implemented can include, for example, the following of walls, corridors or the avoidance of obstacles. However, usually the existing fuzzy logic methods use only the distance between the robot and the obstacle in order to avoid conflicts and so deal with static environments.

We propose a fuzzy model for the navigation in dynamic and uncertain environments based on"the nature" and "the velocity" of obstacles. For example, if there is an obstacle in front of the robot, it is more logical to reason about its velocity. In fact, the distance between the robot and the obstacle allows it to immediately change the trajectory without taking into account the obstacle nature (mobile or static) and whether the obstacle is going in the same direction of the robot or not. In contrast, the velocity allows the robot to predict if there is conflict in the future. In the reminder of this section, we present robot behaviour modules implemented as fuzzy logic controllers. Each behaviour module receives data input that describes the situation and produces an output to be addressed to the actuators. Our fuzzy architecture is composed of two fuzzy controllers. The first controller is in charge of determining the angular velocity of the two wheels. This controller allows the robot to change its angular orientation based on the prediction of the nature of the obstacle, in order to avoid conflict and to move closer to its target. The second controller uses the velocity of the obstacle. It is in charge of determining whether the robot's speed should be increased (accelerate its speed if there is a free space) or decreased (minimise its speed if there is an obstacle in its trajectory).

4.4.1 Obstacle nature prediction: Time collision concept

Typically ultrasonic sensors calculate the distance between the robot and the obstacle and therefore they do not provide information about the obstacle's nature (mobile or static) and its position in the future state. However, this kind of information is necessary to predict conflict situations in the future. In order to overcome this problem, in our work, the robot uses this distance to calculate new information called "Time Collision" between the robot and the obstacle.

The Time Collision (TC) represents the time left for the robot before a collision occurs with an obstacle. In order to predict the nature and the position of the obstacle in the future, the robot operates as follows. At time Ti, the robot should observe its environment (using the perception model). If it detects an obstacle, it calculates a Time Collision Needed (TCN) representing the time required to collide with this obstacle in the future while keeping the same velocity (see figure 3). At time Ti+1, it repeats the same procedure to recalculate a Time Collision Remaining (TCR) representing the time required to collide the same obstacle in the future while keeping the same velocity (see figure 3). The TC can be obtained by applying the formula 2. Di represents the distance between the robot and the obstacle at time Ti, and VRi represents the velocity of the robot.

$$TC = D_i / VR_i \tag{2}$$

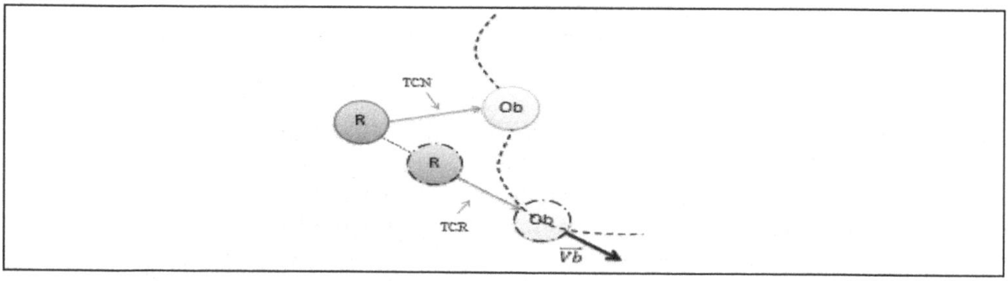

Fig. 3. Illustration of "the Time Collision".

From these two values, the robot can calculate a Difference Time Collision (DTC) with this obstacle by applying formula 3. If the DTC is equal to zero, it means that the obstacle in the trajectory of the robot is static. In this situation, the robot should immediately change its direction in order to avoid local minimum. If the DTC is positive, then the obstacle is considered as a mobile one. In this situation, the robot can expect that the trajectory will be free after a period of time and then it does not need to change its direction. If the DTC is negative, then obstacle is mobile and it is going towards the robot. In this case, the robot should minimise its velocity in order to avoid conflict with this obstacle, and then decide to change the direction completely if it is possible (moves to another subspace, which is safe), or wait until this mobile object moves away (the current subspace will be free), if there is no other free subspace.

$$DTC=TCN-TCR \quad (3)$$

The advantage of this information is that it allows the robot not only to avoid collision and local minimum at real time with obstacles around it, but also to anticipate their natures and their positions in the future so as to predict conflict situations.

We present in figure 4 the FLC of angular velocity. We describe four linguistic variables that are: the DTCL, DTCF, DTCR, and TO. These variables represent respectively: the difference time collision on the left, the difference time collision in the front, the difference time collision on the right and the current target orientation between the robot and the goal. Each

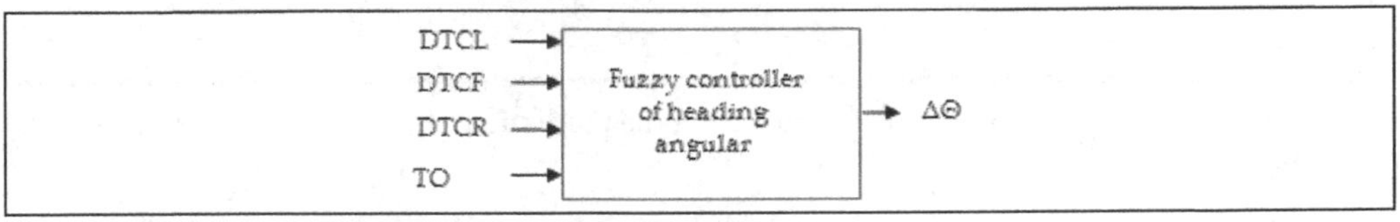

Fig. 4. Fuzzy controller of heading angular

DTC have two membership functions: Fixed Obstacle and Moving Obstacle. The DTCL is used to control the left area. We present in algorithm 1 the reasoning process of the robot that is used to predict the nature of the nearest obstacle in this area. Thus, if the DTCL is

```
if (DTCi) ≤ the time interval Δ then
    The obstacle is static
else
    The obstacle is mobile
    Free space
end if
```

Algorithm 1. The robot's behaviour is based on the prediction of the nature of the obstacles in left and right areas

smaller or equal to Δ (the time interval between two successive perceptions) it means that the obstacle is static; else it is considered as mobile. We can obtain the value of this variable from the formula 4. The same reasoning process is applied to the right area based on the DTCR that can be obtained from the formula 5. The value of DTCF variable can be obtained from the formula 6. Algorithm 2 describes the reasoning mechanism of the robot that is used to predict the nature of the nearest obstacle in the front area. For example, if the DTCF is equal to zero, this means that the obstacle is static. Therefore, the robot should change its trajectory immediately in order to avoid collision. Otherwise, if the DTCF is equal to infinity, this means that the space is free. In order to move toward a specific goal, the robot

```
if (DTCF) =0 then
    The obstacle is static. The robot should change its trajectory
else
    if ((DTCF)>0) and ((DTCF) < + ∞) then
The obstacle is mobile and it is moving away from the robot.
    else
if ((DTCF) <0) and ((DTCF)> −∞) then
The obstacle is mobile and it is going toward the robot.
else
if (DTCF) =- ∞ then
The space is considered as free at t_i
else
if (INF (DTCF) =+ ∞ then
No obstacle in front of the robot. Free space
end if
end if
            end if
        end if
end if
```

Algorithm 2. The robot's behaviour is based on the prediction of the nature of the obstacles in front area

should have an idea about the area in which the target exists. For this purpose, the TO which gives an idea about this position is combined with the DTC for each area in order to reach a compromise between "avoid conflict" and "reach goal". The TO has three membership functions that are: L: Left, F: Front, R: Right. The output of this controller is the angular velocity, it has five membership functions: LLT: Large Left Turn, SLT: Small Left Turn, NT: No Turn, SRT: Small Right Turn, LRT: Large Right Turn. Note that figure 5 describes the input of the fuzzy sets involved in the definition of the heading angular. Figure 6 describes the output variable for the angular velocity.

$$DTCL = Sup\ (DTCL_i)\ \text{where}\ i=3, 4 \tag{4}$$

$$DTCR = Sup\ (DTCR_i)\ \text{where}\ i=5, 6 \tag{5}$$

$$DTCF = Inf\ (DTCF_i)\ \text{where}\ i=1, 2, 7, 8 \tag{6}$$

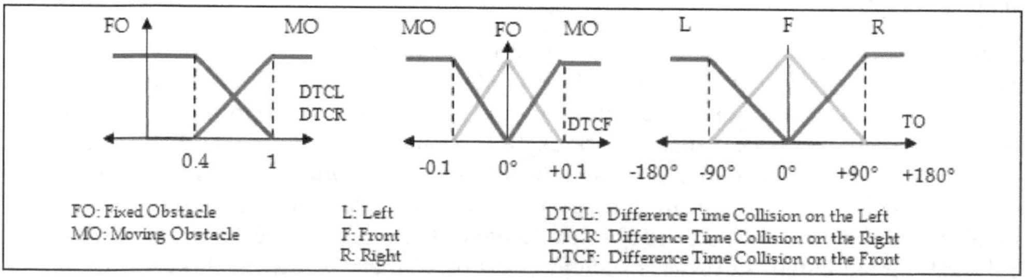

Fig. 5. The definition of angular velocity input membership functions.

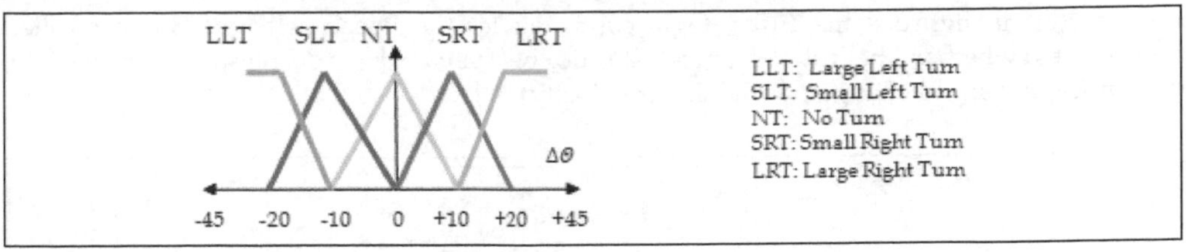

Fig. 6. The definition of angular velocity output membership functions.

4.4.2 Linear velocity prediction: Obstacle velocity concept

The fuzzy controller for angular velocity uses the DTC that gives information about the nature of an obstacle and its trajectory. So, the robot is able to predict the future situation of the environment if it keeps moving in the same direction. While the second fuzzy controller for linear velocity uses the obstacle's velocity in the front area. This information determines whether the robot's speed should be increased or decreased to respond to the nearest situation detected.

Hence, the obstacle's velocity can be calculated according to formula 7 where D represents the distance travelled by the obstacle during the time interval Δ.

$$\vartheta = D/\Delta \tag{7}$$

The distance D is calculated according to formula 8 where X_c and Y_c represent the coordinates of the obstacle at time t_i and X'_c and Y'_c represent the coordinates of the obstacle at $t_i+\Delta$ (see figure 7). The coordinates are obtained according to formula 9.

$$D=\sqrt{(X_c-Y_c)^2+(X'_c-Y'_c)^2} \tag{8}$$

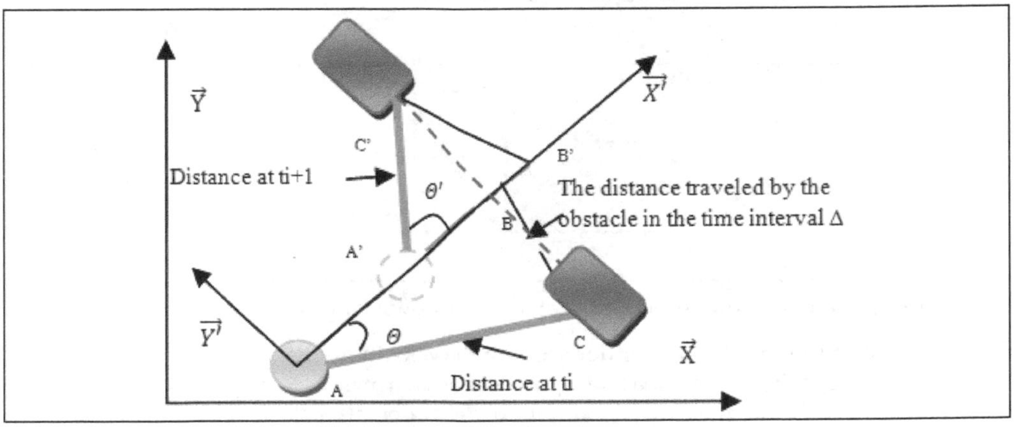

Fig. 7. Illustration of distance traversed by the obstacle.

$$\text{Cord}= \begin{cases} X_c=X_a+(X_b-X_a)\cos\theta-(Y_b-Y_a)\sin\theta \\ Y_c=Y_a+(Y_b-Y_a)\cos\theta+(X_b-X_a)\sin\theta \end{cases} \tag{9}$$

We present in figure 8 the fuzzy logic controller for the linear velocity. This controller determines whether the robot's speed should be increased or decreased. It is used to determine the speed change $\Delta\vartheta$ (the linear velocity).

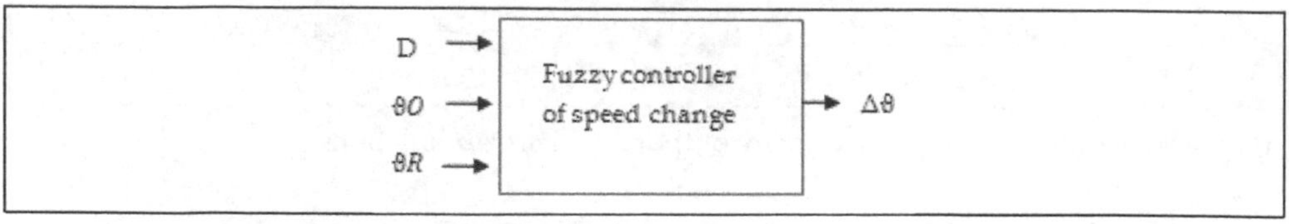

Fig. 8. Fuzzy controller of speed change

The D, ϑO, and the ϑR represent respectively the distance between the robot and the nearest obstacle in the front area, the velocity of the nearest obstacle in the front area and the current velocity of the robot. The input and output membership functions of the fuzzy controller for linear velocity are shown respectively in figure 9 and figure 10.

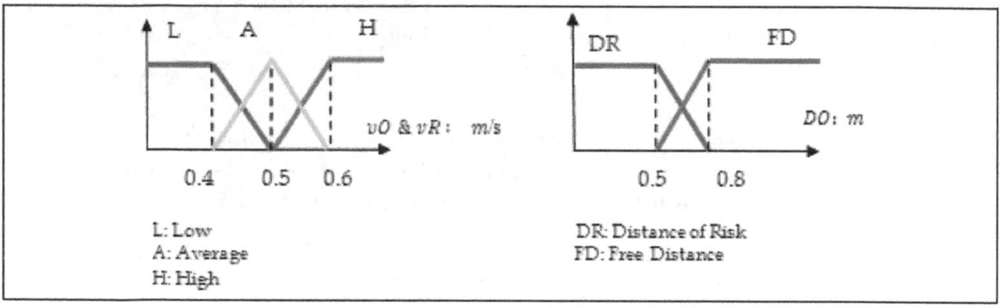

Fig. 9. Input membership functions for the linear velocity

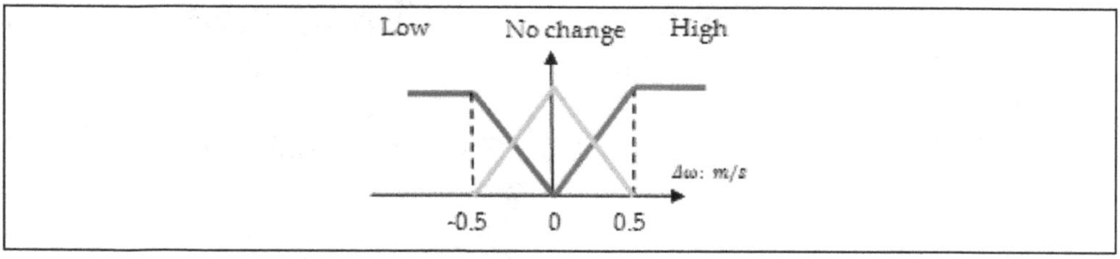

Fig. 10. Output membership functions. Velocity up/down control.

So, the speed of the robot can be influenced by the velocity of the obstacle in the front area. We present in algorithm 3 the reasoning process of the robot used to predict the velocity of the nearest obstacle in this area in order to determine its velocity. For example, if the velocity of the obstacle is smaller than the velocity of the robot, then the robot decreases its speed in order to avoid collision in the near future. If the velocity of the obstacle is larger than the velocity of the robot, the robot keeps the same speed if its velocity and the velocity of the obstacle are close. However, if there is a difference between the speed of the robot and the obstacle, it accelerates.

```
if (VO) =0 then
    The obstacle is static
    The robot must minimise its acceleration.
else
    if (VO < VA) then
        Reduce the velocity of the robot in order to avoid conflict in the near future.
    else
if (VO > VA) then
        Keep the same speed if the velocity of the robot and the velocity of the
obstacle
        are close or accelerate if they are different.
else
            Keep the same speed.
        end if
    end if
end if
```

Algorithm 3. The velocity of the robot according to the concept of the velocity of the obstacle

4.4.3 Fuzzy system rules

Robot navigation is actually governed by rules in the fuzzy controller which represent qualitatively the human driving heuristics. The fuzzy logic controllers are applied to coordinate and to combine the various behaviours of the robot in order to choose a suitable decision. The fuzzification converts the input variables into input grades by means of the membership functions shown in figure 5 and figure 9. The inference and aggregation generate a resultant output membership function with respect to fuzzy rules. The defuzzification gives the output membership function shown in figure 6 and in figure 10. Therefore, the total number of rules to determine the angular velocity is N= 24. We can reduce this number into 9 if-else rules by eliminating redundancies (rules that have different values of input variables and the same value of the output variable can be replaced by one rule). We present in table 2 the fuzzy rules for the angular velocity. For example, if DTCF=M (the obstacle in the front area is moving) and TO=F (the target is in front area) then $\Delta\theta$=NT (the robot do not change its trajectory) whatever the value of DTCL and DTCR. Likewise, for the linear velocity there are N=18 rules, but we can reduce them into 9 if-else rules.

The use of a small number of rules is required in dynamic environments, this is especially important when a fast response to the changes is required (short execution and high frequency of sensor readings). We use the method of Center of Gravity Defuzzification to

DTCF		F				M			
DTCL		F		M		F		M	
DTCR		F	M	F	M	F	M	F	M
TO	F	LRT	LRT	SLT	SLT	NT	NT	NT	NT
	L	LRT	LRT	LLT	LLT	NT	NT	LLT	LLT
	R	LRT	LRT	LRT	LRT	NT	LRT	NT	LRT

Table 2. The fuzzy rules for the angular velocity.

calculate the output for the two controllers. The constants of output membership function for the first controller (angular velocity) are shown in figure 6 and the values of the constants of output membership function for the second controller (linear velocity) are shown in figure 10. During simulation work, the robot's normal velocity VN is set to 0.5m/s.

5. Experimentation results

The robot calculates a "TC" with each obstacle around it. The value of "TC" allows it to predict the nature of each obstacle (static or dynamic). Then, the robot can calculate the velocity of each obstacle in order firstly to anticipate the state of the environment in the future and therefore to adjust its speed in order to avoid collisions.
In order to prove the efficiency of the proposed model, we proceed as follows. In a first time, we test our method in static environments (composed of U-shape and T-shape obstacles). Having obtained valid results, we achieve, in a second time, tests in dynamic environments (including moving robots). We have used the Simbad simulation platform.

5.1 The parameters of the robot and the controller in simulation

In simulation, the parameters of the robot are: the maximum linear velocity $\vartheta max = 1m/s$ and the maximum angular velocity $\theta max = 45/s$. The height of the mobile robot is 0.78 m, the diameter is 0.4 m, the weight is 50 kg and the radius of the wheel is 0.05m. The Δ time interval between two successive perceptions is 0.4s. This interval time is used to calculate the TC.
In addition, we have defined a variety of basic scenarios for the experimentation stage. In the reminder of this section we detail some of them. For each scenario, the robot is initially positioned at the centre of the environment and starts to work without having any preliminary information. The simulation environment is a square area composed of multiple traps like U-shape and T-shape forming the static obstacle, and moving agents (moving objects and other robots).

5.2 Experimentation results in static environments

Test results in static environments are presented in figure 11. Each scenario shows the robot's environment with the robot's path covered and the detected obstacles.
Scenario 1 The robot is initially positioned in an environment composed of multiple "dead cycle" as shown in figure 11.a. The robot calculates for each trap a DTC in order to predict the nature of the obstacle. The robot's decision depends on this information (e.g. if the DTC is equal to zero then, the obstacle is static and the robot should change its direction). In region 'a' it detects that the DTCF is equal to zero, which means that the robot should change immediately its direction to find a free path towards its target. In region 'b', it changes its behaviour and Turns Right. In region 'c', the robot detects a new local minimum (DTCF=0, DTCR<ΔT and DTCL<ΔT). In this situation, the robot should find a free trajectory in order to move away from this obstacle. When it leaves this local minimum, the robot adjusts its trajectory to minimise the distance between the robot and the target. For each step, the robot reasons about the nature, velocity and the orientation angle in order to choose a direction that will be free in the future (predict the future state using the nature and the velocity of obstacle in order to find a suitable path with fewer constraints). However, in (Wang & Liu, 2005) based on trial-return behaviour phenomenon, the robot searches the nearest exit. Therefore, when the nearest exit is blocked by along wall, the

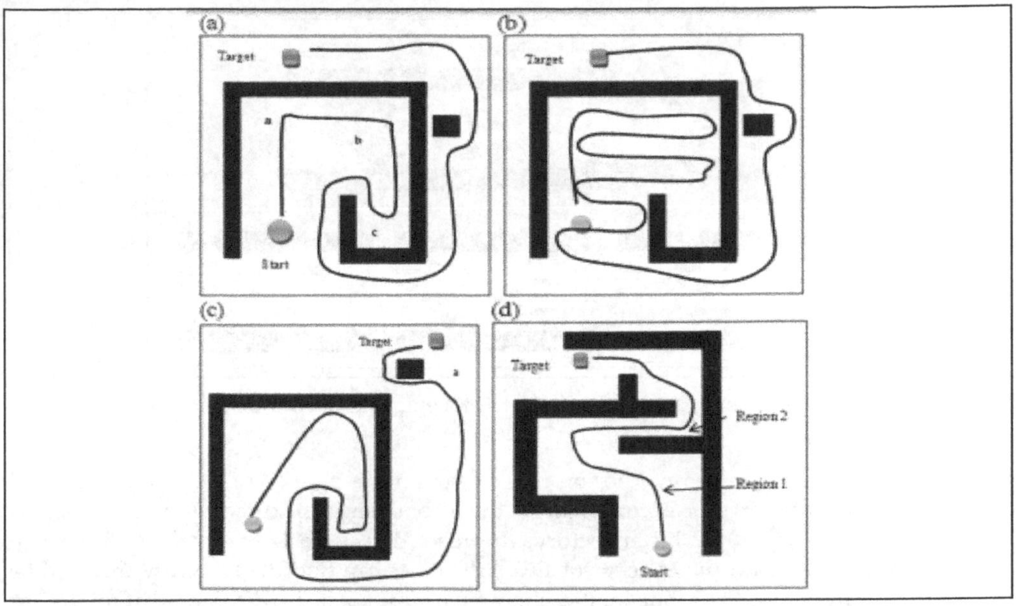

Fig. 11. Experimental results in static environments

nearest exit will be the opening at the right hand side where the wall ends (see figure 11.b). This method takes a large time to find the free end of the wall. However, the problem with trial-return motion is high power consumption and the time spent from start point to target is long when the obstacle is composed of complex traps like U-shape and T-shape. In (Zhu & Yang, 2004), the dead cycle problem is resolved by using a state memory strategy (see figure 11.c). Therefore, there is a situation where the robot cannot go straight toward the target. It has to keep turning around the obstacle (point "a") if the distance between the robot and the obstacle is shorter than the distance between the robot and the target.

Scenario 2 We test our method in an environment composed of corridors (wide and narrow). As shown in figure 11.d the robot moves in a corridor in order to reach its target. The problem is how to ensure the motion's stability without oscillation. In our architecture, there is a compromise between "reach target" and "avoid obstacles". In region 1, the robot crosses a large corridor; it changes its direction when it detects a static obstacle. In region 2, the robot crosses a narrow corridor; in this situation the $DTCR<\Delta T$ and the $DTCL<\Delta T$. Thus, the robot cannot change its trajectory, but it should only adjust its velocity with the velocity of the nearest obstacle in the front area. This reasoning allows the robot to reach its target easily without oscillation.

5.3 Experimentation results in dynamic environments

Scenario 1 Generally it is difficult to deal with narrow passage cases when there are moving actors because there may be oscillation in the trajectory between multiple obstacles. However, we can show, in figure 12, the effectiveness of our method when dealing with this case. The robot navigates in an environment composed of corridors with a moving obstacle. At the first time, the corridor is narrow. In this setting the robot adjusts its velocity to avoid collision with the obstacle without changing the direction because the goal is in the front

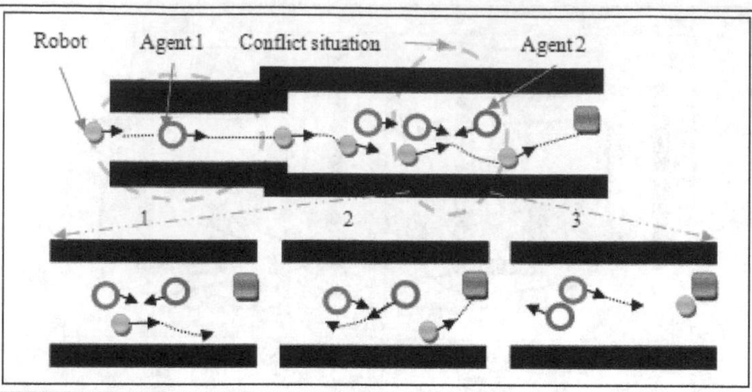

Fig. 12. Experimental results in a dynamic environment: case of conflict situations in narrow and large corridors.

area and there is no free area neither on the left nor on the right (DTC<ΔT on the Left and the Right). When the corridor becomes large, the robot changes its direction because there is a free area on the right. In this trajectory, the robot detects that there is a conflict space (DTCF is negative). With the concept of DTC, the robot can find the trajectory that will be free in the future. Thus, the robot should choose the most adequate direction that minimises states of conflict. So, in our model the robot can go through narrow passages successfully. However, in (Lazkano et al., 2007) a reactive navigation method based on Bayesian Networks is proposed. This method only takes into account the door-crossing behaviour. Its drawback is that it becomes slow when the environment becomes more dynamic because it requires a long computing time.

Scenario 2 We test our method in a dynamic environment composed of multiple traps as shown in figure 13. The robot adjusts its velocity in order to avoid collision with moving

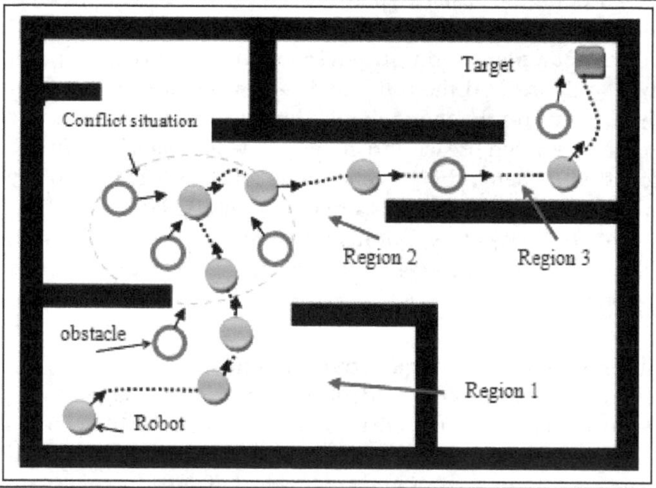

Fig. 13. Experimental results in a dynamic environment: case of conflict situations in a dynamic environment composed of local traps.

obstacles and changes its orientation if it detects a static obstacle. In region 1, it moves away from a U-shape obstacle because it detects that the DTC is equal to zero. In region 2, the robot is in a large corridor with moving obstacles, it calculates the DTC relative to every moving obstacle in order to predict conflict situations. In this setting, the DTC of every obstacle around the robot decreases, which means that the obstacles are going toward the robot. At the first step, the robot calculates the velocity of the nearest moving obstacle in order to adjust its velocity. At the second step, it reasons about the trajectory that it has to choose (minimise interactions with other agents).

According to a large number of simulation experiments, we can conclude that:

1. The robot uses a simple model of perception composed of ultrasonic sensors that provides information at real time.
2. The robot can provide a suitable trajectory in dynamic environments and there are no local minimum encountered.
3. The results provided by our method are satisfactory with the robot's dynamic constraints.
4. The robot can easily predict the nature of obstacles around it, and it uses this kind of information to predict conflict situations with other agents sharing the same resources.
5. The robot can also predict the velocity of each obstacle, using its simple model of perception, to adjust its linear velocity in order to avoid collision with other agents.
6. Our method can be adapted to different cases in dynamic environments.

6. Conclusion and perspectives

In this paper, a new method based on the sensors' information to combine reactivity and anticipation in order to predict conflict situations between robots has been proposed. We have used fuzzy logic to develop a control system that enables the robot to navigate at real time and to minimise its interaction with other agents. Basing on the simulation experiments, we can conclude that our method operates in different cases of dynamic environments. The robot can easily predict the nature of obstacles in order to anticipate conflict situations. It can also predict the velocity of each obstacle and adjust its linear velocity in order to avoid collision. As perspectives for this work, we attend to improve the robot's behaviour by introducing both the ability to learn and the concept of communication between robots in order to explore the environment and to share knowledge.

7. References

Arkin, R. (1990). Integrating behavioural, perceptual and world knowledge in reactive navigation. *Robots and Autonomous Systems*, Vol.6, pp. 105-122

Bonarini, A., Invernizzi, G. & Labella, T. H. (2003). An architecture to coordinate fuzzy behaviours to control an autonomous robot. *Proc. Fuzzy sets Systems*, pp.101-115

Canny, J., Rege, A. & Reif, J. (1990). An exact algorithm for kino dynamic planning in the plane. *In ACM Symp.On Computational Geometry*,pp. 271-280.

Darwiche, A. (2009). Modelling and Reasoning with Bayesian Networks. *New York, NY: Cambridge University Press*

Ferber, J. (1995). Les systèmes multi-agents vers une intelligence collective.*InterEditions*, Paris.

Fiorini, P. & Shiller, Z. (1998). Motion planning in dynamic environments using velocity obstacles. *International.Journal.Robot. Res*, Vol.17 (7), pp. 760-772

Fox, D., Burgard, W. & Thrun, S. (1997). The dynamic window approach to collision avoidance. *IEEE Robotics and Automation Magazine*, pp.23-33

Fulgenzi, C. (2009). Autonomous navigation in dynamic uncertain environment using probabilistic models of perception and collision risk prediction, PhD thesis, INRIA Rhône-Alpes, Grenoble France

Fulgenzi, C., Tay, C., Spalanzani, A. & Laugier, C. (2008). Probabilistic navigation in dynamic environment using Rapidly-exploring Random Trees and Gaussian Processes. *Proc.IEEE/RSJ International Conference on Intelligent Robotsand Systems*, pp.508-513

Huang, W.H., Fajen, R., Fink, J. & Warren, W.H. (2006). Visual navigation and obstacle avoidance using a steering potential function, *Proc. Robotics and Autonomous Systems*, pp. 288-299

Huq, R., Mann, G. K. I. & Gosine, R. G. (2008). Mobile robot navigation usingmotors schema and fuzzy context dependent behaviour modulation. *Appl. Soft. Comput.*, pp. 422-436

Innocenti, B. (2007). A multi-agent architecture with cooperative fuzzy control for a mobile robot. *Proc. Robotics and Autonomous Systems*, Vol. 55, PP. 881-891

Jing, X., Wang, Y. & Tan, D. (2003). Cooperative motion behaviours using biology-modelling behaviour decision-making rules. *Cont. Theory. Appl*, Vol.20 (3), pp. 407-410

Khatib, O. (1986). Real-Time Obstacle Avoidance for Manipulators and Mobile Robots. *In International Journal of Robotics Research*, pp.90-98

Klir, G. J., Yuan, B. & Clair, U. S. (1997). Fuzzy Set Theory: Foundations and Applications. *Prentice-Hall, Upper Saddle River*

Langer, D., Rosenblatt, J. & Hebert, M. (1994). A behaviour-based system for off-road navigation. *Proc.IEEE Transaction. Robot. Automation*, Vol.10, pp. 776-783

Lazkano, E., Sierra, B., Astigarraga, A. & Martnez-Otzeta, J.M. (2007). On the use of Bayesian Networks to develop behaviours for mobile robots. *Robotics and Autonomous Systems*, Vol.55, pp. 253-265

Mamdani, E. H. (1974). Applications of fuzzy algorithms for simple dynamic plants.*Proc.IEEE, 121(12)*, pp. 1585-1588

Minguez, J. & Montano, L. (2000). Nearness diagram navigation: a new real time collision avoidance approach. *Proc. IntelligentRobots and Systems*, pp.2094-2100

Minguez, J., Montano, L., Siméon, N. & Alami, R. (2002). Global nearness diagram navigation. *Proc. IEEE International Conference on Robotics and Automation.*

Nilsson, N. J. (1980). Principles of Artificial Intelligence. *Morgan Kauf-mann Ed, Los Altos, CA*

Omid, R.E.M., Tang, S.H. & Napsiah, I. (2009). Development of a new minimum avoidance system for a behaviour-based mobile robot. Proc.*Fuzzy Sets and Systems*

Ono, Y. (2004). A mobile robot for corridor navigation: A multi-agent approach. *Proc. ACM Southeast Regional Conference, ACM Press*, pp. 379-384

Ordonez, C., Collins, E. G., Selekwa, M.F. & Dunlap, D. D. (2008). The virtual wall approach to limit cycle avoidance for unmanned ground vehicles. *Proc. Robotic and Autonomous System*, vol. 56, pp.645-657

Oreback, A. & Christensen, H. I. (2003). Evaluation of architectures for mobile robotics. *Autonomous Robots*, Vol.14, pp. 33-49

Petti, S. & Fraichard, T. (2005). Safe motion planning in dynamic environments. *IEEE IROS*.

Posadas, J. L., Poza, J.L., Sim, J.E., Benet, G. & Blanes, F. (2008). Agent-based distributed architecture for mobile robot control. *Engineering Applications of Artificial Intelligence*, Vol.21, pp. 805-823

Pruski, A. & Rohmer, S. (1997). Robust path planning for non-holonomic robots.*J. Intell. Robot.Syst: Theory Appl*, 18(4), pp. 329-350

Quinlan, S., Khatib, O. (1993). Elastic bands: Connecting path planning and control. *Proc. IEEE International Conference on Robotics and Automation*, pp. 802-807

Ros, R. (2005). A CBR system for autonomous robot navigation. *Proc. Frontiers in Artificial Intelligence and Applications*, vol.131, pp. 299-306

Sahota, M. K. (1994). Reactive Deliberation: An Architecture for Real Time Intelligent Control in Dynamic Environments. *Proceedings of the AAAI*, pp. 1303–1308.

Sajotti, A., Konolige, K. & Ruspini, E. H. (1995). A multi valued-logic approach to integrating planning and control. *Artificial Intelligence*, pp. 481-526

Schwartz, J. T. & Sharir, M. (1983). General techniques for computing topological properties of algebraic manifolds. *Advances in Applied Mathematics*, 12, 1983.

Selekwa, M. F., Dunlap, D., Shi, D. & Collins, E. (2008). Robot navigation in very cluttered environments by preference-based fuzzy behaviours. *Proc.Robotics and Autonomous Systems*, pp.231-246

Simmons, R. (1996). The curvature-velocity method for local obstacle avoidance. *Proc. International Conference onRobotics and Automation*, pp.2737-2742

Stachniss, C. & Burgard, W. (2002). An integrated approach to goal-directed obstacle avoidance under dynamic constraints for dynamic environments. *Proc. IEEE International Conference on Intelligent Robots and Systems*, pp.508-513

Stentz, A. (1995). The focussed d* algorithm for real-time re-planning. *Proceedings of the International Joint Conference onArtificialIntelligence*.

Thrun, S. (1998). Bayesian landmark learning for mobile robot navigation. *Machine Learning*,Vol.33 (1), pp. 41-76

Ulrich, I. & Borenstein, J. (2000). VFH: Local obstacle avoidance with look-ahead verification. *Proc. IEEE International Conference on Robotics and Automation*, pp.2505-2511

Wang, M. & Liu, J. N. K. (2005). Fuzzy logic based robot path planning in unknown environments. *Proc. Internat. Conf.on Mach. Learn. And Cybernet*, pp.818-818

Yang, X., Moallem, M. & Patel, R. V. (2006). A layered goal-oriented fuzzymotion planning strategy for mobile robot navigation. *IEEE Trans.Syst, Man Cyber Part B: Cybernetics*, 35(6), pp. 1214-1224.

Zadeh, L. A. (1973). Outline of a New Approach to the Analysis of Complex Systems and Decision-Making Approach. *IEEE Trans. on Systems, Man, and Cybernetics SME-3(1)*, pp. 28-45

Zhu, A. & Yang, S. X.(2004). A fuzzy logic approach to reactive navigation of behaviour-based mobile robots. *Proc. IEEE International Conference on Robotics and Automation*, pp.5045-5050

Part 2

Control

9

Position Control and Trajectory Tracking of the Stewart Platform

Selçuk Kizir and Zafer Bingul
Mechatronics Engineering, Kocaeli University
Turkey

1. Introduction

Demand on high precision motion systems has been increasing in recent years. Since performance of today's many mechanical systems requires high stiffness, fast motion and accurate positioning capability, parallel manipulators have gained popularity. Currently, parallel robots have been widely used several areas of industry such as manufacturing, medicine and defense. Some of these areas: precision laser cutting, micro machining, machine tool technology, flight simulators, helicopter runway, throwing platform of missiles, surgical operations. Some examples are shown in Figure 1. Unlike open-chain serial robots, parallel manipulators are composed of closed kinematic chain. There exist several parallel kinematic chains between base platform and end moving platform. Serial robots consist of a number of rigid links connected in serial so every actuator supports the weight of the successor links. This serial structure suffers from several disadvantages such as low precision, poor force exertion capability and low payload-to-weight-ratio. The parallel robot architecture eliminates these disadvantages. In this architecture, the load is shared by several parallel kinematic chains. This superior architecture provides high rigidity, high payload-to-weight-ratio, high positioning accuracy, low inertia of moving parts and a simpler solution of the inverse kinematics equations over the serial ones. Since high accuracy of parallel robots stems from load sharing of each actuator, there are no cumulative joint errors and deflections in the links. Under heavy loads, serial robots cannot perform precision positioning and oscillate at high-speeds. Positioning accuracy of parallel robots is high because the positioning error of the platform cannot exceed the average error of the legs positions. They can provide nanometer-level motion performance. But they have smaller workspace and singularities in their workspace.

The most widely used structure of a parallel robot is the Stewart platform (SP). It is a six degrees of freedom (DOF) positioning system that consists of a top plate (moving platform), a base plate (fixed base), and six extensible legs connecting the top plate to the bottom plate.

SP was invented as a flight simulator by Stewart in 1965 (Stewart, 1965). This platform contained three parallel linear actuators. Gough had previously suggested a tire test machine similar to Stewart's model (Bonev, 2003). In the test machine, six actuators were used as a mechanism driven in parallel. Gough, the first person, developed and utilized this type parallel structure. Therefore, SP is sometimes named as Stewart-Gough platform in the literature. Stewart's and Gough's original designs are shown in Figure 2.

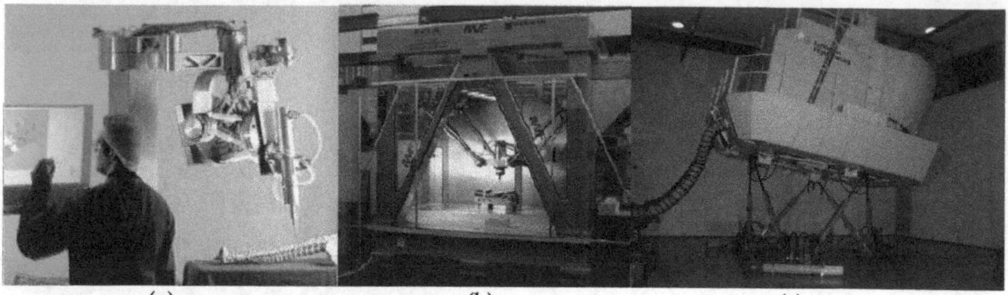

(a) (b) (c)

Fig. 1. Applications of the Stewart Platform: medical, manufacturing and flight simulator (Niesing, 2001; Merlet, 2006; Wikipedia)

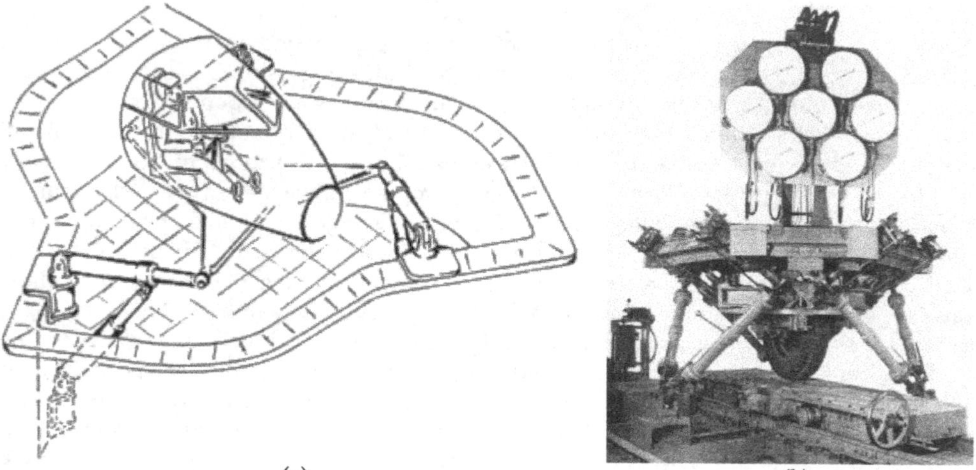

(a) (b)

Fig. 2. Stewart (a) and Gough (b) original design (Bonev, 2003)

SP was not attracted attention during the first 15 years since the first invention. Then, Hunt indicated the advantages of parallel robots. After 1983, researchers realized their high load carrying capacity and high positioning ability of these robots. Researchers were then started to study a detailed analysis of these structures. The widely used structure of SP, where top platform is connected to base platform using 6 linear axis with universal joints, was then developed (Hunt, 1983).

It is a well known fact that the solution of the forward kinematics problem is easier than the inverse kinematics problem for serial robot manipulators. On the other hand, this situation is the just opposite for a parallel robot. Inverse kinematics problem of parallel robot can be expressed as follows: position vector and rotation matrix in Cartesian space is given, and asked to find length of each link in joint space. It is relatively easy to find the link lengths because the position of the connecting points and the position and orientation of the moving platform is known. On the other hand, in the forward kinematics problem, the rotation matrix and position vector of the moving platform is computed with given the link lengths. Forward kinematic of the SP is very difficult problem since it requires the solution of many non-linear equations. In the literature, solutions of the forward (Chen & Song, 1994; Liao et al., 1993; Merlet, 1992; Nauna et al., 1990) and the inverse (Fitcher, 1986; Kim & Chung, 1999; Sefrioui & Gosselin, 1993) kinematics has been given in detail (Kizir et al., 2011).

In this study, design and development stages were given about position control and trajectory tracking of a 6 DOF-Stewart platform using Matlab/Simulink® and DS1103 real time controller. Matlab® (Mathworks Inc.) is a well known and one of the most popular technical computing software package that it is used in a wide area of applications from financial analysis to control designs. Matlab/Simulink® allows easiest way of programming and technical computing to its users. It enables simulations and real time applications of various systems. Third party co-developers improve its abilities allowing using hundreds of hardware. Dspace® company is one of the third party participate of Matlab® that produces rapid control prototyping and hardware-in-the-loop simulation units. DS1103 is a powerful real time controller board for rapid control prototyping (Dspace Inc.).

This chapter is organized in the following manner. System components and real-time controller board are introduced in section 2 and 3, respectively. Position and trajectory tracking control with PID and sliding mode controllers are described in section 4. Finally, experimental results are given in detail.

2. Stewart platform system

The system components are two main bodies (top and base plates), six linear motors, controller, space mouse, accelerometer, gyroscope, laser interferometer, force/torque sensor, power supply, emergency stop circuit and interface board. They are shown in Figure 3.

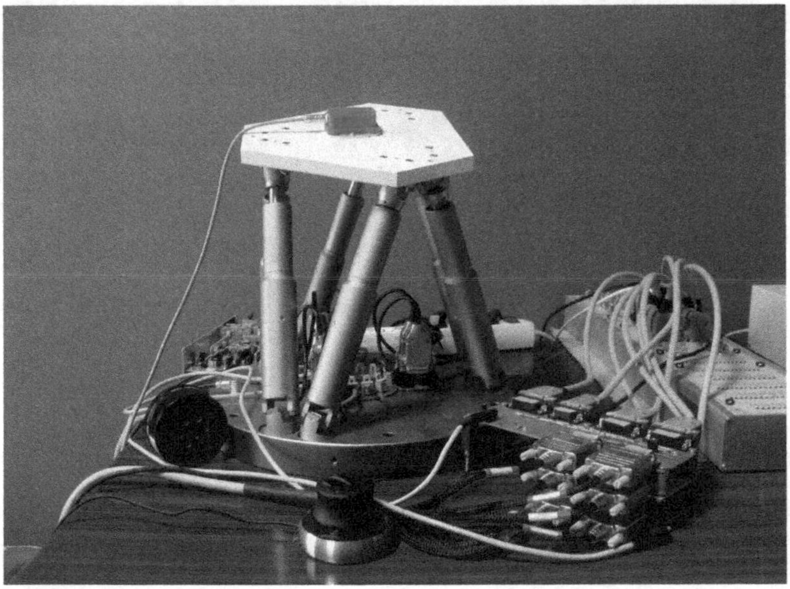

Fig. 3. Stewart platform system

A simple emergency stop circuit was designed to protect the motors, when they move to out of the limits. This circuit controls the power supply which gives the energy to the motors based on the signal of hall-effect sensors on each motor. A switch-mode 150W power supply with inhibit input and EMI filter is used to supply required energy. Also, an interface board was designed between controller and motors.

The Dspace DS1103 real time controller is used to implement control algorithms. DS1103 is a rapid prototyping controller that developed for designing and analyzing complex and difficult control applications. It has various inputs and outputs such as digital, analog digital converter, digital analog converter, serial interface, can-bus, pulse width modulation (PWM) channels and encoders in order to be used lots of peripheral unit like actuators and sensors. DS1103 has a real time interface (RTI) that allows fully programmable from the Simulink® block diagram environment. A dspace toolbox will be added to Simulink® after installing RTI, so it can be configured all I/O graphically by using RTI. You can implement your control and signal processing algorithms on the board quickly and easily. A general DS1103 controller board system is shown in Figure 4. It consists of a DS1103 controller card in expansion box, CLP1103 input-output connector and led panel, DS817 link card and a computer.

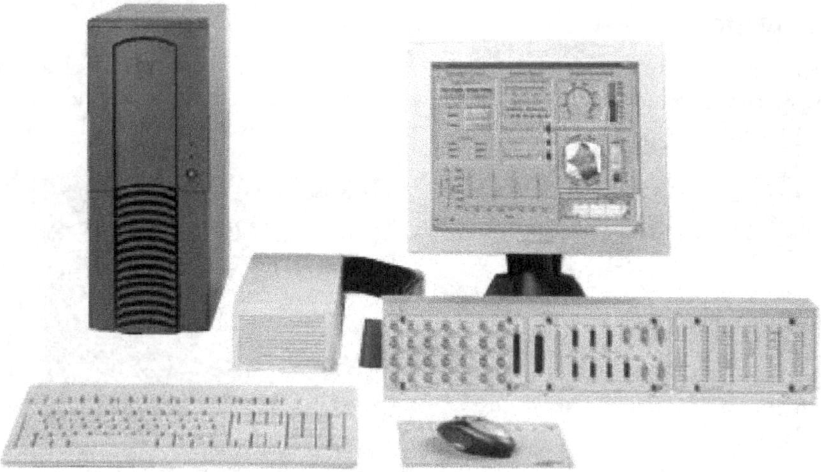

Fig. 4. A general DS1103 set-up

3. DSPACE tool box and design procedures

While Dspace offers various development tools, it is needed at least Control Desk and RTI block set software packages in order to develop projects under Simulink. After installing software, a tool box shown in Figure 5 is added to Simulink®. It can be also opened dspace library shown in Figure 5 typing 'rti' in the matlab command window. Block sets in the library are divided into two categories: master processor and slave dsp. While master have blocks such as ADC, serial, encoder, digital I/O, slave has such as PWM, ADC and digital I/O.

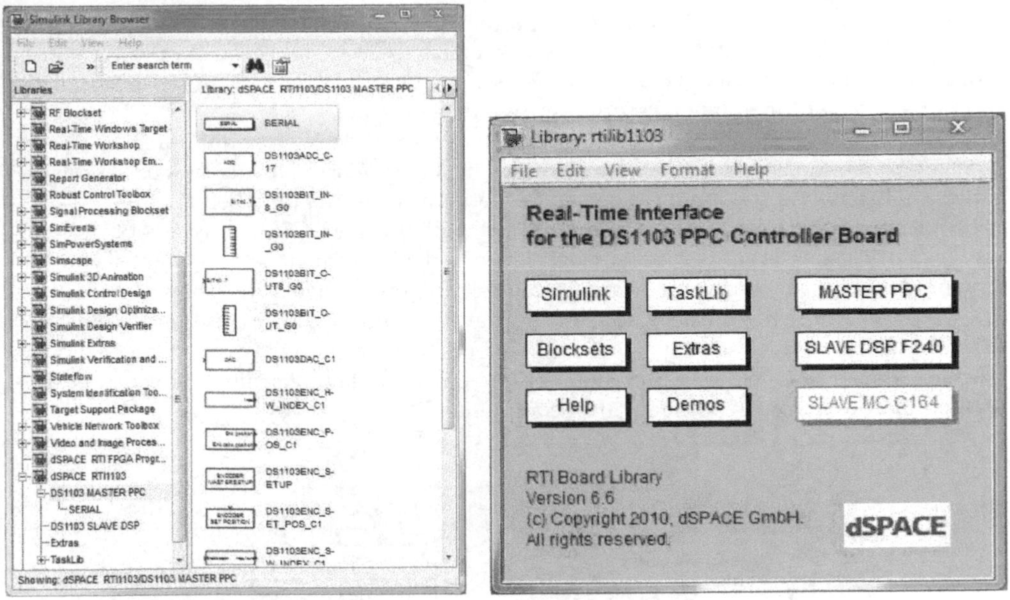

Fig. 5. Dspace toolbox and library

Another software component is the control desk (interface is shown in Figure 6) which allows downloading applications, doing experiments, easily creating graphical user interface and data acquisition.

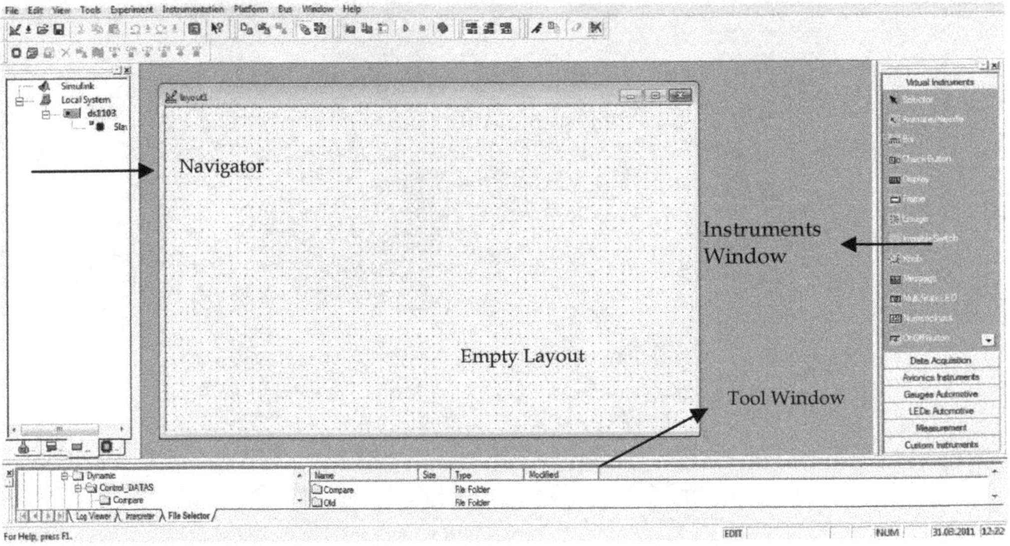

Fig. 6. Control desk interface

As can be seen from Figure 6, panel on the left side is called "Navigator" and it has four tabs: experiment, instrumentation, platform and test automation. All files written for

conducting experiment are listed in the experiment tab. Instrumentation tab allows building instrument panels in order to change and monitor the variables of a model. Supported simulations and connected boards are shown in platform tab. Test automation tab has functions about automation tasks, other software solution of Dspace.
Bottom side is called "Tool Window" having log viewer, file selector, interpreter and open experiment tabs. It is seen errors and warnings in the log viewer tab and files under selected folder are listed where an application can be loaded by drag and drop action.
In order to create a GUI for an experiment, it should be opened an empty layout from file-new layout click. A lot of instruments are listed in the instrument selector right side on the control desk. Virtual instruments and data acquisition elements are shown in Figure 7 below. An instrument can be placed on the layout plotting with mouse left clicked after selecting from virtual instruments panel. Its position and size can be changed and its properties such as color, text, precision and etc can be settled according to needs. It can be saved and added to an experiment after completing GUI.

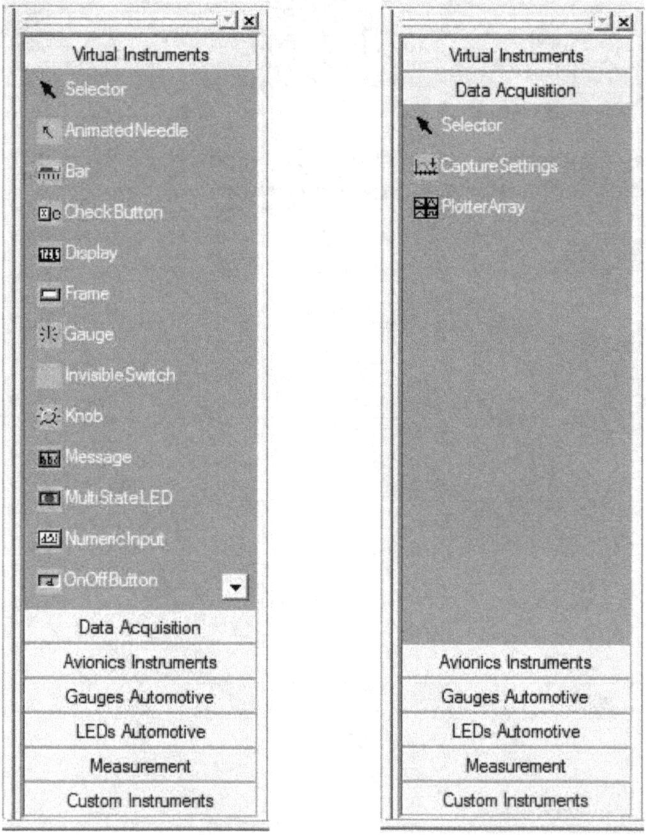

Fig. 7. Virtual instruments and data acquisition in the instruments selector window

All development steps can be illustrated basically in the Figure 8 below. This figure is illustrated for DS1103 in an expansion box. It should be finished all connections before this procedure.

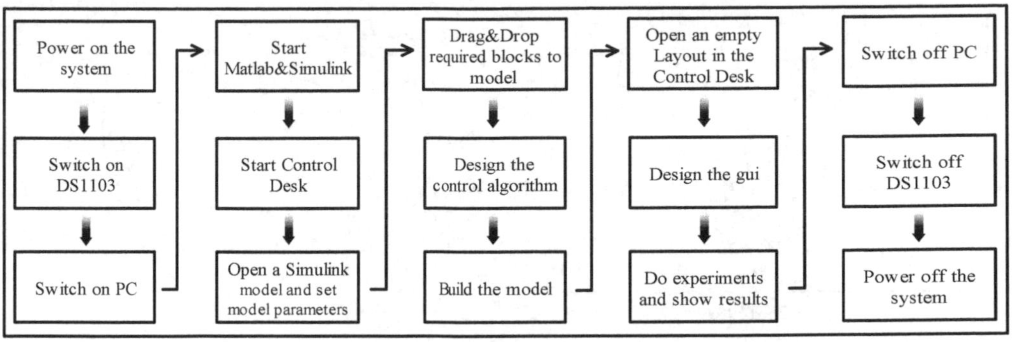

Fig. 8. Basic flow diagram for developing a project

4. Control

A controller is needed to move top platform from initial position to desired position and orientation. It will generate required forces for each motor. Position and trajectory control of the platform can be reduced to leg position control after inverse kinematic and path planning algorithms. A PID (proportional-integrator-derivative) and sliding mode position controllers were developed and implemented. Control algorithms designed in Simulink environment and embedded in the Dspace DS1103 real time controller.

All robots are electro-mechanic devices consisting of actuators, sensors and mechanical structure. In order to control the robots for desired motions kinematic and dynamic equations of the system should be known. Firstly kinematic solution should be computed before controller design. A schematic model of the SP for kinematic solution is illustrated in Figure 9. In the figure, base $B=\{X,Y,Z\}$ and top $T=\{x,y,z\}$ coordinate systems are placed and base and top joint points are labeled as B_i (i=1,2, .. 6) and T_i (i=1,2, .. 6).

It is needed to find leg lengths to reach the moving platform to its desired position and orientation according to fixed platform (inverse kinematics). Required leg vectors (L_i) for given position vector P and orientation matrix R are obtained by using the following equation. Finally, norm of the vectors (L_i) are leg lengths (l_i) (Fitcher, 1986; Kim & Chung, 1999; Sefrioui & Gosselin, 1993).

$$L_i = R_{XYZ}T_i + P - B_i \qquad i:1,2,...,6 \tag{1}$$

In order to have T_i and B_i position vectors based on robot structure, an m-file is written and a Simulink model is designed to obtain inverse kinematic solution by using Equation 1. The model shown in Figure 10 uses the m-file to get required variables and takes the desired position (x, y, z) and orientation (φ, θ, ψ) of the top platform. It outputs the leg lengths. Desired block in this model is shown in Figure 11. The reference inputs can be entered in this block.

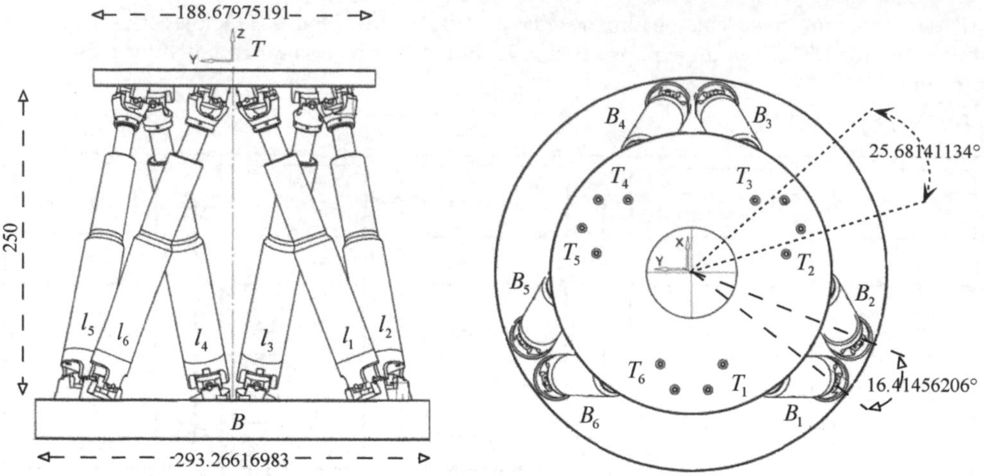

Fig. 9. Schematic diagram of the Stewart platform

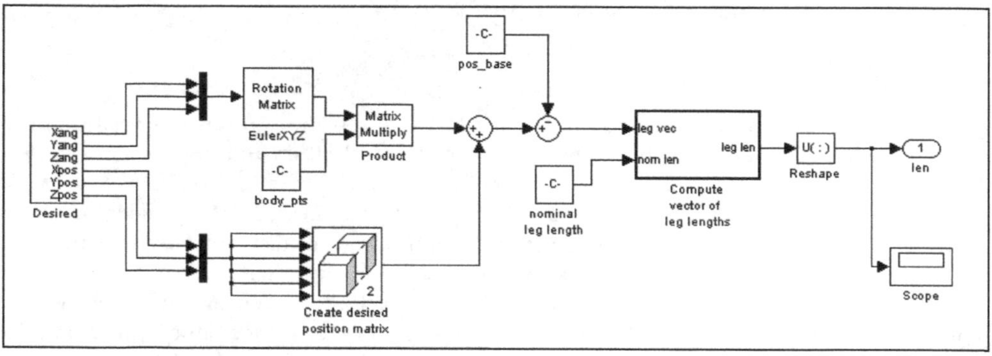

Fig. 10. Simulink model for inverse kinematic solution

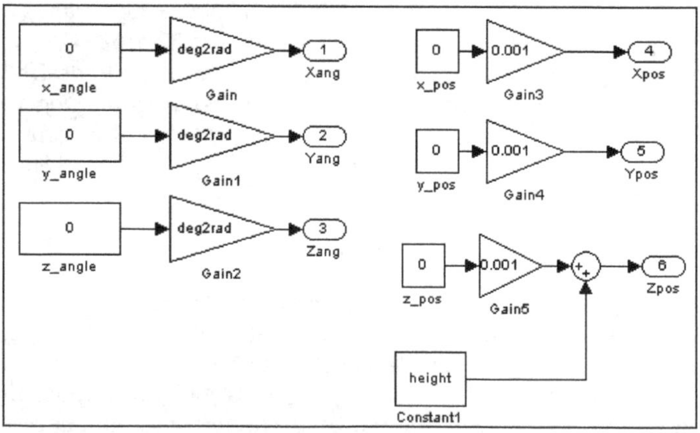

Fig. 11. A subsystem for desired references

4.1 Leg model

The leg system is basically composed of dc motor, precision linear bearing & ball screw and coupling elements. Dc motor model is given below (Küçük & Bingül, 2008).

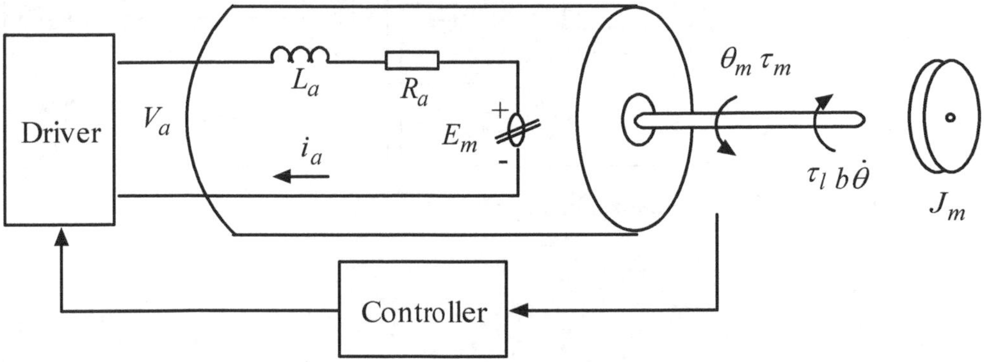

Fig. 12. DC motor model

The symbols represent the following variables here; θ_m is the motor position (*radian*), τ_m is the produced torque by the motor (*Nm*), τ_l is the load torque, V_a is the armature voltage (*V*), L_a is the armature inductance (*H*), R_a is the armature resistance (*Ω*), E_m is the reverse EMF (*V*), i_a is the armature current (*A*), K_b is the reverse EMF constant, K_m is the torque constant.

$$L_a \frac{di_a}{dt} + R_a i_a = V_a - E_m \quad (2)$$

$$E_m = K_b \frac{d\theta_m}{dt} \quad (3)$$

$$\tau_m = K_m i_a \quad (4)$$

$$\tau_m - \tau_l = J_m \frac{d^2\theta_m}{dt^2} \quad (5)$$

4.2 Startup algorithm

Before controller design a startup algorithm is needed to get robot to its home position. The position of each motor is controlled after startup. Motors have incremental encoders therefore firstly they must be brought their zero or home position. When SP system is energized, an index search algorithm looks what the position of the each leg is. The algorithm simply searches index signal of the leg and when it is found encoders are reset by hardware. Designed Simulink model for this purpose is illustrated in Figure 13. Movements to the home position for possible two situations (from upper and lower sides to zero) are shown in Figure 14.

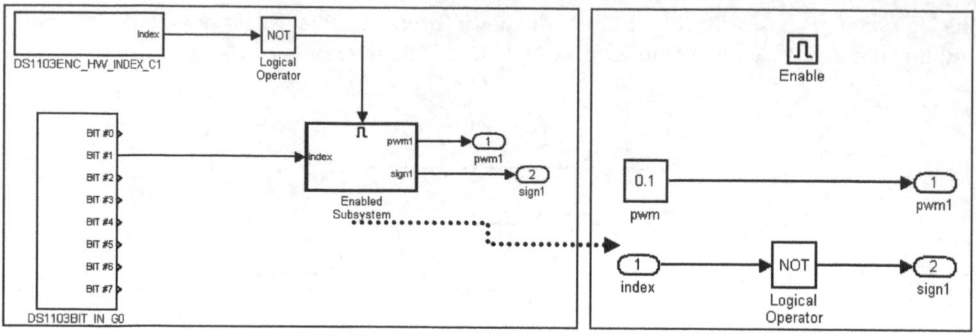

Fig. 13. Simulink model for initialization algorithm

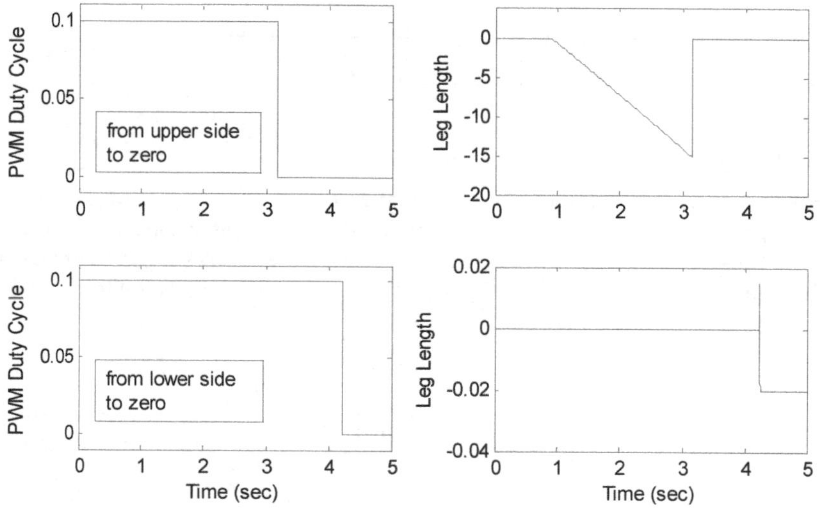

Fig. 14. Initialization routine

4.3 Trajectory generation

For step inputs, the leg lengths obtained from inverse kinematics solution input to independent position control system for each motor. This movement is defined in the joint space. In order to move robot along a straight line, a trajectory planning algorithm is developed. Thus, it can be determined start and stop times of the motion besides desired position and orientation inputs. Also, motors are synchronized each other during the motion. If classical polynomial-based trajectory equations are examined, the following deficiencies are determined: i) there are need many initial and finish values in order to find the polynomial coefficients ii) the acceleration values, especially initially require high levels, iii) the coefficients need to be calculated again each time when the conditions changes. Kane's transition function is used to resolve these shortcomings and it is given the following equation (Reckdahl, 1996).

$$y(t) = y_0 + (y_f - y_0)\frac{t - t_0}{t_f - t_0} - \frac{y_f - y_0}{2\pi}\sin\left(2\pi\frac{t - t_0}{t_f - t_0}\right) \quad (6)$$

where $y(t)$ is the position function, y_0 is the initial position, y_f is the finish position, t is time, t_0 is initial time and t_f is the finish time. An example of the trajectory generation is given below (Figure 15) using Equation 6. As can be seen from figure, position, velocity and acceleration curves are given for two seconds and velocity and acceleration start and finish with zero.

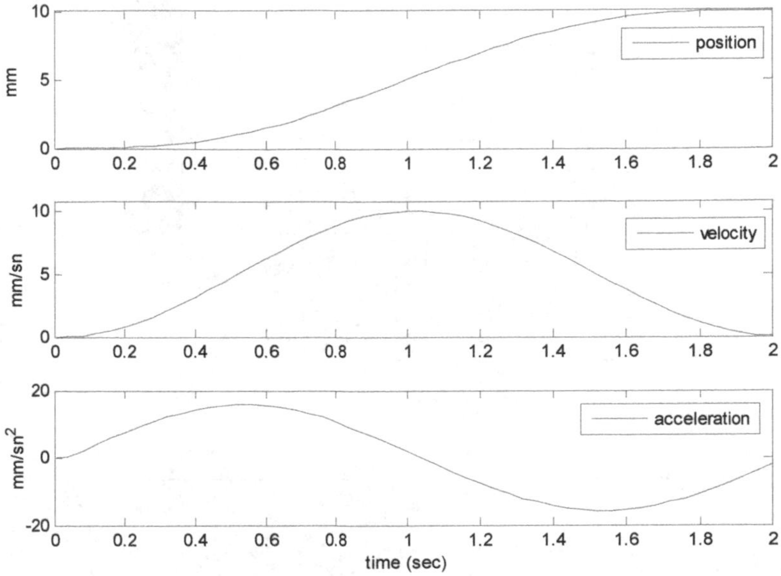

Fig. 15. An example of trajectory generation with Kane function

Equation 6 was embedded in Simulink block. Only end position and path period is given to this structure and other parameters are automatically calculated. This form of motion starts and finishes with zero velocity and zero acceleration. Position, velocity and acceleration values are quite soft changes. This is very important for motors to start and stop more softly. Main model of the trajectory planning with reference inputs is shown in Figure 16 below. Detail of the blocks named 'path' in the Figure 16 is shown in Figure 17 which implements the Equation 6.

4.4 PID controller

PID control is one of the classical control methods and widely used in the industrial applications. The difference between the set point and the actual output is represented by the $e(t)$ error signal. This signal is applied to a PID controller, control signal, $u(t)$ is as follows.

$$u(t) = K_P e(t) + K_I \int e(t)dt + K_D \frac{d}{dt} e(t) \quad (7)$$

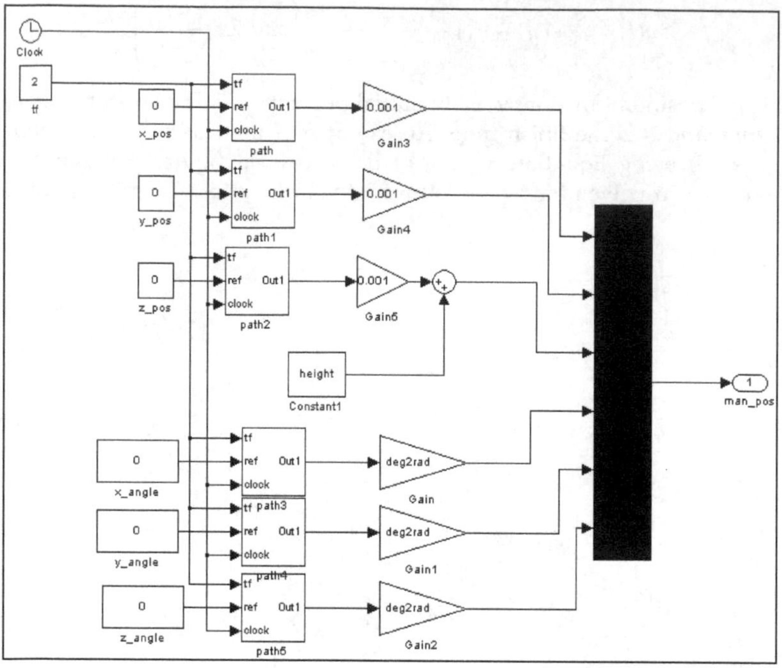

Fig. 16. Reference inputs with trajectory planning

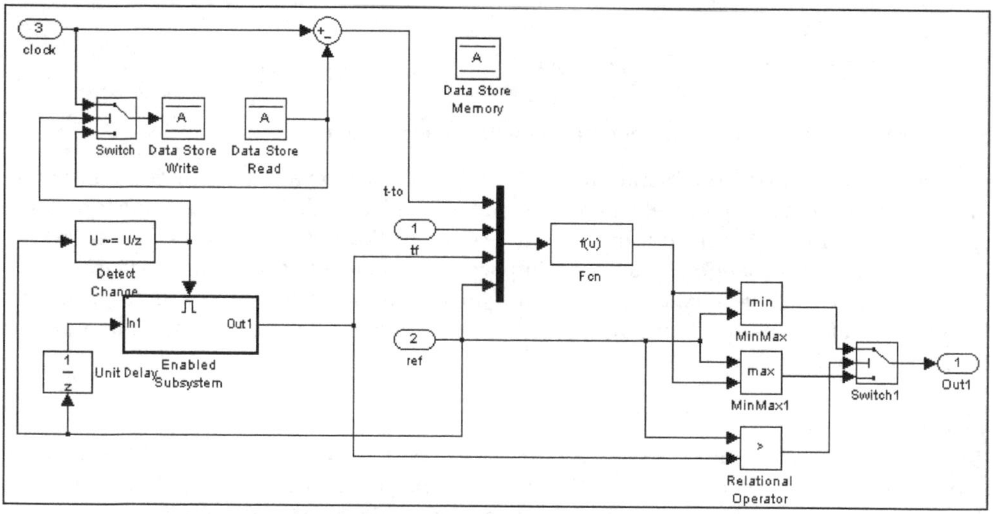

Fig. 17. Implementing Equation 6 in Simulink model

General control schema is summarized as follows: initialization, reference input, inverse kinematic, measurement, closed loop controller, input/outputs and additional blocks for safety reasons. Main PID Simulink model diagram is given Figure 18 below. The model

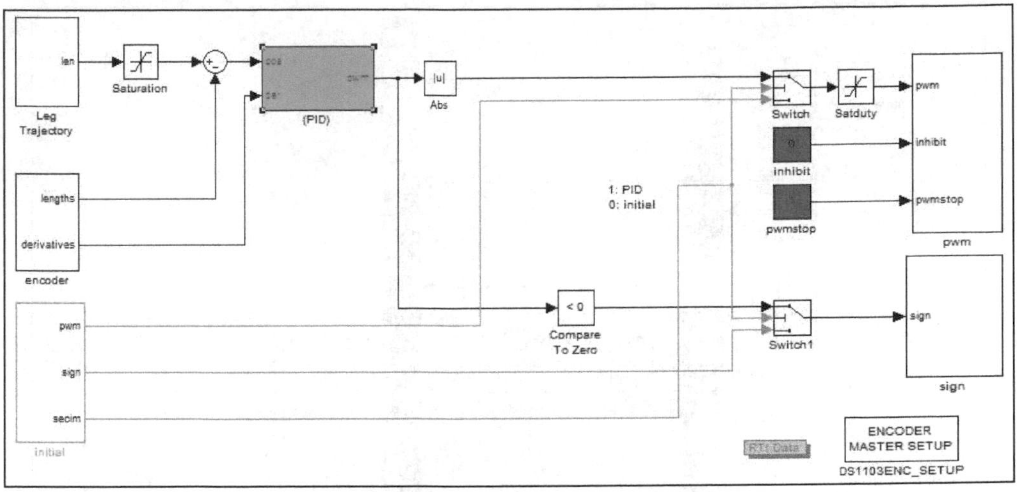

Fig. 18. Main PID controller Simulink model

contains some subsystems such as leg trajectory, encoder, initial, PID, pwm and sign. These subsystems perform the tasks mentioned above.
Project has a Control Desk interface for experiments and it is shown below. All system information can be entered through this interface. It contains variables that can be used in the development phase. Reference input values can be easily entered through the interface.
Leg trajectory subsystem detail was given before, so it will be continued giving other blocks. Firstly, encoder subsystem given in Figure 20 is described. Each encoder is read using an encoder block and then pulses are converted to metric unit (mm) and scaled. Round subsystem shown in Figure 21 is used to eliminate errors less than 500 nanometer.

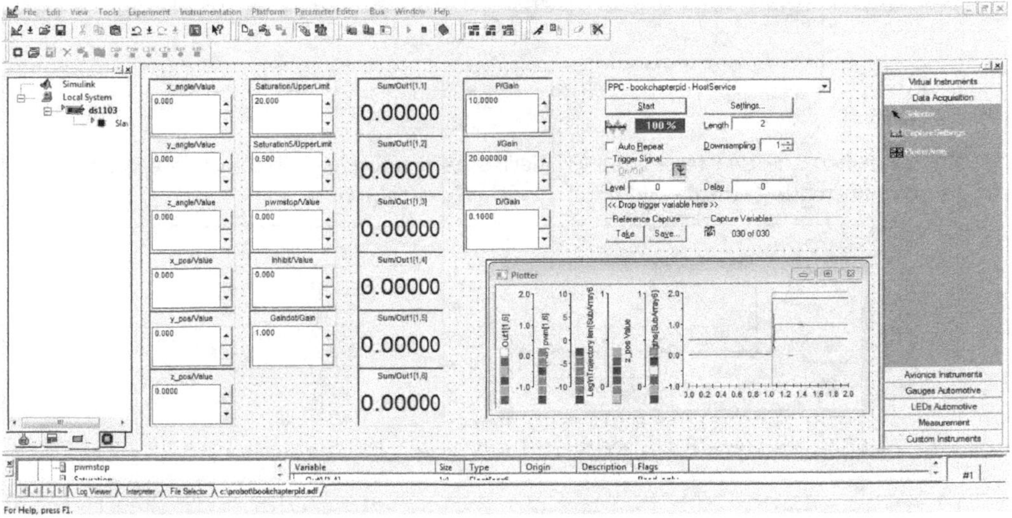

Fig. 19. Control Desk GUI for data acquisition and parameter update

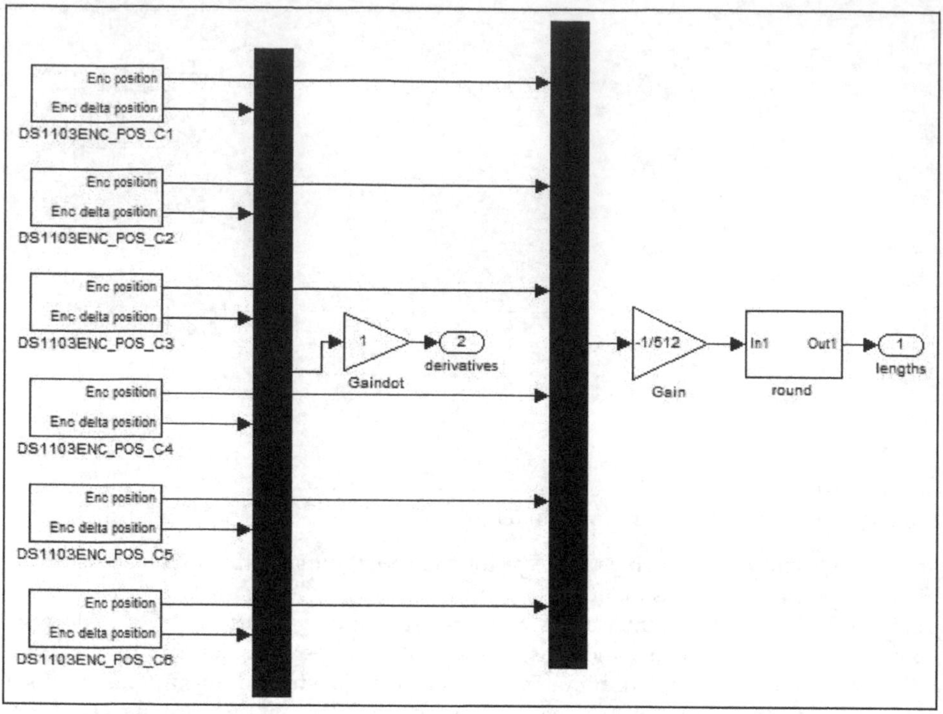

Fig. 20. Encoder subsystem

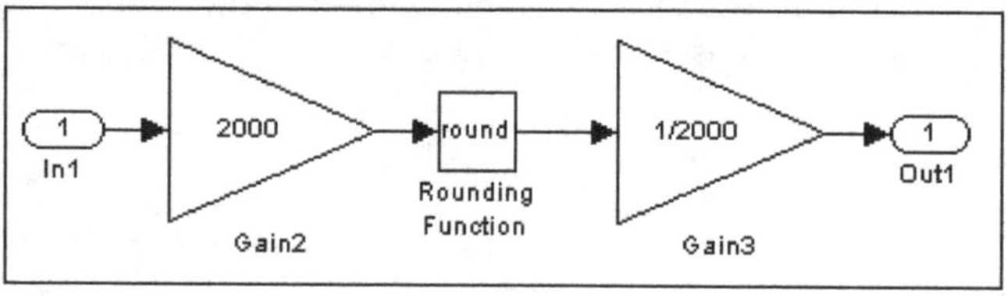

Fig. 21. Round subsystem

A PID controller is added for each leg. These subsystems are shown in figure below. It is a classic way of creating simple and efficient closed control loops.

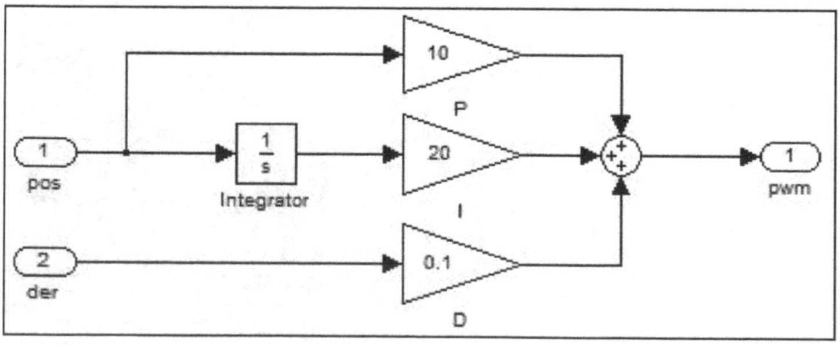

Fig. 22. PID subsystem in main model and simple PID structure

Initial subsystem model is shown in Figure 23 and its details were given before. It takes the index signals of the motors and produces a predefined duty cycle value for PWM generation, motor direction signals and status of the initialization routine.

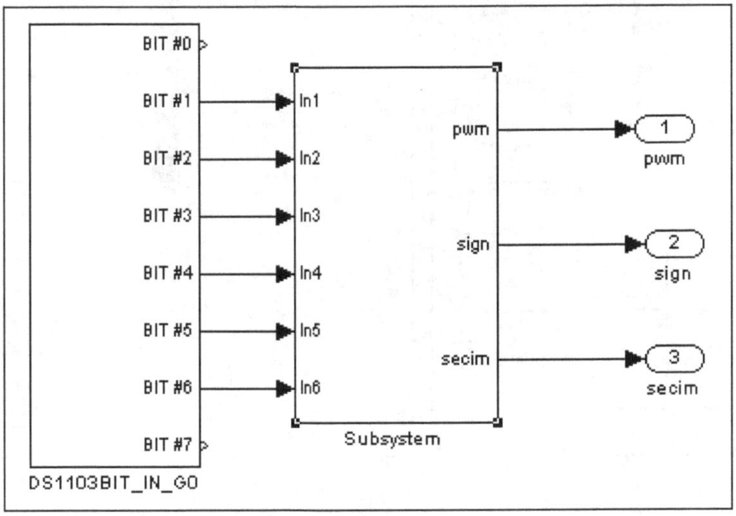

Fig. 23. Initial subsystem

Motors are controlled using 25 KHz PWM signals for amplitude and sign signals for motor directions according to controller outputs. In order to generate PWM signals slave dsp of the DS1103 is used. PWM block in the main model is given below. Saturation blocks are used to determine upper limit of the PWM duty cycles. In Figure 24, PWM and sign subsystems in the main model are shown.

4.4.1 Position control

After creating controller model and user interfaces, experiments can be done and results can be observed. Firstly, PID parameters were determined via trial-error experiments in order to obtain desired responses. A lot of experiments can be done simply and system responses can be taken easily. Output data can be saved '*.mat' extension in order to analyze in detail in Matlab. Some real time responses are given below. For 1 *mm* motion along the *z* direction, responses of the legs with references and control outputs are shown in Figure 25.
An orientation step response in *x* direction and a 500 *nm* step response in *z* direction were given in Figure 26. As can be seen from figures, PID parameters were selected to obtain fast rising time, no steady state error and smaller overshoot. PID controller works in a wide range between 500 *nm* to 100 *mm* with good control behavior.

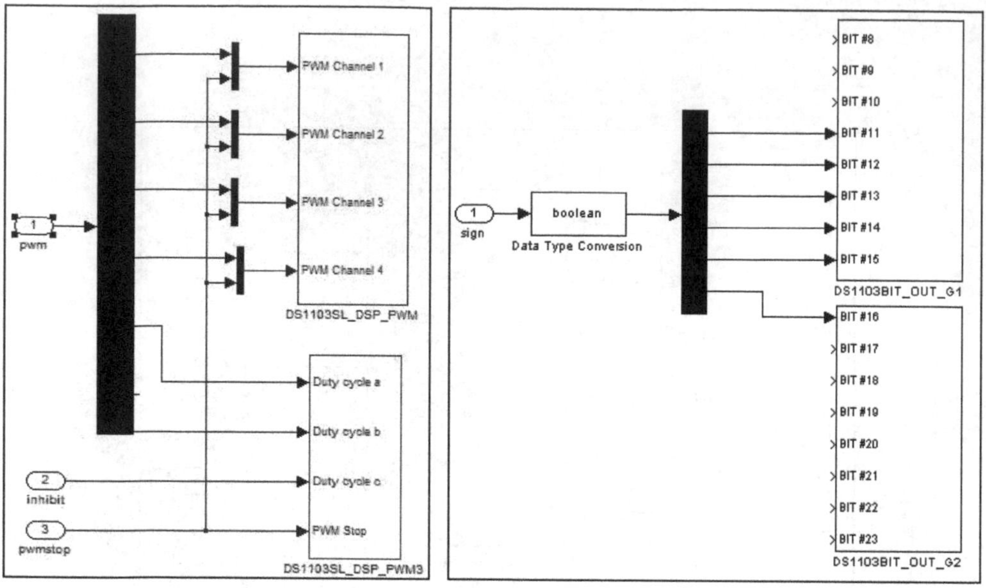

Fig. 24. PWM and Sign subsystems shown in the main model

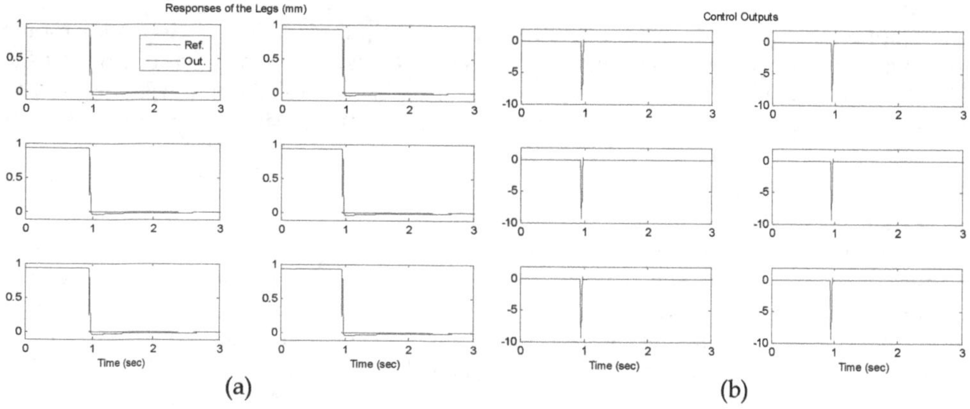

Fig. 25. (a) 1 *mm* step response of the top position in *z* direction (b) PID controller outputs

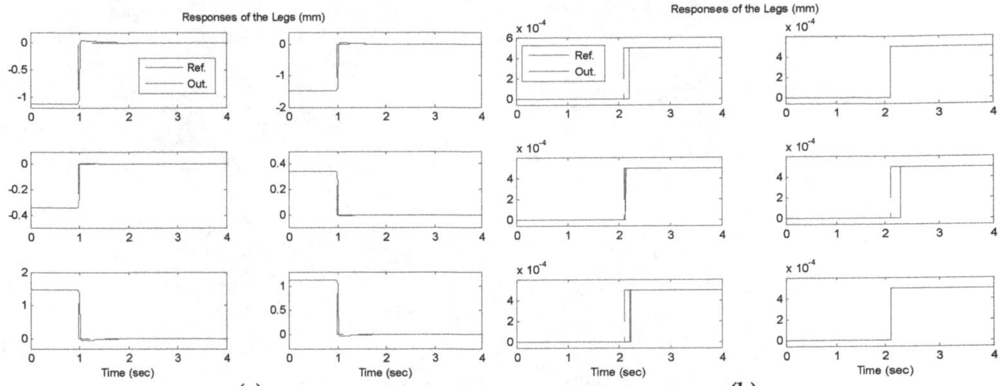

(a) (b)

Fig. 26. (a) 1° rotation in x direction (b) 500 nm linear motion in z direction

4.4.2 Trajectory control

Several trajectory experiments were performed using trajectory generation algorithm mentioned in section 4.3. If motion time and end points are entered in to the user interface, the trajectory is created automatically and motion starts. In order to show the performance of trajectory tracking, some examples are given below. For these examples, references, trajectories and leg errors are illustrated in Figure 27-31. Several cases are examined in the results.

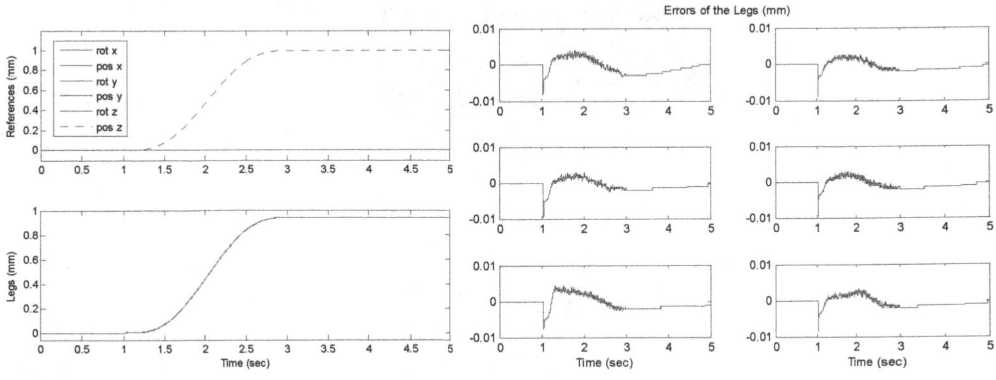

(a) (b)

Fig. 27. (a) Trajectory tracking in z direction (b) Errors of the legs

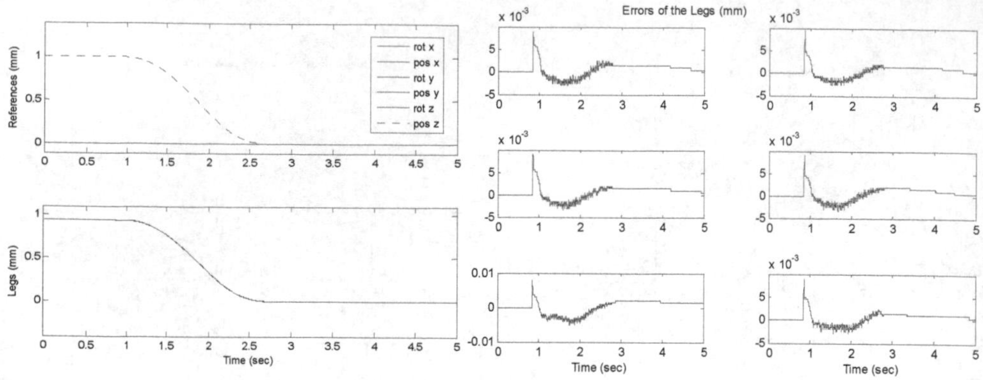

(a) (b)

Fig. 28. (a) Trajectory tracking in z direction (b) Leg errors

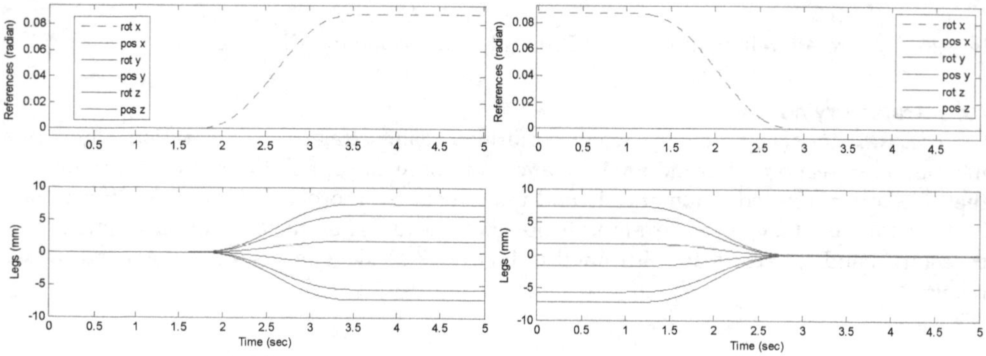

(a) (b)

Fig. 29. (a) Rotation in x direction (b) Rotation in x direction

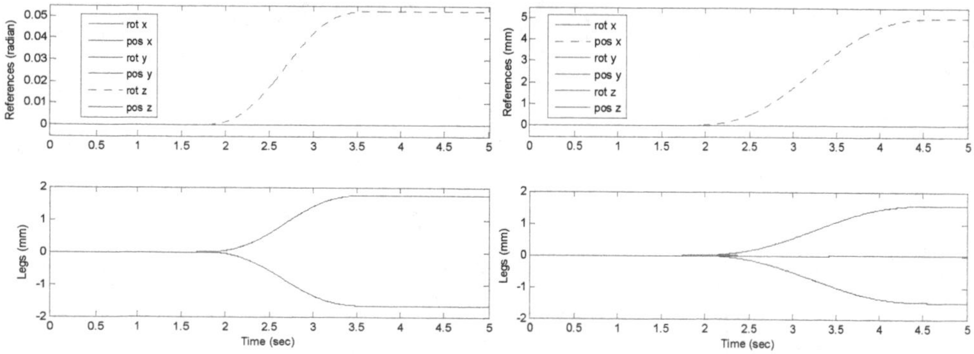

(a) (b)

Fig. 30. (a) Rotation in z direction (b) Linear motion in x direction

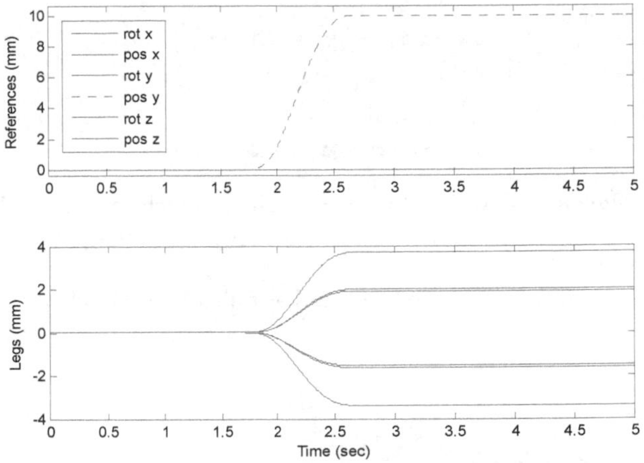

Fig. 31. Linear motion in *y* direction

4.5 Sliding mode controller

Nonlinear control has a very important role in the robot control applications. An important reason for this situation is to create knowledge-based systems which simplify the modeling of complex dynamics in the robot control. SMC is one of the suitable methods used in the knowledge-based systems (Küçük & Bingül, 2008).

SMC is a special case of the variable structure control systems. Variable structure control systems have a structure using feedback control laws and decision making laws together. Decision-making rule called as switching function selects a special feedback control structure according to the behavior of the system. Variable structure control system is designed to force the system states to slide a special surface called the sliding surface in the state space. Once the sliding surface is reached, the SMC tries to keep the states very close to the sliding surface (Küçük & Bingül, 2008).

If it is considered the change in the position error, second order equation of motion given with Equation 2, 3, 4 and 5 can be written as follows according to $x_1 = \dot{\theta}_r - \dot{\theta}_m$ (Utkin, 1993).

$$\begin{aligned} \dot{x}_1 &= x_2 \\ \dot{x}_2 &= -a_1 x_1 - a_2 x_2 + f(t) - bu \end{aligned} \tag{8}$$

a_1, a_2 and b are the positive parameters and *f(t)* is a function depending on load torque, reference input and their derivatives in Equation 8.

For discontinuous control:

$$u = u_0 sign(s), \qquad s = cx_1 + x_2 \tag{9}$$

and as Equation 10 is linear and doesn't depend on *f(t)*, sliding mode on the s = 0 line allows to decreasing error exponentially.

$$cx_1 + \dot{x}_1 = 0 \tag{10}$$

Derivative of the sliding surface is:

$$\dot{s} = cx_2 - a_1x_1 - a_2x_2 + f(t) - bu_0 sign(s) \tag{11}$$

From equation,

$$bu_0 > \left| cx_2 - a_1x_1 - a_2x_2 + f(t) \right| \tag{12}$$

s and $\dot{s}$ have opposite signs and the state $s = 0$ will approach the sliding line after a while. Inequality given with Equation 12 determines the required voltage to force the system to sliding mode (Utkin, 1993).

A candidate Lyapunov function can be selected as follows for stability analysis (Kassem & Yousef, 2009),

$$v = \frac{1}{2}\sigma^2 \tag{13}$$

The stability condition from Lyapunov's second theorem,

$$\frac{1}{2}\frac{d\sigma^2}{dt} = \sigma\dot{\sigma} \le -K|\sigma| \tag{14}$$

where K is a positive constant.

After theoretical steps, sliding mode controller was designed in Simulink similar to PID controller. Main model is shown in Figure 32. Details of some subsystems different from using subsystems in PID will be given only.

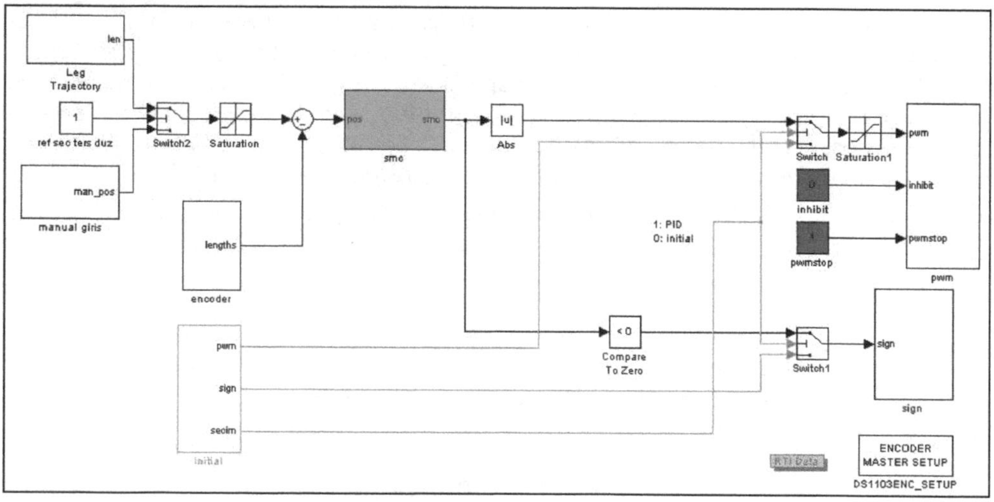

Fig. 32. Main sliding mode controller model

"smc" subsystem is shown in the Figure 33. This model contains sliding mode control and integrator algorithm for one leg. In order to reduce the chattering, a rate limiter block was added to output. In order to eliminate the steady state error, an integrator was added to the

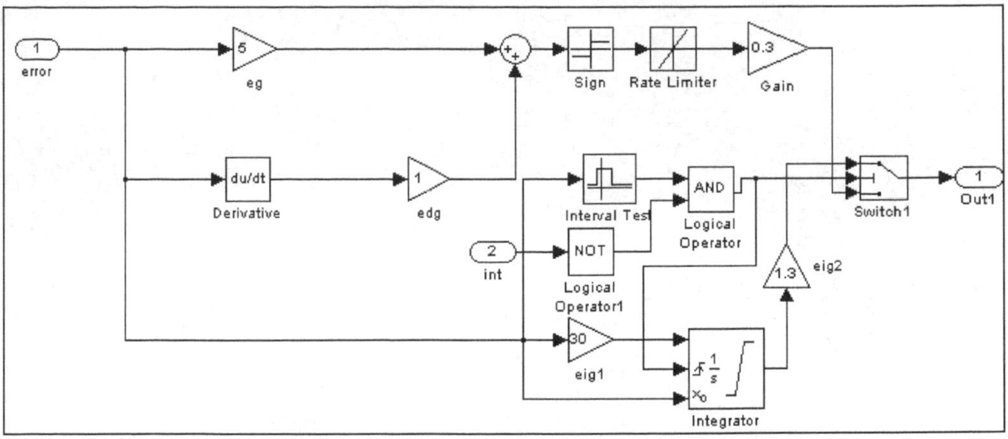

Fig. 33. Simulink model of SMC with integrator for one leg

controller. This integrator is switched on only small limited range between ± 0,008 mm. This integrator does not affect the performance of sliding mode controller. It does not slow down the system response. Also, integrator is disabled during trajectory tracking by switching.

4.5.1 Position control

Some real time responses for position control are given below. In Figure 34, errors of the legs were shown in the motions linear and rotation in the x and z direction with 15 *mm* and 20° inputs. As can be seen from the figures, overshoot and steady state error are very small. But, system response is slower.

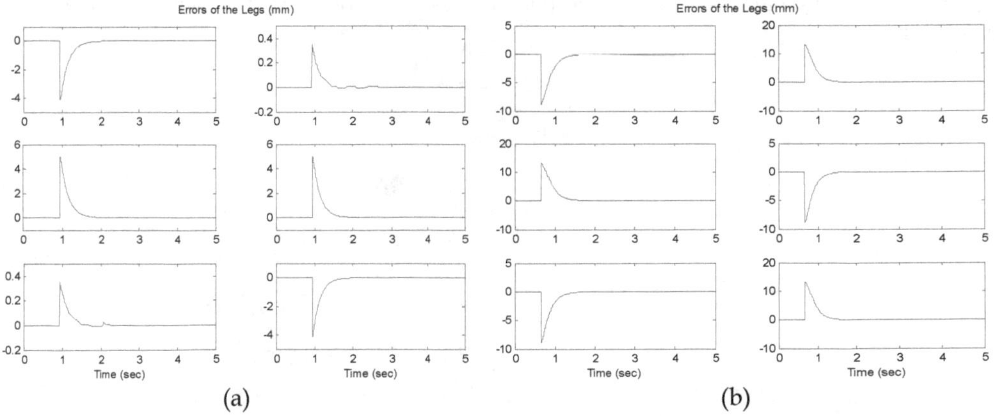

Fig. 34. (a) Linear motion in x direction with 15 *mm* reference (b) Rotation in z direction with 20° reference

Figure 35 shows a phase diagram of the system with SMC. In the phase diagram, the states of the system are leg position and leg velocity. As can be seen from the figure, SMC pushes states to sliding line and the states went to the desired values along sliding line when 5 *mm* step input along the z axis in Cartesian space was applied to the system.

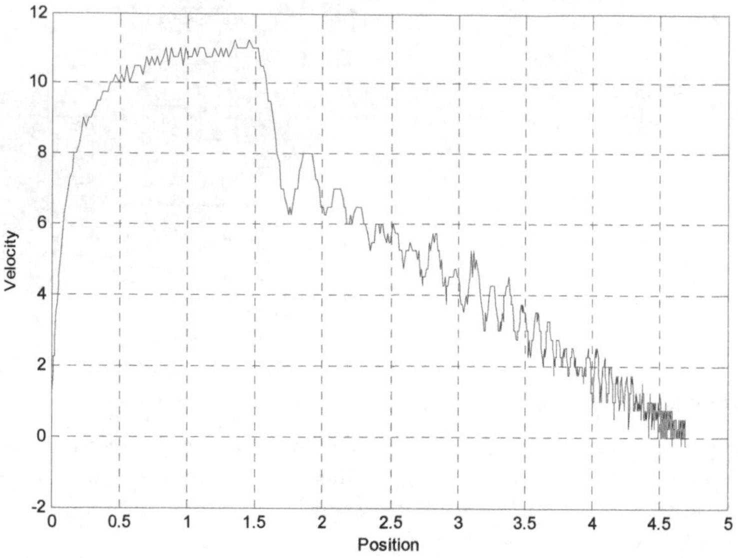

Fig. 35. Phase diagram of the SMC position control

4.5.2 Trajectory control

Different situations in trajectory control are considered in this section. These are shown in Figure 36-38. As can be seen from the figures legs followed the desired trajectories synchronous.

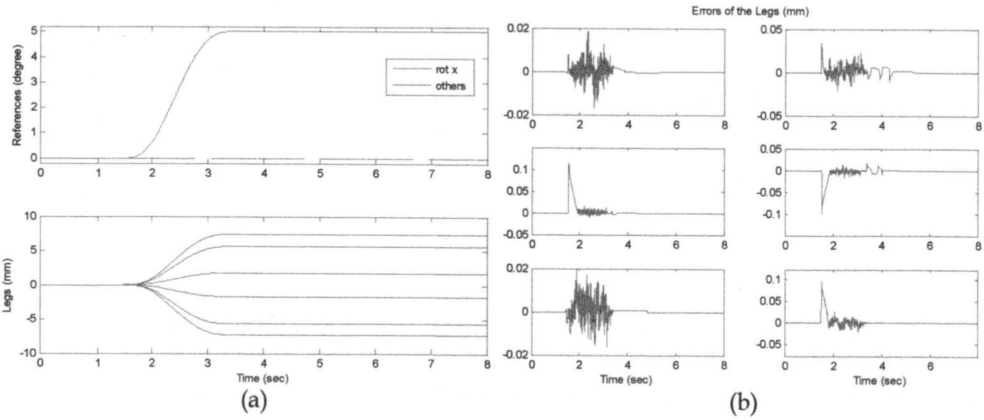

Fig. 36. (a) Rotation in x direction (b) Errors of the legs

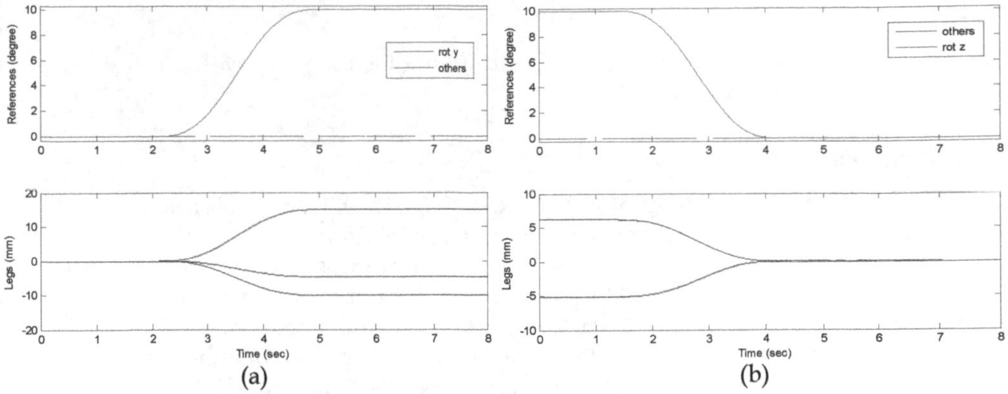

Fig. 37. (a) Rotation in y and (b) z direction

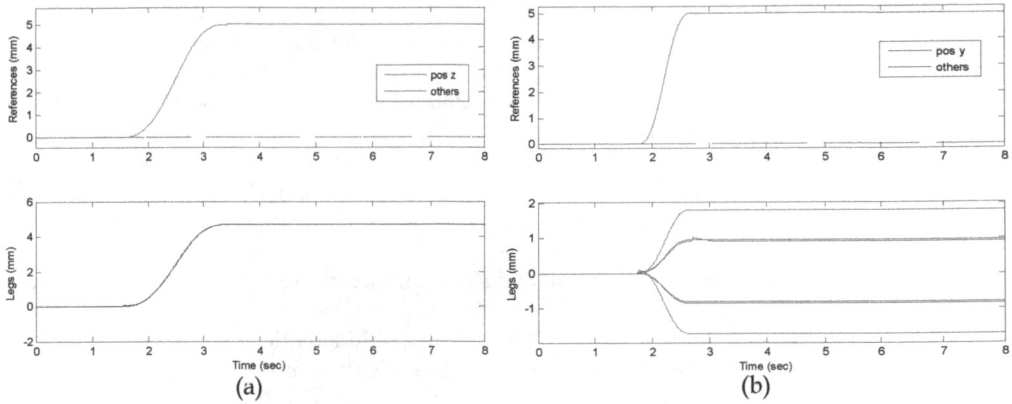

Fig. 38. (a) Linear motion in z and (b) in y direction

5. Conclusion

In this study, a high precision 6 DOF Stewart platform is controlled by a PID and sliding mode controller. These controllers were embedded in a Dspace DS1103 real time controller which is programmable in the Simulink environment. Design details and development stages of the PID and SMC are given from subsystems to main model in Simulink. This study can be a good example to show how a real time controller can be developed using Matlab/Simulink and Dspace DS1103. In order to test the performance of the controllers, several position and trajectory tracking experiments were conducted. Step inputs are used for position control and Kane transition function is used to generate trajectory. In the position experiments using both controllers, there is no steady state error and moving plate of the SP is positioned to the desired target with an error less than 0.5 μm. Sliding mode controller is better performance in terms of overshoot than PID but PID has faster response due to high gain. In the tracking experiments, PID and SMC have similar responses under no load. If nonlinear external forces are applied to moving platform, control performance of the SMC will be better than PID.

6. Acknowledgment

This work is supported by The Scientific and Technological Research Council of Turkey (TUBITAK) under the Grant No. 107M148.

7. References

Bonev, I. (2003). The True Origins of Parallel Robots, 06.04.2011, Available from: http://www.parallemic.org/Reviews/Review007.html

Chen, N.X.; Song, S.M. (1994). Direct position analysis of the 4-6 Stewart Platform, *ASME J. of Mechanical Design,* Vol. 116, No. 1, (March 1994), pp. (61-66), ISSN 1050-0472

Dspace Inc., 06.04.2011, Available from: http://www.dspaceinc.com/en/inc/home.cfm

Hunt, K.H. (1983). Structural kinematics of in-parallel-actuated robot-arms, *ASME J. Mech., Trans. Automat. Des.*, Vol. 105, No. 4, (December 1983), pp. (705–712), ISSN 0738-0666

Fitcher, E.F. (1986). A Stewart Platform-Based Manipulator: General Theory and Practical Construction, *Int. J. of Robotics Research,* Vol. 5, No. 2, (June 1986), pp. (157-182), ISSN 0278-3649

Kassem, A.M.; Yousef, A. M. (2009). Servo DC Motor Position Control Based on Sliding Mode Approach, *5th Saudi Technical Conference and Exhibition STCEX 2009,* January 2009

Kim, D.; Chung, W. (1999). Analytic Singularity Equation and Analysis of Six-DOF Parallel Manipulators Using Local Structurization Method, *IEEE Transactions on Robotics and Automation,* Vol. 15, No. 4, (August 1999), pp. (612-622), ISSN 1042-296X

Kizir, S. ; Bingül, Z. ; Oysu, C. ; Küçük, S. (2011). Development and Control of a High Precision Stewart Platform, *International Journal of Technological Sciences,* Vol. 3, No. 1, pp. (51-59)

Küçük, S.; Bingül, Z. (2008). *Robot Dinamiği ve Kontrolü,* Birsen press, ISBN 978-975-511-516-0, İstanbul

Liao, Q.; Seneviratne, L.D.; Earles, S.W.E. (1993). Forward kinematic analysis for the general 4-6 Stewart Platform, *IEEE/RSJ International Conference on Intelligent Robots and Systems IROS'93,* ISBN 0-7803-0823-9, Yokohama, Japan, July 1993

Mathworks Inc., 06.04.2011, Available from: http://www.mathworks.com/

Merlet, J.P. (1992). Direct kinematics and assembly modes of parallel manipulators, *Int. J. of Robotics Research,* Vol. 11, No. 2, (April 1992), pp. (150-162), ISSN 0278-3649

Merlet, J.P. (Ed(s).). (2006). *Parallel Robots,* Springer, ISBN-10 1-4020-4133-0, Netherlands

Nauna, P.; Waldron, K.J.; Murthy, V. (1990). Direct kinematic solution of a Stewart Platform, *IEEE Trans. Robotics Automat.,* Vol. 6, No. 4, (August 1990), pp. (438-444), ISSN 1042-296X

Niesing, B. (2001). Medical Engineering, *Fraunhofer Magazine,* Vol. 2

Reckdahl, K.J. (1996). Dynamics and control of mechanical systems containing closed kinematic chains, *Phd Thesis,* Stanford University

Sefrioui, J.; Gosselin., C.M. (1993). Singularity analysis and representation of planar parallel manipulators, *Robot. Autom. Syst.,* Vol. 10, No. 4, pp. (209-224)

Stewart, D. (1965). A Platform with Six Degrees of Freedom, *Proceedings of the Institute of Mechanical Engineering,* Vol. 180, Part 1, No. 5, pp. (371-386)

Utkin, V.I., (1993). Sliding mode control design principles and applications to electric drives, *IEEE Transactions on Industrial Electronics,* Vol. 40, No. 1, pp. (23–36), ISSN 1042-296X

Wikipedia, 06.04.2011, Available from: http://en.wikipedia.org/wiki/Stewart_platform

10

Obstacle Avoidance for Redundant Manipulators as Control Problem

Leon Žlajpah and Tadej Petrič
Jožef Stefan Institute, Ljubljana
Slovenia

1. Introduction

One of the goals of robotics research is to provide control algorithms that allow robotic manipulators to move in an environment with objects. The contacts with these objects may be part of the task, e.g. in the assembly operations, or they may be undesired events. If the task involves some contacts with the environment it is necessary to control the resulting forces. For that purpose, different control approaches have been proposed like hybrid position/force control (Raibert & Craig, 1981) or impedance control (Hogan, 1985), which have also been applied to redundant manipulators (Khatib, 1987; Park et al., 1996; Woernle, 1993; Yoshikawa, 1987). However, in most cases the contact is supposed to occur between the end-effector or the handling object and the object in the workspace. Except in some special cases all other contacts along the body of the robot manipulator are not desired and have to be avoided. In the case when the contacts are not desired, the main issue is how to accomplish the assigned task without any risk of collisions with the workspace objects. A natural strategy to avoid obstacles would be to move the manipulator away from the obstacle into a configuration where the manipulator is not in contact with the obstacle. Without changing the motion of the end-effector, the reconfiguration of the manipulator into a collision-free configuration can be made only if the manipulator has redundant degrees-of-freedom (DOF). The flexibility depends on the degree-of-redundancy, i.e., on the number of redundant DOF. A high degree-of-redundancy is important, especially when the manipulator is working in an environment with many potential collisions with obstacles.

Generally, the obstacle-avoidance (or collision-avoidance) problem can be solved with two classes of strategies: global (planning) and local (control). The global ones, like high-level path planning, guarantee to find a collision-free path from the initial point to the goal point, if such a path exists. They often operate in the configuration space into which the manipulator and all the obstacles are mapped and a collision-free path is found in the unoccupied portion of the configuration space (Lozano-Perez, 1983). However, these algorithms are very computationally demanding and the calculation times are significantly longer than the typical response time of a manipulator. This computational complexity limits their use for practical obstacle avoidance just to simple cases. Furthermore, as global methods do not usually rely on any sensor feedback information, they are only suitable for static and well-defined environments. On the other hand, local strategies treat obstacle avoidance as a control problem. Their aim is not to replace the higher-level, global, collision-free path planning but to make use of the capabilities of low-level control, e.g., they can use the sensor information to change the path if the obstacle appears in the workspace or it moves. Hence, they are suitable when the obstacle position is not known in advance and must be detected in real-time during the task execution. A significant advantage of local methods is that they are less computationally demanding and more flexible. These characteristics make local methods good candidates for on-line collision avoidance, especially in unstructured environments. However, the drawback is that they may cause globally suboptimal behavior or may even get stuck when a collision-free path cannot be found from the current configuration.

With the problem of the collision avoidance of redundant manipulators, there have been different approaches to the local methods proposed by many researchers in the past (Brock et al., 2002; Colbaugh et al., 1989; Glass et al., 1995; Guo & Hsia, 1993; Khatib, 1986; Kim & Khosla, 1992; Maciejewski & Klein, 1985; McLean & Cameron, 1996; Seraji & Bon, 1999; Volpe & Khosla, 1990). The approach proposed by Maciejewski and Klein (Maciejewski & Klein, 1985) is to assign to the critical point an avoiding task space vector, with which the point is directed away from the obstacle. Colbaugh, Glass and Seraji (Colbaugh et al., 1989; Glass et al., 1995) used configuration control and defined the constraints representing the obstacle-avoidance. The next approach is based on potential functions, where a repulsive potential is assigned to the obstacles and an attractive potential to the goal position (Khatib, 1986; Kim & Khosla, 1992; McLean & Cameron, 1996; Volpe & Khosla, 1990). The fourth approach uses the optimization of an objective function maximizing the distance between the manipulator and the obstacles (Guo & Hsia, 1993). Most of the proposed methods solve the obstacle-avoidance problem at the kinematic level. Velocity null-space control is an appropriate way to control the internal motion of a redundant manipulator. Some of the control strategies are acceleration based or torque based, considering also the manipulator dynamics (Brock et al., 2002; Khatib, 1986; Newman, 1989; Xie et al., 1998). However, it is established that certain acceleration-based control schemes exhibit instabilities (O'Neil, 2002). An alternative approach is the augmented Jacobian, as introduced in Egeland (1987), where the secondary task is added to the primary task so as to obtain a square Jacobian matrix that can be inverted. The main drawback of this technique are the so-called algorithmic singularities. They occur when the secondary task causes a conflict with the primary task. Khatib investigated in depth the use of the second-order inverse kinematic, either at the torque or acceleration level, starting from Khatib (1987) to recent task-prioritised humanoid applications (Mansard et al., 2009; Sentis et al., 2010).

Here, we want to present some approaches to on-line obstacle-avoidance for the redundant manipulators at the kinematic level and some approaches, which are considering also the dynamics of the manipulator.

Like in most of the local strategies that solve the obstacle-avoidance problem at the kinematic level (Colbaugh et al., 1989; Glass et al., 1995; Guo & Hsia, 1993; Kim & Khosla, 1992; Maciejewski & Klein, 1985; Seraji & Bon, 1999), the aim of the proposed strategies is to assign each point on the body of the manipulator, which is close to the obstacle, a motion component in a direction away from the obstacle. The emphasis is given to the definition of the avoiding motion. Usually, the avoiding motion is defined in the Cartesian space. As obstacle avoidance is typically a one-dimensional problem, we use a one-dimensional operational space for each critical point. Consequently, some singularity problems can be avoided when not enough "redundancy" is available locally. Additionally, we propose an approximative calculation of the motion that is faster that the exact one. Another important issue addressed in this paper is how the obstacle avoidance is performed when there are more simultaneously active

obstacles in the neighborhood of the manipulator. We propose an algorithm that considers all the obstacles in the neighborhood of the robot.
Most tasks performed by a redundant manipulator are broken down into several subtasks with different priorities. Usually, the task with the highest priority, referred to as the main task, is associated with the positioning of the end-effector in the task space, and other subtasks are associated with the obstacle avoidance and other additional tasks (if the degree-of-redundancy is high enough). However, in some cases it is necessary (e.g., for safety reasons) that the end-effector motion is not the primary task. As, in general, task-priority algorithms do not always allow simple transitions or the changing of priority levels between the tasks (Sciavicco & Siciliano, 2005), we propose a novel formulation of the primary and secondary tasks, so that the desired movement of the end-effector is in fact a secondary task. The primary task is now the obstacle avoidance and it is only active if we approach a pre-defined threshold, i.e., the critical distance to one of the obstacles. While far from the threshold, our algorithm allows undisturbed control of the secondary task. If we approach the threshold, the primary task smoothly takes over and only allows joint control in the null-space of the primary task. A similar approach can also be found in Sugiura et al. (2007), where they used a blending coefficient for blending the end-effector motion with the obstacle-avoidance motion, and in Mansard et al. (2009) where they proposed a generic solution to build a smooth control law for any kind of unilateral constraints.
Next, we propose strategies that are also considering the dynamics. In this case, it is reasonable to define the obstacle avoidance at the force level, i.e., the forces are supposed to generate the motion that is necessary to avoid the obstacle. We discuss three approaches regarding the sensors used to detect the obstacles: no sensors, tactile sensors and proximity sensors or vision. First of all, we want to investigate what happens if the manipulator touches an obstacle, especially how to control the contact forces and how to avoid the obstacle after the contact. Therefore, we propose a strategy that utilizes the self-motion caused by the contact forces to avoid an obstacle after the collision. The main advantage of this strategy is that it can be applied to the systems without any contact or force sensors. However, a prerequisite for this strategy to be effective is that the manipulator is backdrivable. As an alternative for stiff systems (having high-ratio gears, high friction, etc.), we propose using tactile sensors. Finally, we deal with proximity sensors and we propose a virtual forces strategy, where a virtual force component in a direction away from the obstacle is assigned to each point on the body of the manipulator, which is close to an obstacle. Like other classical methods for obstacle avoidance this one prevents any part of the manipulator touching an obstacle. Also here we address the problem of multiple obstacles in the workspace, which have to be simultaneously avoided.
The computational efficiency of all the proposed algorithms (at the kinematic level and considering the dynamics) allows the real-time application in an unstructured or time-varying environment. The efficiency of the proposed control algorithms is illustrated by simulations of highly redundant planar manipulator moving in an unstructured and time-varying environment and by experiments on a real robot manipulator.

2. Background

The robotic systems under study are serial manipulators. We consider redundant systems, i.e., the dimension of the joint space n exceeds the dimension of the task space m. The difference between n and m will be denoted as the degree of redundancy r, $r = n - m$. Note that in this definition the redundancy is not only a characteristics of the manipulator itself but also of the task. This means that a nonredundant manipulator may also become a redundant manipulator for a certain task.

2.1 Modelling

Let the configuration of the manipulator be represented by the n-dimensional vector $\boldsymbol{q}$ of joint positions, and the end-effector position (and orientation) by the m-dimensional vector $\boldsymbol{x}$ of the task positions (and orientations). Then, the kinematics can be described by the following equations

$$\boldsymbol{x} = \boldsymbol{f}(\boldsymbol{q}) \qquad \dot{\boldsymbol{q}} = \mathbf{J}^{\#}\dot{\boldsymbol{x}} + \mathbf{N}\dot{\boldsymbol{q}} \qquad \ddot{\boldsymbol{q}} = \mathbf{J}^{\#}(\ddot{\boldsymbol{x}} - \dot{\mathbf{J}}\dot{\boldsymbol{q}}) + \mathbf{N}\ddot{\boldsymbol{q}} \tag{1}$$

where $\boldsymbol{f}$ is an m-dimensional vector function representing the manipulator's forward kinematics, $\mathbf{J}$ is the $m \times n$ manipulator's Jacobian matrix, $\mathbf{J}^{\#}$ is the generalized inverse of the Jacobian matrix $\mathbf{J}$ and $\mathbf{N}$ is a matrix representing the projection into the null space of $\mathbf{J}$, $\mathbf{N} = (\mathbf{I} - \mathbf{J}^{\#}\mathbf{J})$.

For the redundant manipulators the static relationship between the m-dimensional generalized force in the task space $\boldsymbol{F}$, and the corresponding n-dimensional generalized joint space force $\boldsymbol{\tau}$ is

$$\boldsymbol{\tau} = \mathbf{J}^T\boldsymbol{F} + \mathbf{N}^T\boldsymbol{\tau} \tag{2}$$

where $\mathbf{N}^T$ is a matrix representing the projection into the null space of $\mathbf{J}^{T\#}$.

Assuming the manipulator consists of rigid bodies the joint space equations of motion can be written in the form

$$\boldsymbol{\tau} = \mathbf{H}\ddot{\boldsymbol{q}} + \boldsymbol{h} + \boldsymbol{g} - \boldsymbol{\tau}_F \tag{3}$$

where $\boldsymbol{\tau}$ is an n-dimensional vector of control torques, $\mathbf{H}$ is an $n \times n$ inertia matrix, $\boldsymbol{h}$ is an n-dimensional vector of Coriolis and centrifugal forces, $\boldsymbol{g}$ is an n-dimensional vector of gravity forces, and the vector $\boldsymbol{\tau}_F$ represents torques due to external forces acting on the manipulator.

2.2 Kinematic control

For velocity control the following kinematic controller can be used

$$\dot{\boldsymbol{q}} = \mathbf{J}^{\#}\dot{\boldsymbol{x}}_c + \mathbf{N}\dot{\boldsymbol{\varphi}} \tag{4}$$

where $\dot{\boldsymbol{x}}_c$ and $\dot{\boldsymbol{\varphi}}$ represent the task space control law and the arbitrary joint velocities, respectively. The task space control $\dot{\boldsymbol{x}}_c$ can be selected as

$$\dot{\boldsymbol{x}}_c = \dot{\boldsymbol{x}}_e + \mathbf{K}_p\boldsymbol{e} \tag{5}$$

where $\boldsymbol{e}$, $\boldsymbol{e} = \boldsymbol{x}_d - \boldsymbol{x}$, is the tracking error, $\dot{\boldsymbol{x}}_e$ is the desired task space velocity, and $\mathbf{K}_p$ is a constant gain matrix. To perform the additional subtask, the velocity $\dot{\boldsymbol{\varphi}}$ is used. Let p be a function representing the desired performance criterion. Then, to optimize p we can select $\dot{\boldsymbol{\varphi}}$ as

$$\dot{\boldsymbol{\varphi}} = \mathbf{K}_p\nabla p \tag{6}$$

Here, ∇p is the gradient of p and $\mathbf{K}_p$ is a gain.

2.3 Torque control

To decouple the task space and null-space motion we propose using a controller given in the form

$$\boldsymbol{\tau}_c = \mathbf{H}(\mathbf{J}^{\#}(\ddot{\boldsymbol{x}}_c - \dot{\mathbf{J}}\dot{\boldsymbol{q}}) + \mathbf{N}(\boldsymbol{\phi} - \dot{\mathbf{N}}\dot{\boldsymbol{q}})) + \boldsymbol{h} + \boldsymbol{g} \tag{7}$$

where $\ddot{x}_c$ and ϕ represent the task space and the null-space control law, respectively. The task space control $\ddot{x}_c$ can be selected as

$$\ddot{x}_c = \ddot{x}_d + \mathbf{K}_v \dot{e} + \mathbf{K}_p e \tag{8}$$

where e, $e = x_d - x$, is the tracking error, $\ddot{x}_d$ is the desired task space acceleration, and $\mathbf{K}_v$ and $\mathbf{K}_p$ are constant gain matrices. The selection of $\mathbf{K}_v$ and $\mathbf{K}_p$ can be based on the desired task space impedance. To perform the additional subtask, the vector ϕ is given in the form

$$\phi = \mathbf{N}\ddot{\varphi} + \dot{\mathbf{N}}\dot{\varphi} + \mathbf{K}_n \dot{e}_n, \qquad \dot{e}_n = \mathbf{N}(\dot{\varphi} - \dot{q}) \tag{9}$$

where $\dot{\varphi}$ is the desired null space velocity and $\mathbf{K}_n$ is an $n \times n$ diagonal gain matrix. The velocity $\dot{\varphi}$ is defined by the subtask. E.g., let p be a function representing the desired performance criterion. To optimize p we can select $\dot{\varphi}$ as (6).

3. Obstacle-avoidance strategy

The obstacle-avoidance problem is usually defined as how to control the manipulator to track the desired end-effector trajectory while simultaneously ensuring that no part of the manipulator collides with any obstacle in the workspace of the manipulator. To avoid any obstacles the manipulator has to move away from the obstacles into the configuration where the distance to the obstacles is larger (see Fig. 1). Without changing the motion of the end-effector, the reconfiguration of the manipulator into a collision-free configuration can be done only if the manipulator has redundant degrees-of-freedom (DOF). Note that the flexibility or the *degree-of-redundancy* of the manipulator does not depend only on the number of redundant DOF but also on the "location" of the redundant DOF. Namely, it is possible that the redundant manipulator cannot avoid an obstacle, because it is in a configuration where the avoiding motion in the desired direction is not possible. A high degree-of-redundancy is important, especially when the manipulator is working in an environment with many potential collisions with obstacles.

A good strategy for obstacle avoidance is to identify the points on the robotic arm that are near obstacles and then assign to them a motion component that moves those points away from the obstacle, as shown in Fig. 1. The motion of the robot is perturbed only if at least one part of the robot is in the critical neighborhood of an obstacle, i.e., the distance is less than a prescribed minimal distance. We denote the obstacles in the critical neighborhood as the *active obstacles* and the corresponding closest points on the body of a manipulator as the *critical points*. Usually, it is assumed that the motion of the end effector is not disturbed by any obstacle. Otherwise, the task execution has to be interrupted and the higher-level path planning has to recalculate the desired motion of the end-effector. If the path-tracking accuracy is not important control algorithms which move the end-effector around obstacles on-line, can be used.

As the obstacle avoidance is supposed to be done on-line, it is not necessary to know the exact position of the obstacles in advance. Of course, to allow the manipulator to work in an unstructured and/or dynamic environment, some sensors have to be used to determine the position of the obstacles or measure the distance between the obstacles and the body of the manipulator. There are different types of sensor systems that can be used to detect objects in the neighborhood of the manipulator. They can be tactile or proximity sensors. The tactile sensors, like artificial skin, can detect the obstacle only if they touch it. On the other hand, the proximity sensors can sense the presence of an obstacle in the neighborhood. We have

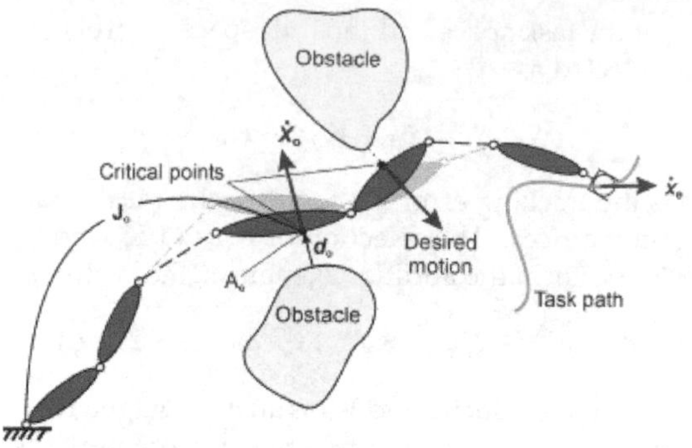

Fig. 1. Manipulator motion in presence of some obstacles

compared the capabilities of a manipulator equipped with both types of sensors. Actually, we have also investigated if and how the manipulator can avoid obstacles without any sensors for the detection of obstacles. As in the case of tactile sensors and when no sensors are used the manipulator has to "touch" the obstacles, we allow a collision with an obstacle. However, after the collision the manipulator should move away from the obstacle and the collision forces should be kept as low as possible.

4. Obstacle avoidance using kinematic control

The proposed velocity strategy considers the obstacle-avoidance problem at the kinematic level. Let $\dot{x}_e$ be the desired velocity of the end-effector, and A_o be the critical point in the neighborhood of an obstacle (see Fig. 1). To avoid a possible collision, one possibility is to assign to A_o such a velocity that it moves away from the obstacle, as proposed in Maciejewski & Klein (1985). Hence, the motion of the end-effector and the critical point can be described by the equations

$$\mathbf{J}\dot{q} = \dot{x}_e \qquad \mathbf{J}_o\dot{q} = \dot{x}_o \tag{10}$$

where $\mathbf{J}_o$ is a Jacobian matrix associated with the point point A_o. There are some possibilities to find a common solution for both equations.

4.1 Exact solution

Let $\dot{x}_c$ in (4) equal $\dot{x}_e$. Then, combining (4) and (10) yields

$$\dot{\varphi} = (\mathbf{J}_o\mathbf{N})^{\#}(\dot{x}_o - \mathbf{J}_o\mathbf{J}^{\#}\dot{x}_e) \tag{11}$$

Now, using this $\dot{\varphi}$ in (4) gives the final solution for $\dot{q}$ in the form

$$\dot{q} = \mathbf{J}^{\#}\dot{x}_c + (\mathbf{J}_o\mathbf{N})^{\#}(\dot{x}_o - \mathbf{J}_o\mathbf{J}^{\#}\dot{x}_e) \tag{12}$$

because $\mathbf{N}$ is both hermitian and idempotent (Maciejewski & Klein, 1985; Nakamura et al., 1987). The meaning of the terms in the above equation can be easily explained. The first term $\mathbf{J}^{\#}\dot{x}_c$ guarantees the joint motion necessary for the desired end-effector velocity. Also, $\dot{x}_c$ is used in (12) instead of $\dot{x}_e$ to indicate that a task space controller can be used to compensate for any task space tracking errors, e.q.

$$\dot{x}_c = \dot{x}_d + \mathbf{K}e. \tag{13}$$

where $\dot{\boldsymbol{x}}_d$ is the desired task space velocity, $\mathbf{K}$ is an $m \times m$ positive-definite matrix and $\boldsymbol{e}$ is the task position error, defined as

$$\boldsymbol{e} = \boldsymbol{x}_d - \boldsymbol{x}, \tag{14}$$

where $\boldsymbol{x}_d$ is the desired task space position. The second term in (12), i.e., the homogeneous solution $\dot{\boldsymbol{q}}_h$, represents the part of the joint velocity causing the motion of the point A_o. The term $\mathbf{J}_o\mathbf{J}^{\#}\dot{\boldsymbol{x}}_e$ is the velocity in A_o due to the end-effector motion. The matrix $\mathbf{J}_o\mathbf{N}$ is used to transform the desired critical point velocity from the operational space of the critical point into the joint space. Note that the above solution guarantees that we achieve exactly the desired $\dot{\boldsymbol{x}}_o$ only if the degree of redundancy of the manipulator is sufficient.

4.2 Exact solution using a reduced operational space

The matrix $\mathbf{J}_o\mathbf{N}$ combines the kinematics of the critical point A_o and the null-space matrix of the whole manipulator. Hence, its properties define the flexibility of the system for avoiding the obstacles. We want to point out that the properties of the matrix $\mathbf{J}_o\mathbf{N}$ do not depend only on the position of the point A_o but also on the definition of the operational space associated with the critical point. Usually, it is assumed that all critical points belong to the Cartesian space. Hence, the velocity $\dot{\boldsymbol{x}}_o$ is a 3-dimensional vector and the dimension of the matrix $\mathbf{J}_o\mathbf{N}$ is $3 \times n$. This also implies that 3 degrees-of-redundancy are needed to move one point from an obstacle. Consequently, it seems that a manipulator with 2 degrees-of-redundancy is not capable of avoiding obstacles, and of course, this is not true. For example, consider a planar 3 DOF manipulator that is supposed to move along a line, as shown in Fig. 2. As this is a planar case, the task space is 2-dimensional (e.g. x and y) and the manipulator has 1 degree-of-redundancy. Defining the velocity $\dot{\boldsymbol{x}}_o$ in the same space as the end-effector velocity, i.e., as a 2-dimensional vector, reveals the matrix $\mathbf{J}_o\mathbf{N}$ to have the dimension 2×3. Furthermore, due to 1 degree-of-redundancy the components of the velocity vector $\dot{\boldsymbol{x}}_o$ are not independent. Hence, the rank of $\mathbf{J}_o\mathbf{N}$ is 1, and the pseudoinverse $(\mathbf{J}_o\mathbf{N})^{\#}$ does not exist.
As the obstacle-avoidance strategy only requires motion in the direction of the line connecting the critical point with the closest point on the obstacle, this is a one-dimensional constraint and only one degree-of-redundancy is needed to avoid the obstacle, generally. Therefore, we propose using a reduced operational space for the obstacle avoidance and define the Jacobian $\mathbf{J}_o$ as follows.
Let $\boldsymbol{d}_o$ be the vector connecting the closest points on the obstacle and the manipulator (see Fig. 1) and let the operational space in A_o be defined as one-dimensional space in the direction of $\boldsymbol{d}_o$. Then, the Jacobian, which relates the joint space velocities $\dot{\boldsymbol{q}}$ and the velocity in the direction of $\boldsymbol{d}_o$, can be calculated as

$$\mathbf{J}_{d_o} = \boldsymbol{n}_o^T \mathbf{J}_o \tag{15}$$

where $\mathbf{J}_o$ is the Jacobian defined in the Cartesian space and $\boldsymbol{n}_o$ is the unit vector in the direction of $\boldsymbol{d}_o$, $\boldsymbol{n}_o = \frac{\boldsymbol{d}_o}{\|\boldsymbol{d}_o\|}$. Now, the dimension of the matrix $\mathbf{J}_{d_o}$ is $1 \times n$, and the velocities $\dot{\boldsymbol{x}}_o$ and $\mathbf{J}_{d_o}\mathbf{J}^{\#}\dot{\boldsymbol{x}}_e$ become scalars. Consequently, the computation of $(\mathbf{J}_{d_o}\mathbf{N})^{\#}$ is faster. For example, when calculating the Moore-Penrose pseudoinverse of $(\mathbf{J}_{d_o}\mathbf{N})$ defined as

$$(\mathbf{J}_{d_o}\mathbf{N})^{\#} = (\mathbf{J}_{d_o}\mathbf{N})^T \left(\mathbf{J}_{d_o}\mathbf{N}\,(\mathbf{J}_{d_o}\mathbf{N})^T\right)^{-1} = \mathbf{N}\mathbf{J}_{d_o}^T(\mathbf{J}_{d_o}\mathbf{N}\mathbf{J}_{d_o}^T)^{-1} \tag{16}$$

we do not have to invert any matrix because the term $(\mathbf{J}_{d_o}\mathbf{N}\mathbf{J}_{d_o}^T)$ is a scalar. Going back to our example in Fig. 2, the pseudoinverse $(\mathbf{J}_{d_o}\mathbf{N})^{\#}$ exists and the manipulator can perform the primary task and simultaneously avoid the obstacle, as shown in Fig. 2.

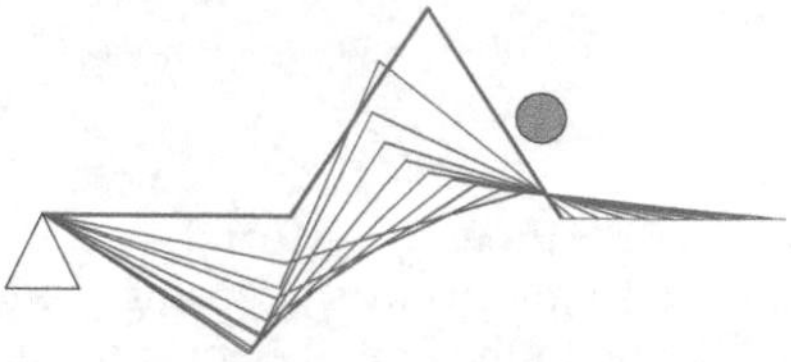

Fig. 2. Planar 3DOF manipulator: tracking of a line and obstacle avoidance using the Jacobian $\mathbf{J}_{d_o}$

4.3 Selection of avoiding velocity

The efficiency of the obstacle-avoidance algorithm also depends on the selection of the desired critical point velocity $\dot{x}_o$. We propose changing $\dot{x}_o$ with respect to the obstacle distance

$$\dot{x}_o = \alpha_v v_o \tag{17}$$

where v_o is the nominal velocity and α_v is the obstacle-avoidance gain defined as

$$\alpha_v = \begin{cases} \left(\frac{d_m}{\|\boldsymbol{d}_o\|}\right)^2 - 1 & \text{for} \quad \|\boldsymbol{d}_o\| < d_m \\ 0 & \text{for} \quad \|\boldsymbol{d}_o\| \geq d_m \end{cases} \tag{18}$$

where d_m is the critical distance to the obstacle (see Fig. 3). If the obstacle is too close ($\|\boldsymbol{d}_o\| \leq d_b$) the main task should be aborted. The distance d_b is subjected to the dynamic properties of the manipulator and can also be a function of the relative velocity $\dot{\boldsymbol{d}}_o$. To assure smooth transitions it is important that the magnitude of $\dot{\boldsymbol{x}}_o$ at d_m is zero. Special attention has to be given to the selection of the nominal velocity v_o. Large values of v_o would cause unnecessarily high velocities and consequently the manipulator would move far from the obstacle. Such motion may cause problems if there are more obstacles in the neighborhood of the manipulator. Namely, the manipulator may bounce between the obstacles. On the other hand, too small value of v_o would not move the critical point of the manipulator away from the obstacle.

For a smoother motion, was is proposed in Maciejewski & Klein (1985) to change the amount of the homogenous solution to be included in the total solution

$$\dot{\boldsymbol{q}}_{EX} = \mathbf{J}^{\#}\dot{\boldsymbol{x}}_c + \alpha_h (\mathbf{J}_{d_o}\mathbf{N})^{\#}(\dot{\boldsymbol{x}}_o - \mathbf{J}_{d_o}\mathbf{J}^{\#}\dot{\boldsymbol{x}}_e) \tag{19}$$

We have selected α_h as

$$\alpha_h = \begin{cases} 1 & \text{for} \quad \|\boldsymbol{d}_o\| \leq d_m \\ \frac{1}{2}(1 - \cos(\pi \frac{\|\boldsymbol{d}_o\| - d_m}{d_i - d_m})) & \text{for} \quad d_m < \|\boldsymbol{d}_o\| < d_i \\ 0 & \text{for} \quad d_i \leq \|\boldsymbol{d}_o\| \end{cases} \tag{20}$$

where d_i is the distance where the obstacle influences the motion. From Fig. 3 it can be seen that in the region between d_b and d_m the complete homogenous solution is included in the motion specification and the avoidance velocity is inversely related to the distance. Between d_m and d_i the avoidance velocity is zero and only a part of the homogenous solution is included. As the homogenous solution compensates for the motion in the critical point due to the end-effector motion, the relative velocity between the obstacle and the critical point

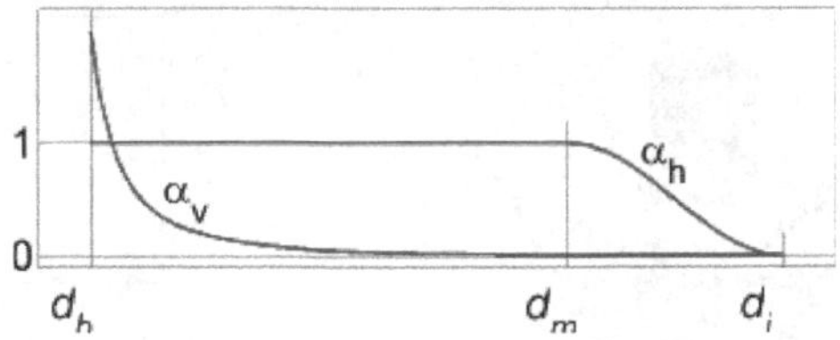

Fig. 3. The obstacle avoidance gain α_v and the homogenous term gain α_h versus the distance to the obstacle

decreases when approaching from d_i to d_m, if the obstacle is not moving, of course. With such a selection of α_v and α_h smooth velocities can be obtained.

The control law (19) can be used for a single obstacle. When more than one obstacle is active at one time then the worst-case obstacle (nearest) has to be used, which results in discontinuous velocities and may cause oscillations in some cases. Namely, when switching between active obstacles the particular homogenous solutions are not equal and a discontinuity in the joint velocities occurs. To improve the behavior we propose to use a weighted sum of the homogenous solution of all the active obstacles

$$\dot{\boldsymbol{q}}_{EX} = \mathbf{J}^{\#}\dot{\boldsymbol{x}}_c + \sum_{i=1}^{n_o} w_i \alpha_{h,i} \dot{\boldsymbol{q}}_{h,i} \tag{21}$$

where n_o is the number of active obstacles, and w_i, $\alpha_{h,i}$ and $\dot{\boldsymbol{q}}_{h,i}$ are the weighting factor, the gain and the homogenous solution for the i-th active obstacle, respectively. The weighting factors w_i are calculated as

$$w_i = \frac{d_i - \|\boldsymbol{d}_{o,i}\|}{\sum_{i=1}^{n_o}(d_i - \|\boldsymbol{d}_{o,i}\|)} \tag{22}$$

Although the actual velocities in the critical points differ from the desired ones, using $\dot{\boldsymbol{q}}_{EX}$ improves the behavior of the system and when one point is much closer to the obstacle than another, then its weight approaches 1 and the velocity in that particular point is close to the desired value.

As an illustration we present a simulation of a planar manipulator with 4 revolute joints. The primary task is to move along a straight line from point P_1 to point P_2, and the motion is obstructed by an obstacle. The task trajectory has a trapezoid velocity profile with an acceleration of $4ms^{-2}$ and a max. velocity of $0.4ms^{-1}$. We chose the the critical distance $d_m = 0.2m$. The simulation results using the exact velocity controller EX (19) are presented in Fig. 4(a).

4.4 Approximate solution

Another possible solution for $\dot{\varphi}$ is to calculate joint velocities that satisfy the secondary goal as

$$\dot{\varphi} = \mathbf{J}_{d_o}^{\#}\dot{x}_o \tag{23}$$

without compensating for the contribution of the end-effector motion and then substitute $\dot{\varphi}$ into (4), which yields

$$\dot{\boldsymbol{q}}_{AP} = \mathbf{J}^{\#}\dot{\boldsymbol{x}}_c + \mathbf{N}\mathbf{J}_{d_o}^{\#}\dot{x}_o \tag{24}$$

This approach avoids the singularity problem of $(\mathbf{J}_{d_o}\mathbf{N})$ (Chiaverini, 1997). The formulation (24) does not guarantee that we will achieve exactly the desired $\dot{x}_o$ even if the degree of redundancy is sufficient because $\mathbf{J}_{d_o}\mathbf{N}\mathbf{J}_{d_o}^{\#}\dot{x}_o$ is not equal to $\dot{x}_o$, in general.

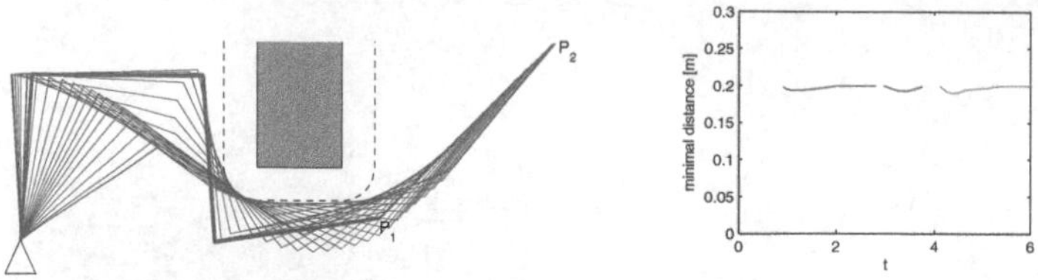

a) EX controller

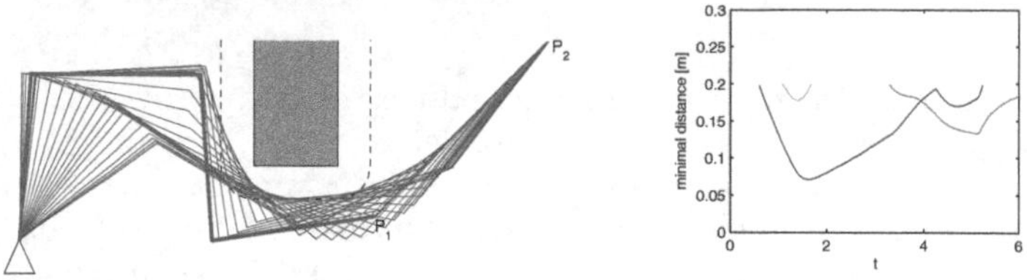

b) AP controller

Fig. 4. Planar 4DOF manipulator tracking a line while avoiding obstacles using velocity control (the dotted line indicates the critical distance)

To avoid the obstacle the goal velocity in A_o is represented by the vector $\dot{\boldsymbol{x}}_o$. Using the original method (12) the velocity in A_o is exactly $\dot{\boldsymbol{x}}_o$. The joint velocities $\dot{\boldsymbol{q}}_{EX}$ assure that the component of the velocity at point A_o (i.e., $\mathbf{J}_o\dot{\boldsymbol{q}}_{EX}$) in the direction of $\dot{\boldsymbol{x}}_o$ is as required. The approximate solution $\dot{\boldsymbol{q}}_{AP}$ gives, in most cases, a smaller magnitude of the velocity in the direction of $\dot{\boldsymbol{x}}_o$ (see $\mathbf{J}_o\dot{\boldsymbol{q}}_{AP}$). Therefore, the manipulator moves closer to the obstacle when $\dot{\boldsymbol{q}}_{AP}$ is used. This is not so critical, because the minimal distance also depends on the nominal velocity v_o, which can be increased to achieve larger minimal distances. Additionally, the approximate solution possesses certain advantages when many active obstacles have to be considered. The joint velocities can be calculated as

$$\dot{\boldsymbol{q}}_{AP} = \mathbf{J}^{\#}\dot{\boldsymbol{x}}_c + \mathbf{N}\sum_{i=1}^{n_o}\mathbf{J}^{\#}_{d_o,i}\dot{x}_{o,i} \tag{25}$$

where n_o is the number of active obstacles, and therefore, the matrix $\mathbf{N}$ has to be calculated only once. Of course, pseudoinverses $\mathbf{J}^{\#}_{o,i}$ have to be calculated for each active obstacle. For the same system and task as before, the simulation results using the approximate velocity controller AP (24) are presented in Fig. 4(b). We can see that in the case of the AP controller, the links are coming closer to the obstacle as in the case of the EX controller. However, when changing the desired critical point velocity $\dot{x}_o$, i.e., using a higher order in (18), the minimal distance can be increased.

4.5 Obstacle avoidance as a primary task

For a redundant system multiple tasks can be arranged in priority. Let us consider two tasks, T_a and T_b

$$\boldsymbol{x}_a = \boldsymbol{f}_a(\boldsymbol{q}) \qquad\qquad \boldsymbol{x}_b = \boldsymbol{f}_b(\boldsymbol{q}) \tag{26}$$

For each of the tasks, the corresponding Jacobian matrices can be defined as $\mathbf{J}_a$ and $\mathbf{J}_b$, and their corresponding null-space projections as $\mathbf{N}_a$ and $\mathbf{N}_b$. Assuming that task T_a is the primary task, Eq. (4) can be rewritten as

$$\dot{\boldsymbol{q}} = \mathbf{J}_a^{\#}\dot{\boldsymbol{x}}_a + \mathbf{N}_a\mathbf{J}_b^{\#}\dot{\boldsymbol{x}}_b \tag{27}$$

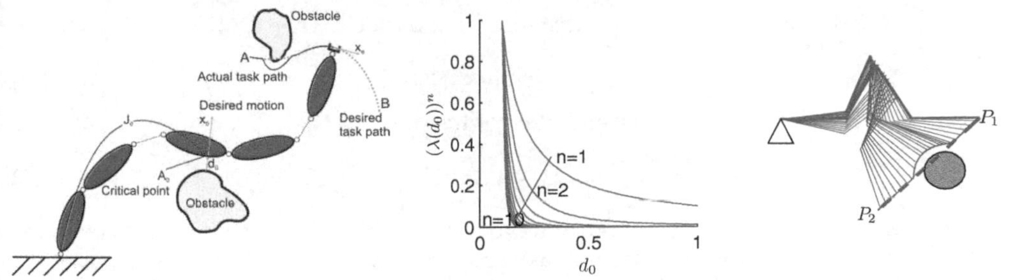

Fig. 5. Robot motion in the presence of some obstacle (left figure). Obstacle-avoidance proximity gain to the power of n (middle figure) and the planar 4DOF manipulator tracking line while avoiding the obstacle (right figure).

In many cases it would be of benefit to have the possibility to change the priority of particular subtasks. Using formulation (27) this cannot be done in a smooth way. Therefore, we propose a new definition of the velocity $\dot{\boldsymbol{q}}$. The velocity $\dot{\boldsymbol{q}}$ is now defined as

$$\dot{\boldsymbol{q}} = \mathbf{J}_a^{\#}\dot{\boldsymbol{x}}_a + \mathbf{N}'_a\mathbf{J}_b^{\#}\dot{\boldsymbol{x}}_b, \tag{28}$$

where the matrix $\mathbf{N}'_a$ is given as

$$\mathbf{N}'_a = \mathbf{I} - \lambda(\boldsymbol{x}_a)\mathbf{J}^{\#}\mathbf{J}, \tag{29}$$

where $\lambda(\boldsymbol{x}_a)$ is a scalar measure of how "active" is the primary task T_a, scaling the vector $\boldsymbol{x}_a$ to the interval $[0,1]$. When the primary task T_a is active the $\lambda(\boldsymbol{x}_a) = 1$ and the value when the task T_a is not active $\lambda(\boldsymbol{x}_a) = 0$.

The proposed algorithm allows a smooth transition in both ways, i.e., between observing the task T_a and the task T_b in null-space of the task T_a or just the unconstrained movement of the task T_b. It can be used for different robotic tasks. When applied to obstacle avoidance, task T_a is the obstacle avoidance and the end-effector motion is the task T_b. Before, we have assumed that the end-effector motion is not disturbed by an obstacle. Now, it is assumed that the motion of the end effector can be disturbed by any obstacle (see Fig. 5). If such a situation occurs, usually the task execution has to be interrupted and higher-level path planning has to be employed to recalculate the desired motion of the end effector. However, if the end-effector path tracking is not essential, we can use the proposed control (28). Consequently, no end-effector path recalculation or higher-level path planning is needed.

Let the primary task T_a be the motion in the direction $\boldsymbol{d}_0$ and the motion of the end-effector be the task T_b. Using the reduced operational space yields

$$\mathbf{J}_a = \mathbf{J}_o, \tag{30}$$

$$\mathbf{J}_b = \mathbf{J}. \tag{31}$$

Next, let the avoiding velocity $\dot{\boldsymbol{x}}_{d_0}$ be defined as

$$\dot{\boldsymbol{x}}_{d_0} = \lambda\left(\boldsymbol{d}_0\right)\boldsymbol{v}_0, \tag{32}$$

where $\boldsymbol{v}_0$ is a nominal velocity and $\lambda\left(\boldsymbol{d}_0\right)$ is defined as

$$\lambda(\boldsymbol{d}_0) = \begin{cases} \left(\dfrac{d_m}{||\boldsymbol{d}_0||}\right)^n, \; n = 1,2,3... & ||\boldsymbol{d}_0|| \geq d_m \\ 1 & ||\boldsymbol{d}_0|| < d_m \end{cases} \tag{33}$$

where $n = 1, 2, 3...$ and d_m is the critical distance to the obstacle. Then eq. (28) can be rewritten in the form

$$\dot{q} = \mathbf{J}_o^{\dagger}\lambda\left(\boldsymbol{d}_0\right)\boldsymbol{v}_0 + \mathbf{N}_0'\mathbf{J}^{\#}\dot{\boldsymbol{x}}_c. \tag{34}$$

Here, $\dot{\boldsymbol{x}}_c$ is the task controller for the end-effector tracking and $\mathbf{N}_0'$ is given by

$$\mathbf{N}_0' = \mathbf{I} - \lambda\left(\boldsymbol{d}_0\right)\mathbf{J}_o^{\dagger}\mathbf{J}_o. \tag{35}$$

Formulation (34) allows unconstrained joint movement, while $\lambda(\boldsymbol{d}_0)$ is close to zero ($\lambda(\boldsymbol{d}_0) \approx 0$). Thus, the robot can track the desired task space path, while it is away from the obstacle. On the other hand, when the robot is close to the obstacle ($\lambda(\boldsymbol{d}_0) \approx 1$), the null space in (35) takes the form $\mathbf{N}_0' = \mathbf{N}_0$, and only allows movement in the null space of the primary task, i.e., the obstacle-avoidance task. In this case, we can still move the end effector, but the tracking error can increase due to the obstacle-avoiding motion.

4.6 Singular configurations

An important issue in the control of redundant manipulators is singular configurations where the associated Jacobian matrices lose the rank. Usually, only the configuration of the whole manipulator is of interest, but in obstacle avoidance we have also to consider the singularities of two manipulator substructures defined by the critical point A_o: (a) the part of the manipulator between the base and the point A_o and (b) the part between A_o and the end-effector. Although it can be assumed that the end-effector Jacobian $\mathbf{J}^{\#}$ is not singular along the desired end-effector path (otherwise the primary task can not be achieved), this is not always true for the matrix $\mathbf{J}_o\mathbf{N}$. Namely, when part (a) is in the singular configuration then $\mathbf{J}_o$ is not of full rank and when part (b) is in the singular configuration then $\mathbf{J}_o$ retains the rank but $\mathbf{J}_{d_o}\mathbf{N}$ becomes singular. Hence, when approaching either singular configuration the values of $\dot{q}_h$ become unacceptably large. As the manipulator is supposed to move in an unstructured environment it is practically impossible to know when $\mathbf{J}_o\mathbf{N}$ will become singular. Therefore, a very important advantage of the proposed Jacobian $\mathbf{J}_{d_o}$ compared to $\mathbf{J}_o$ is that the system has significantly fewer singularities when $\mathbf{J}_{d_o}$ is used.

5. Obstacle avoidance using forces

Impedance control approaches for obstacle-avoidance were first introduced by Hogan (Hogan, 1985). These approaches make use of the additive property of impedances to supplement an impedance controller with additional disturbance forces to avoid the obstacles. The disturbance forces are generated from the artificial potential field. Lee demonstrated this approach with his reference adaptive impedance controller and gave promising results for a simulated 2-DOF robot (Lee et al., 1997).

The advantage of these approaches is that the dynamic behavior of the manipulator as it interacts with obstacles is adjustable through the gains in the obstacle-induced disturbing force. However, this approach is tightly coupled with the control scheme and requires the use of a compliance controller, which may not be desirable, depending on the task.

5.1 Obstacle avoidance without any sensors

First we want to investigate what happens if the manipulator collides with an obstacle. We are especially interested in how to control the manipulator so that it avoids the obstacle after the collision and how to minimize the contact forces. When the task itself includes contacts with the environment (e.g. assembly), the manipulator is equipped with an appropriate sensor

to measure the contact forces and the controller includes a force control loop. But when the contacts between the manipulator and an obstacle can take place anywhere on the body of the manipulator, it is questionable whether the forces arising from such contacts can be measured. Assuming that the contact forces are not measurable and no other sensors to detect the obstacles are present, the contact forces have to be considered in the controller as disturbances. Now the problem is, how to generate a motion that would move a part of the manipulator away from the obstacle. To solve this problem we propose to follow a very basic principle: *an action causes a reaction*. In other words, if a manipulator acts with a force on an obstacle then the obstacle acts with a force in the opposite direction on the manipulator and our idea is to take advantage of this reaction force to move the manipulator away from the obstacle. To make such a motion possible, the control must not force the manipulator to oppose the reaction force (e.g. by preserving the configuration). This means that the system should be compliant to these external forces. A prerequisite for such an approach is that the manipulator is *backdrivable*, meaning that any force at the manipulator is immediately felt at the motors, so the manipulator reacts rapidly, drawing back from the source of the force.
To decouple the task space and the null-space motion we propose to use the controller (7). Actually, from the obstacle-avoidance point of view, any controller for the redundant manipulators could be used provided that the controller outputs are the joint torques. However, this is not enough to decouple the task space and null-space motion. Namely, torques applied through the null-space of $\mathbf{J}^{T\#}$, i.e., $\mathbf{N}^T\boldsymbol{\tau}$, can affect the end-effector acceleration, depending on the choice of the generalized inverse. It turns out for redundant systems that only the so-called "dynamically consistent" generalized inverse $\bar{\mathbf{J}}$ defined as

$$\bar{\mathbf{J}} = \mathbf{H}^{-1}\mathbf{J}^T(\mathbf{J}\mathbf{H}^{-1}\mathbf{J}^T)^{-1} \tag{36}$$

is the unique generalized inverse that decouples the task space and the null-space motion, i.e., assures that the task space acceleration is not affected by any torques applied through the null space of $\bar{\mathbf{J}}^T$, and that the end-effector forces do not produce any accelerations in the null-space of $\mathbf{J}$ (Featherstone & Khatib, 1997).
Using the inertia weighted generalized inverse in Eq. (7) yields

$$\boldsymbol{\tau} = \mathbf{H}(\bar{\mathbf{J}}(\ddot{\boldsymbol{x}}_c - \dot{\mathbf{J}}\dot{\boldsymbol{q}}) + \bar{\mathbf{N}}(\boldsymbol{\phi} - \dot{\bar{\mathbf{N}}}\dot{\boldsymbol{q}})) + \boldsymbol{h} + \boldsymbol{g} \tag{37}$$

Combining (3) and (37), and considering Eqs. (8) and (9) yields

$$\mathbf{H}\ddot{\boldsymbol{q}} + \boldsymbol{h} + \boldsymbol{g} - \boldsymbol{\tau}_F = \mathbf{H}(\bar{\mathbf{J}}(\ddot{\boldsymbol{x}}_d + \mathbf{K}_v\dot{\boldsymbol{e}} + \mathbf{K}_p\boldsymbol{e} - \dot{\mathbf{J}}\dot{\boldsymbol{q}}) + \bar{\mathbf{N}}(\ddot{\boldsymbol{\varphi}} + \dot{\bar{\mathbf{N}}}\dot{\boldsymbol{\varphi}} + \mathbf{K}_n\dot{\boldsymbol{e}}_n - \dot{\bar{\mathbf{N}}}\dot{\boldsymbol{q}})) + \boldsymbol{h} + \boldsymbol{g} \tag{38}$$

which simplifies to

$$\bar{\mathbf{J}}(\ddot{\boldsymbol{e}} + \mathbf{K}_v\dot{\boldsymbol{e}} + \mathbf{K}_p\boldsymbol{e}) + \bar{\mathbf{N}}(-\ddot{\boldsymbol{q}} + \ddot{\boldsymbol{\varphi}} + \dot{\bar{\mathbf{N}}}\dot{\boldsymbol{\varphi}} + \mathbf{K}_n\dot{\boldsymbol{e}}_n - \dot{\bar{\mathbf{N}}}\dot{\boldsymbol{q}}) = -\mathbf{H}^{-1}\boldsymbol{\tau}_F \tag{39}$$

where $\boldsymbol{\tau}_F$ represents the influence of all external forces caused by the contacts with obstacles

$$\boldsymbol{\tau}_F = \sum_{i=1}^{a_o} \mathbf{J}_{o,i}^T \boldsymbol{F}_{o,i} \tag{40}$$

Premultiplying (39) with $\mathbf{J}$ yields

$$\ddot{\boldsymbol{e}} + \mathbf{K}_v\dot{\boldsymbol{e}} + \mathbf{K}_p\boldsymbol{e} = -\mathbf{J}\mathbf{H}^{-1}\boldsymbol{\tau}_F \tag{41}$$

Note that the contact forces affect the motion in the task space, which results in the tracking error of the end-effector. Of course, if the obstacle avoidance is successful the contact forces vanish and the task position error converges to zero. The proper choice of $\mathbf{K}_v$ and $\mathbf{K}_p$ assures the asymptotical stability of the homogeneous part of the system, $\ddot{e} + \mathbf{K}_v\dot{e} + \mathbf{K}_p e = 0$. A detailed analysis of the influence of the external forces is given in Žlajpah & Nemec (2003).
Next we analyse the behavior in null-space. Premultiplying (39) with $\bar{\mathbf{N}}$ yields

$$-\bar{\mathbf{N}}\mathbf{H}^{-1}\boldsymbol{\tau}_F = \bar{\mathbf{N}}(-\ddot{q} + \ddot{\varphi} + \dot{\bar{\mathbf{N}}}\dot{\varphi} + \mathbf{K}_n\dot{e}_n - \dot{\bar{\mathbf{N}}}\dot{q}) \tag{42}$$

Rearranging Eq. (42) and using the relations $\ddot{e}_n = \bar{\mathbf{N}}(\ddot{\varphi} - \ddot{q}) + \dot{\bar{\mathbf{N}}}(\dot{\varphi} - \dot{q})$ and $\bar{\mathbf{N}}\mathbf{H}^{-1} = \mathbf{H}^{-1}\bar{\mathbf{N}}^T$, we obtain

$$\bar{\mathbf{N}}(\ddot{e}_n + \mathbf{K}_n\dot{e}_n) = -\bar{\mathbf{N}}\mathbf{H}^{-1}\bar{\mathbf{N}}^T\boldsymbol{\tau}_F = -\mathbf{H}_n^{\ddagger}\boldsymbol{\tau}_F \tag{43}$$

where $\mathbf{H}_n$ is the *null-space effective inertia matrix* describing the inertial properties of the system in the null-space

$$\mathbf{H}_n = \bar{\mathbf{N}}^T\mathbf{H}\bar{\mathbf{N}}$$

and $\mathbf{H}_n^{\ddagger}$ is the generalized inverse of $\mathbf{H}_n$, defined as

$$\mathbf{H}_n^{\ddagger} = \bar{\mathbf{N}}\mathbf{H}^{-1}\bar{\mathbf{N}}^T$$

The selection of the null space dynamic properties is subjected to the subtask that the manipulator should perform. Actually, in most cases it is required that the null space velocity tracks a given desired null space velocity $\dot{\varphi}$. For good null space velocity tracking, the gain matrix $\mathbf{K}_n$ should be high, which means that the system is stiff in the null space. On the other hand, when the manipulator collides with an obstacle, the self-motion is initiated externally by the contact force. As we have already mentioned, in this case the system should be compliant in the null-space. Therefore, the gain matrix $\mathbf{K}_n$ should be low. As both requirements for $\mathbf{K}_n$ are in conflict, a compromise has to made when selecting $\mathbf{K}_n$ or the on-line adaptation of $\mathbf{K}_n$ has to be used. The problem with the second possibility is that we have to be able to detect the current state, i.e., if the manipulator is in contact with an object. If this is possible, then it is easy to set low gains when the collision occurs and high gains when the manipulator is "free". If we cannot detect the contact and the probability of collisions is high, then it is rational to select a low $\mathbf{K}_n$, but we must be aware that too low values of $\mathbf{K}_n$ may cause instability in the null-space. The lower bounds for $\mathbf{K}_n$ can be obtained with a Lyapunov analysis (Žlajpah & Nemec, 2003).
As an illustration we use the same example as before, where the manipulator is supposed to have 4 revolute joints. The primary task is to move along a straight line from point P_1 to point P_2, and the motion is obstructed by an obstacle. The task trajectory has a trapezoid velocity profile with an acceleration of $4ms^{-2}$ and a max. velocity of $0.4ms^{-1}$. The control algorithm is given by Eq. (37) (CF). The task space controller parameters (Eq. 8) are $\mathbf{K}_p = 1000\mathbf{I}s^{-2}$ and $\mathbf{K}_v = 80\mathbf{I}s^{-1}$, which are tuned to ensure good tracking of the task trajectory (stiff task space behavior). The null space controller is a modified version of the controller (9)) $\phi = -\mathbf{K}_n\dot{q}$, i.e., the desired null-space velocity is set to zero. We have compared two sets of null-space gains: (i) Low $\mathbf{K}_n$, which ensures compliant null-space behavior, and (ii) Medium $\mathbf{K}_n$, which makes the null-space more stiff. The simulation results are presented in Figs. 6 and 7.
Although the manipulator has finished the desired task in both cases, we can see that making the null-space more compliant results in decreased contact forces. As expected, the impact force does not depend on the stiffness in the null space and cannot be decreased by increasing

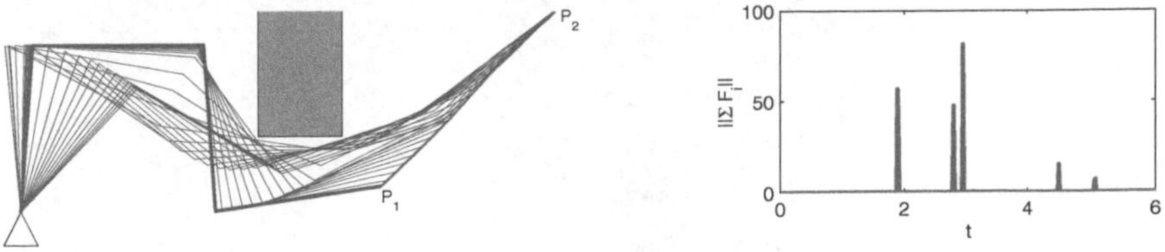

Fig. 6. Planar 4DOF manipulator tracking a line while avoiding obstacles without using any sensors to detect the obstacles (CF); compliant in null space

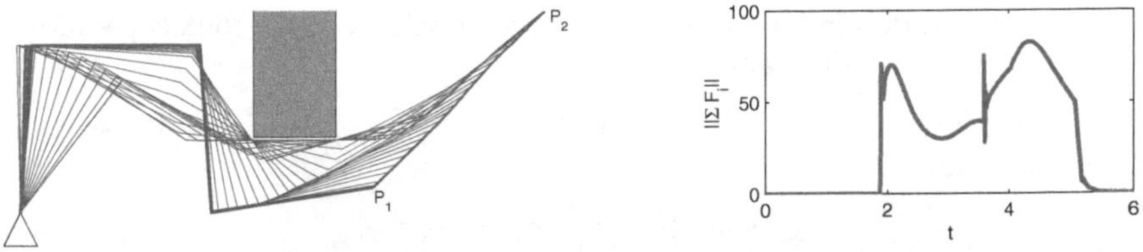

Fig. 7. Planar 4DOF manipulator tracking a line while avoiding obstacles without using any sensors to detect the obstacles (CF); stiff in null space

the compliancy in the null space. Actually, the impact forces are the main problem with this type of control. Namely, the magnitude of the impact forces cannot be controlled with the controller, the controller can decrease the contact forces after the impact. The magnitude of the impact forces depends primarily on the kinetic energy of the bodies that collide and the stiffness of the contact. Therefore, this approach to obstacle avoidance is limited to cases where the manipulator is moving slowly or the manipulator body (or obstacles) is covered with soft material, which reduces the impact forces. Note that the impact forces (the magnitude and the time of occurrence) are not the same in these examples due to the different close-loop dynamics of the system.

5.2 Obstacle avoidance with tactile sensors

As already mentioned, the obstacle-avoidance approach based on contact forces without any contact sensors cannot be applied to manipulators that are not backdrivable, like manipulators with high-ratio gears. The alternative strategy is a sensor based motion control, where a kind of tactile sensors mounted on a manipulator detect the contact with an obstacle. The main advantage of using the tactile sensor information in control is that the obstacle avoidance becomes "active", meaning the avoiding motion is initiated by the controller. So, it can be applied to any manipulator, irrespective of the backdrivability of the manipulator.
With tactile sensors the collisions can be detected. Our approach is to identify the points on the manipulator that are in contact with obstacles and to assign to them additional virtual force components that move the points away from the obstacle (see Fig. 8). We propose that these forces are not included into the close-loop controller, but they are applied as the virtual external forces to the manipulator.
Suppose that $\boldsymbol{F}_o$ is acting at point A_o somewhere on the link i (see Fig. 8).
The static relation between the force $\boldsymbol{F}_o$ and the corresponding joint torques τ_o is

$$\tau_o = \mathbf{J}_o^T \boldsymbol{F}_o \tag{44}$$

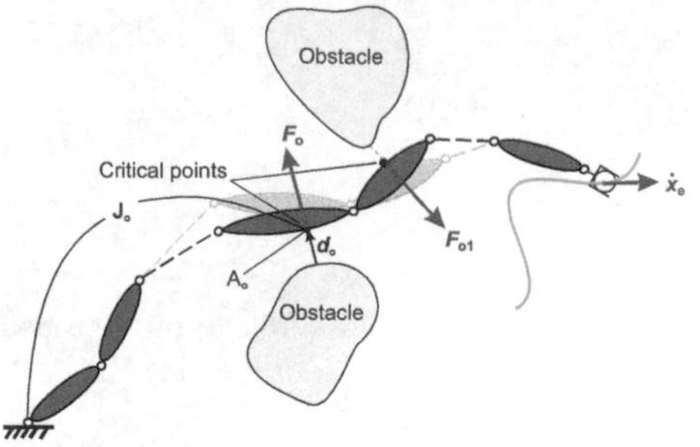

Fig. 8. Manipulator motion in the presence of some obstacles: links in contact are moved away from obstacles by additional virtual forces

where $\mathbf{J}_o$ is a Jacobian matrix associated with the point A_o. Applying $\boldsymbol{\tau}_o$ to the system yields the equation of motion in the form

$$\boldsymbol{\tau} = \mathbf{H}\ddot{\boldsymbol{q}} + \boldsymbol{h} + \boldsymbol{g} - \boldsymbol{\tau}_F - \boldsymbol{\tau}_o \tag{45}$$

which is the same as Eq. (3), except that the torques due to the virtual forces forces $\boldsymbol{\tau}_F$ are added. In general, by applying the torques $\boldsymbol{\tau}_o$ as defined in Eq. (44), the motion of the end-effector and the self-motion of the manipulator are influenced. Note that one of the goals of obstacle avoidance is to disturb the end-effector motion as little as possible. As we are adding virtual forces, it is not necessary that the applied virtual forces correspond to the "real" forces. Therefore, only torques that do not influence the end-effector motion are proposed to be added to generate the obstacle-avoidance motion. In general, the force $\boldsymbol{F}_o$ can be substituted by a force acting in the task space

$$\boldsymbol{F}_{oe} = \mathbf{J}^{T\#}\mathbf{J}_o^T\boldsymbol{F}_o \tag{46}$$

and by joint torques acting in the null-space of $\mathbf{J}^{T\#}$

$$\boldsymbol{\tau}_{oN} = \mathbf{N}^T\mathbf{J}_o^T\boldsymbol{F}_o = \mathbf{N}^T\boldsymbol{\tau}_o \tag{47}$$

Substituting $\boldsymbol{\tau}_{oN}$ for $\boldsymbol{\tau}_o$ in Eq. (45) yields

$$\boldsymbol{\tau} = \mathbf{H}\ddot{\boldsymbol{q}} + \boldsymbol{h} + \boldsymbol{g} - \boldsymbol{\tau}_F - \bar{\mathbf{N}}^T\boldsymbol{\tau}_o \tag{48}$$

The efficiency of the obstacle-avoidance algorithm depends on the selection of the desired force $\boldsymbol{F}_o$. In our approach, the virtual force $\boldsymbol{F}_o$ depends on the location of the critical point A_o, i.e., the locations of the tactile sensors that detect the contact. As the positions of the sensors are known in advance, it is easy to get the corresponding Jacobian matrix $\mathbf{J}_o$ for each sensor. Additionally, each sensor also has a predefined avoiding direction $\boldsymbol{n}_o$. We define the virtual forces $\boldsymbol{F}_o$ as

$$\boldsymbol{F}_o = \alpha_t f_o \boldsymbol{n}_o \tag{49}$$

where α_t is the obstacle-avoidance gain. We propose that α_t depends on the duration of the contact and that the achievedvalue is preserved for some time after the contact between the manipulator and the obstacle is lost (see Fig. 9)

$$\alpha_t = \begin{cases} 0 & \text{for} \quad t < t_s \\ e^{T_i(t-t_s)} & \text{for} \quad t_s \leq t < t_e \\ e^{T_i(t_e-t_s)}e^{-T_d(t-t_e)} & \text{for} \quad t_e \leq t \end{cases} \tag{50}$$

where T_i and T_d are the constants for the increase of the gain value and the delayed action, and t_s and t_e represent the time when the contact occurs and the time when the contact is lost, respectively. With the appropriate selection of T_i and T_d a robust behavior can be obtained.

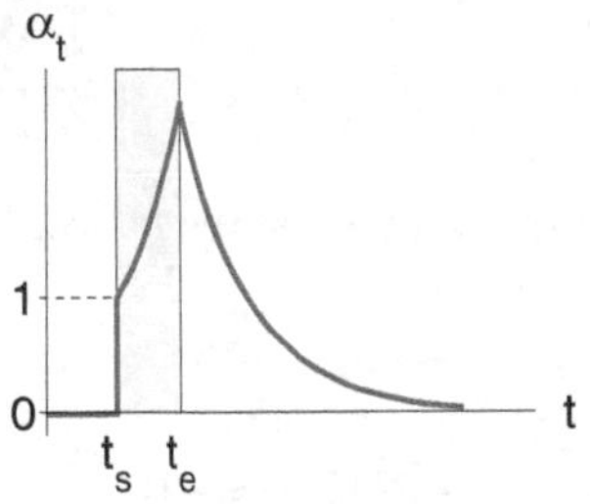

Fig. 9. The obstacle-avoidance gain α_t versus the duration of the contact

The next issue is how the obstacle avoidance is performed when there are more simultaneously active obstacles in the neighborhood of the manipulator. Using the proposed obstacle-avoidance formulation the problem of many simultaneously active obstacles can be solved very efficiently. The additivity of the torques makes it possible to avoid multiple obstacles by using the sum of the torques due to the relevant virtual forces,

$$\tau_o = \sum_{i=1}^{a_o} \mathbf{J}_{o,i}^T \boldsymbol{F}_{o,i} \tag{51}$$

and considering this in Eq. (47) yields

$$\tau_{o,N} = \bar{\mathbf{N}}^T \sum_{i=1}^{a_o} \mathbf{J}_{o,i}^T \boldsymbol{F}_{o,i} \tag{52}$$

where a_o is the number of active obstacles. It is clear that when more than one obstacle is active it is necessary to calculate only the transpose of the Jacobian $\mathbf{J}_{o,i}^T$ for each critical point and not the generalized inverse $\bar{\mathbf{J}}_{o,i}$, as is the case with velocity-based strategies. The redundancy of the manipulator is considered in the term $\bar{\mathbf{N}}^T$, which does not depend on the location of particular critical points $A_{o,i}$. Hence, there is no limitation on a_o regarding the degree-of-redundancy. In the case when a_o is greater than the degree-of-redundancy, the manipulator is pushed into a configuration where the virtual forces compensate each other, i.e., $\tau_o = 0$. Actually, such situations can occur even when a_o is less than the degree-of-redundancy, e.g. when one link is under the influence of more than one obstacle. The force formulation has its advantages computationally, e.g. in Eq. (52) the term $\bar{\mathbf{N}}^T$ is calculated only once.
For the close-loop control the controller (37) is used again. Augmenting (38) with virtual forces the behavior of the system with contact sensors is described by

$$\mathbf{H}\ddot{\boldsymbol{q}} + \boldsymbol{h} + \boldsymbol{g} - \tau_F - \bar{\mathbf{N}}^T\tau_o = \mathbf{H}(\bar{\mathbf{J}}(\ddot{\boldsymbol{x}}_d + \mathbf{K}_v\dot{\boldsymbol{e}} + \mathbf{K}_p\boldsymbol{e} - \dot{\mathbf{J}}\dot{\boldsymbol{q}}) + \bar{\mathbf{N}}(\ddot{\boldsymbol{\varphi}} + \dot{\bar{\mathbf{N}}}\dot{\boldsymbol{\varphi}} + \mathbf{K}_n\dot{\boldsymbol{e}}_n - \dot{\bar{\mathbf{N}}}\dot{\boldsymbol{q}})) + \boldsymbol{h} + \boldsymbol{g} \tag{53}$$

Actually, the term $\bar{\mathbf{N}}^T\tau_o$ is also part of the controller and should be on the right-hand side of Eq. (53), but we put it on the left-hand side to emphasize its role. Eq. (53) simplifies to

$$\bar{\mathbf{J}}(\ddot{\boldsymbol{e}} + \mathbf{K}_v\dot{\boldsymbol{e}} + \mathbf{K}_p\boldsymbol{e}) + \bar{\mathbf{N}}(-\ddot{\boldsymbol{q}} + \ddot{\boldsymbol{\varphi}} + \dot{\bar{\mathbf{N}}}\dot{\boldsymbol{\varphi}} + \mathbf{K}_n\dot{\boldsymbol{e}}_n - \dot{\bar{\mathbf{N}}}\dot{\boldsymbol{q}}) = -\mathbf{H}^{-1}(\bar{\mathbf{N}}^T\tau_o + \tau_F) \tag{54}$$

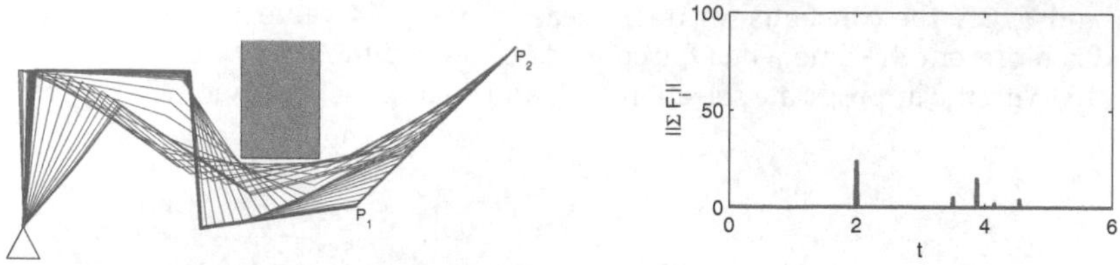

Fig. 10. Planar 4DOF manipulator tracking a line while avoiding obstacles using tactile sensors and virtual forces (TF controller)

Premultiplying (54) with $\mathbf{J}$ yields

$$\ddot{e} + \mathbf{K}_v \dot{e} + \mathbf{K}_p e = -\mathbf{J}\mathbf{H}^{-1}\tau_F \tag{55}$$

which is the same as Eq. (41), and premultiplying (54) with $\bar{\mathbf{N}}$ yields after simplifications

$$\bar{\mathbf{N}}\ddot{e}_n + \bar{\mathbf{N}}\mathbf{K}_n \dot{e}_n = -\mathbf{H}_n^{\ddagger}(\tau_o + \tau_F) \tag{56}$$

From Eqs. (55) and (56) we can see that the task space and the null-space motion are influenced by the contact forces. However, these forces decrease after the impact because of the avoiding motion.

Next, we present the simulation results of the same task as before for the system where the tactile sensors (TF) were used to detect the obstacle. The close loop controller was the same as for the CF case (ii) approach, which assures stiff null space behavior. The avoidance motion was generated using Eqs. (49) – (50) with the parameters $f_o = 100N$, $T_i = 6s$ and $T_d = 8s$. The results, see Fig. 10, show that with tactile sensors we can decrease the contact forces after the contact, but the impact force cannot be decreased. Unfortunately, the magnitude of the impact forces is not controllable and hence this approach also requires low velocities and a soft contact.

5.3 Obstacle avoidance with proximity sensors

When proximity sensors are used to detect obstacles, the collisions between the manipulator and obstacles can be avoided. Also here, our approach is to identify the points on the manipulator that are near obstacles and to assign to them force components that move the points away from the obstacle (see Fig. 8). The strategy is similar to those given in (Brock et al., 2002), although it was developed independently.

Suppose that $\boldsymbol{F}_o$ is acting at point A_o somewhere on the link i (see Fig. 8). Applying τ_o (see Eq. 44) to the system yields the equation of motion in the form

$$\tau = \mathbf{H}\ddot{q} + h + g - \tau_o \tag{57}$$

which is the same as Eq. (3), except that the real external forces τ_F are replaced by the virtual forces τ_o. As we are using virtual forces, only torques that do not influence the end-effector motion are considered

$$\tau_{oN} = \mathbf{N}^T \mathbf{J}_o^T \boldsymbol{F}_o = \mathbf{N}^T \tau_o \tag{58}$$

Substituting τ_{oN} for τ_o in Eq. (57) yields

$$\tau = \mathbf{H}\ddot{q} + h + g - \bar{\mathbf{N}}^T \tau_o \tag{59}$$

The main difference compared to the system with tactile sensors is how the virtual forces $\boldsymbol{F}_o$ are generated. Now, the virtual force $\boldsymbol{F}_o$ depends on the location of the critical point and the closest point on the obstacle. Clearly, the location of all the objects in the workspace of the manipulator has to be known. This information can be provided by a higher control level that has access to sensory data. As sensor systems are not our concern, we will assume that the sensor can detect obstacles in the workspace of the manipulator and that it outputs the direction and the distance to the closest point obstacle, and the position of the critical point on the manipulator.
Let A_o be the critical point and $\boldsymbol{d}$ the vector connecting the closest point on the obstacle and A_o (see Fig. 8). To avoid a possible collision a virtual force $\boldsymbol{F}_o$ is assigned to A_o, defined as

$$\boldsymbol{F}_o = \alpha_f f_o \boldsymbol{n}_o \tag{60}$$

where f_o is a scalar gain representing the nominal force, $\boldsymbol{n}_o$ is the unit vector in the direction of $\boldsymbol{d}$, and α_f is the obstacle-avoidance gain. The gain α_f should depend on the distance to the obstacle, as shown in Fig. 11.

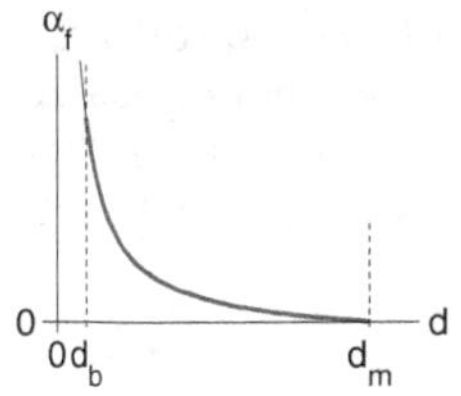

Fig. 11. The obstacle-avoidance gain α_f versus the distance to the obstacle

There are two distances characterizing the changes in the value of the gain: the critical distance d_m and the abort distance d_b. If the distance between the manipulator and the object is greater than d_m then the motion of the manipulator is not perturbed. When the distance is decreasing, the force should increase smoothly (to assure smooth transitions it is important that the magnitude of $\boldsymbol{F}_o$ at d_m is zero). However, if the manipulator is too close to the obstacle (less than d_b) the main task should be, for safety reasons, aborted so that the manipulator can avoid the obstacles (such a situation can occur, especially if moving obstacles are present in the workspace). The distance d_b is subjected to the dynamic properties of the manipulator and can also be a function of the relative velocity between the critical point on the manipulator and the obstacle. The gain α_f can be chosen arbitrarily as long as it has the prescribed form. We propose using the following function

$$\alpha_f = \begin{cases} \left(\frac{d_m}{\|\boldsymbol{d}_o\|}\right)^2 - 1 & \text{for} \quad \|\boldsymbol{d}_o\| < d_m \\ 0 & \text{for} \quad \|\boldsymbol{d}_o\| \geq d_m \end{cases} \tag{61}$$

Special attention has to be given to the selection of the nominal force f_o. Large values of f_o can cause unnecessarily high accelerations and velocities. Consequently, the manipulator could move far from the obstacle. Such a motion may cause problems if there are more obstacles in the neighborhood of the manipulator. Namely, the manipulator may bounce between the obstacles. On the other hand, too small values of f_o would not move the critical point sufficiently away from the obstacle.

When there are more simultaneously active obstacles in the neighborhood of the manipulator, the sum of the torques due to the relevant virtual forces is used,

$$\tau_o = \sum_{i=1}^{a_o} \mathbf{J}_{o,i}^T \boldsymbol{F}_{o,i} \tag{62}$$

and considering this in Eq. (58) yields

$$\tau_{o,N} = \bar{\mathbf{N}}^T \sum_{i=1}^{a_o} \mathbf{J}_{o,i}^T \boldsymbol{F}_{o,i} \tag{63}$$

where a_o is the number of active obstacles.
To decouple the task space and the null-space motion we propose as before the controller (37). Combining (48) and (37), and considering Eqs. (8) and (9) yields

$$\mathbf{H}\ddot{q} + \boldsymbol{h} + \boldsymbol{g} - \bar{\mathbf{N}}^T\tau_o = \mathbf{H}(\bar{\mathbf{J}}(\ddot{\boldsymbol{x}}_d + \mathbf{K}_v\dot{e} + \mathbf{K}_p e - \dot{\mathbf{J}}\dot{q}) + \bar{\mathbf{N}}(\dot{\varphi} + \dot{\bar{\mathbf{N}}}\varphi + \mathbf{K}_n\dot{e}_n - \dot{\bar{\mathbf{N}}}\dot{q})) + \boldsymbol{h} + \boldsymbol{g} \tag{64}$$

As before, the term $\bar{\mathbf{N}}^T\tau_o$ is also part of the controller and should be on the right-hand side of Eq. (64), but we put it on the left-hand side to emphasize its role.
One of the reasons for using the above control law is that the null-space velocity controller (9) can be used for other lower-priority tasks. Hence, to optimize p we can select φ in (9) as

$$\varphi = \mathbf{H}^{-1}k_p\nabla p \tag{65}$$

Note that when the weighted generalized inverse of $\mathbf{J}$ is used, the desired null space velocity has to be multiplied by the inverse of the weighting matrix (in our case $\mathbf{H}^{-1}$) to assure the convergence of the optimization of p (Nemec, 1997).
Next we analyse the behavior of the close-loop system. Premultiplying Eq. (64) by $\mathbf{H}^{-1}$ yields

$$\ddot{q} - \mathbf{H}^{-1}\bar{\mathbf{N}}^T\tau_o = \bar{\mathbf{J}}(\ddot{\boldsymbol{x}}_d + \mathbf{K}_v\dot{e} + \mathbf{K}_p e - \dot{\mathbf{J}}\dot{q}) + \bar{\mathbf{N}}(\dot{\varphi} + \dot{\bar{\mathbf{N}}}\varphi + \mathbf{K}_n\dot{e}_n - \dot{\bar{\mathbf{N}}}\dot{q})$$

and by using (1) the above equation can be rewritten in the form

$$\bar{\mathbf{J}}(\ddot{e} + \mathbf{K}_v\dot{e} + \mathbf{K}_p e) + \bar{\mathbf{N}}(-\ddot{q} + \dot{\varphi} + \dot{\bar{\mathbf{N}}}\varphi + \mathbf{K}_n\dot{e}_n - \dot{\bar{\mathbf{N}}}\dot{q}) = -\mathbf{H}^{-1}\bar{\mathbf{N}}^T\tau_o \tag{66}$$

The dynamics of the system in the task space and in the null-space can be obtained by premultiplying Eq. (66) by $\mathbf{J}$ and $\bar{\mathbf{N}}$, respectively. Premultiplying (66) with $\mathbf{J}$ yields

$$\ddot{e} + \mathbf{K}_v\dot{e} + \mathbf{K}_p e = 0 \tag{67}$$

since $\mathbf{J}\bar{\mathbf{J}} = \mathbf{I}$, $\bar{\mathbf{N}}\mathbf{H}^{-1} = \mathbf{H}^{-1}\bar{\mathbf{N}}^T$, and $\mathbf{J}\bar{\mathbf{N}} = \mathbf{0}$. The proper choice of $\mathbf{K}_v$ and $\mathbf{K}_p$ assures the asymptotical stability of the system. Furthermore, it can be seen that the virtual forces do not affect the motion in the task space. Next, premultiplying (66) by $\bar{\mathbf{N}}$ yields

$$\bar{\mathbf{N}}(-\ddot{q} + \dot{\varphi} + \dot{\bar{\mathbf{N}}}\varphi + \mathbf{K}_n\dot{e}_n - \dot{\bar{\mathbf{N}}}\dot{q}) = -\bar{\mathbf{N}}\mathbf{H}^{-1}\bar{\mathbf{N}}^T\tau_o \tag{68}$$

because $\bar{\mathbf{N}}\bar{\mathbf{J}} = \mathbf{0}$. Rearranging Eq. (68) and using the relation $\ddot{e}_n = \bar{\mathbf{N}}(\dot{\varphi} - \ddot{q}) + \dot{\bar{\mathbf{N}}}(\varphi - \dot{q})$ we obtain

$$\bar{\mathbf{N}}(\dot{\varphi} - \ddot{q} + \dot{\bar{\mathbf{N}}}(\varphi - \dot{q}) + \mathbf{K}_n\dot{e}_n) = -\bar{\mathbf{N}}\mathbf{H}^{-1}\bar{\mathbf{N}}^T\tau_o = -\mathbf{H}_n^{\ddagger}\tau_o \tag{69}$$

From Eq. (69) we can see that the obstacle-avoidance motion is not controlled directly by $\boldsymbol{F}_o$. The force $\boldsymbol{F}_o$ initiates the motion, but the resulting motion depends mainly on the null-space controller (9). Although the null-space controller has in this case the same structure as in the case of the contact forces approach, the gains $\mathbf{K}_n$ can be selected to meet the requirements of the subtask the manipulator should perform, i.e., in most cases to track the desired null space velocity. Hence, the gain matrix $\mathbf{K}_n$ should be high. Consequently, to perform satisfactory obstacle avoidance the nominal force f_o must be higher to predominate over the velocity controller when necessary.

In the following the simulation results for the simple task are shown. The task space controller parameters are the same as in the previous example. To avoid the obstacles the proximity sensor distance was selected as $d_m = 0.2m$. The virtual forces were calculated using Eq. (40) with the parameter $f_o = 800N$. The simulation results, see Fig. 12, clearly show that the obstacle is avoided without a deviation of the end-effector from the assigned task.

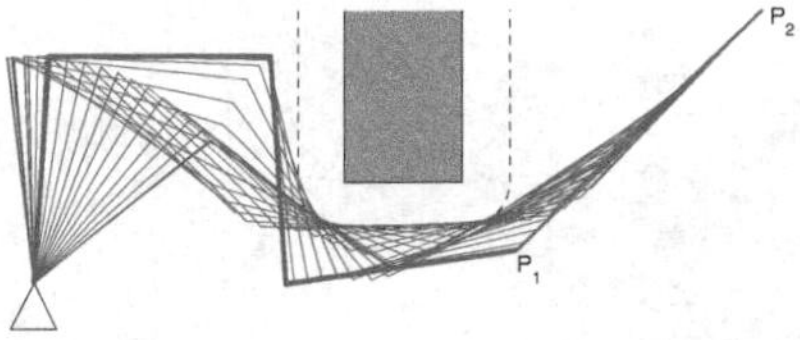

Fig. 12. Planar 4DOF manipulator tracking a line while avoiding obstacles using a virtual forces approach (VF)

6. Experimental results

The proposed algorithms were tested on the laboratory manipulator (see Fig. 13 and two KUKA LWR robots (see Fig. 14). The laboratory manipulator was specially developed for testing the different control algorithms. To be able to test the algorithms for the redundant systems the manipulator has four revolute DOF acting in a plane. The link lengths of the manipulator are $l = (0.184, 0.184, 0.184, 0.203)m$ and the link masses are $m = 0.83, 0.44, 0.18, 0.045)kg$. The manipulator is a part of the integrated environment for the design of the control algorithms and the testing of these algorithms on a real system (Žlajpah, 2001).

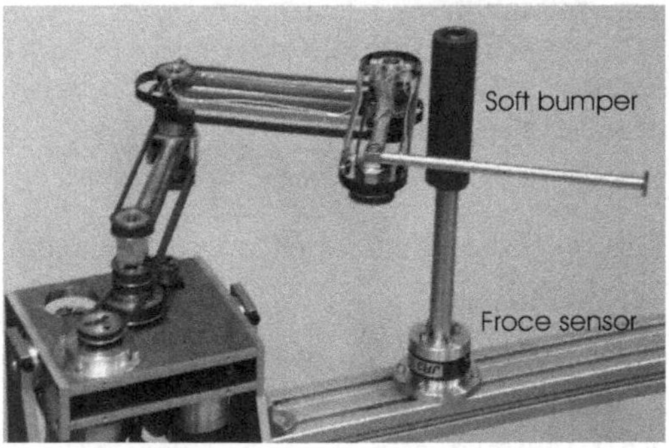

Fig. 13. Experimental manipulator with external force sensor

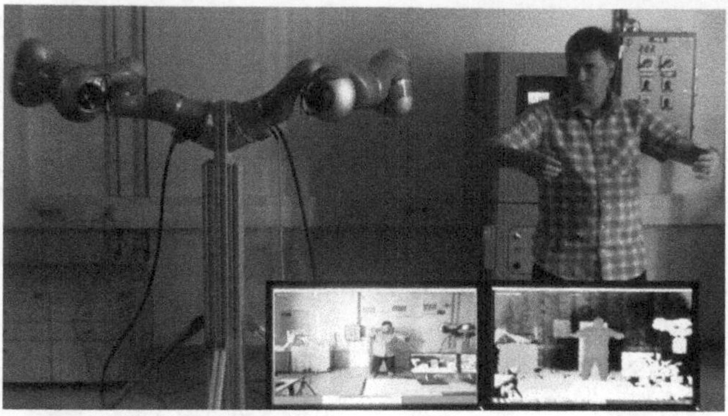

Fig. 14. Experimental setup with two KUKA LWR robots for bimanual movement imitation.

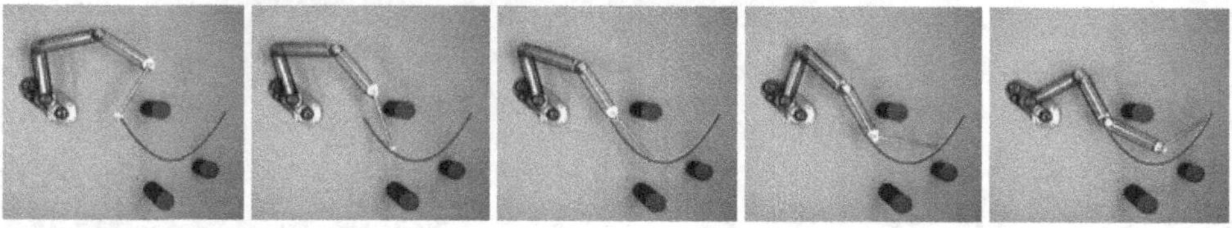

Fig. 15. Planar 4DOF manipulator tracking a path in an unstructured environment with three obstacles using velocity-based control (sampled every 1s)

6.1 Velocity controller

The task in these experiments was tracking the path $x_d = [0.4 - 0.2\sin(2\pi t/8), -0.1 + 0.1\sin(2\pi t/4)]^T$ and the motion of the manipulator was obstructed by three obstacles (see Fig. 15). In the current implementation the vision system is using a simple USB WebCam that can recognize the scene and output the position of all the obstacles in less that 0.04s. To avoid the obstacles the proximity sensor distance was selected as $d_m = 0.08m$. The rate of the velocity controller was 200Hz (the necessary joint velocities for the avoiding motion are calculated in less than 0.5ms). The experimental results are given in Fig. 15. As we can see, the obstacles were successfully avoided.

6.2 Obstacle avoidance as a primary task

We applied our algorithm to two Kuka LWR robots as shown in Fig. 14. Our algorithm is used as a low-level control to prevent self-collision, i.e., a collision between the robots themselves. As mentioned previously, the desired movement of the robot is a secondary task. The task of collision avoidance is only observed if we approach a pre-defined threshold. While far from the threshold the algorithm allows direct control of the separate joints ($\dot{q}_n$). If we approach the threshold, the task of collision avoidance smoothly takes over and only allows joint control in the null space projection of this task. Note that $\dot{q}_n$ is in joint space.
The task for both arms in this experiment was to follow the human demonstrator in real-time. The human motion is captured using the Microsoft Kinect sensor. Microsoft Kinect is based on arange camera developed by PrimeSense, which interprets 3D scene information from a continuously-projected infrared structured light. By processing the depth image, the PrimeSense application programming interface (API) enables tracking of the user's movement in real time. Imitating the motion of the user's arm requires some basic understanding of

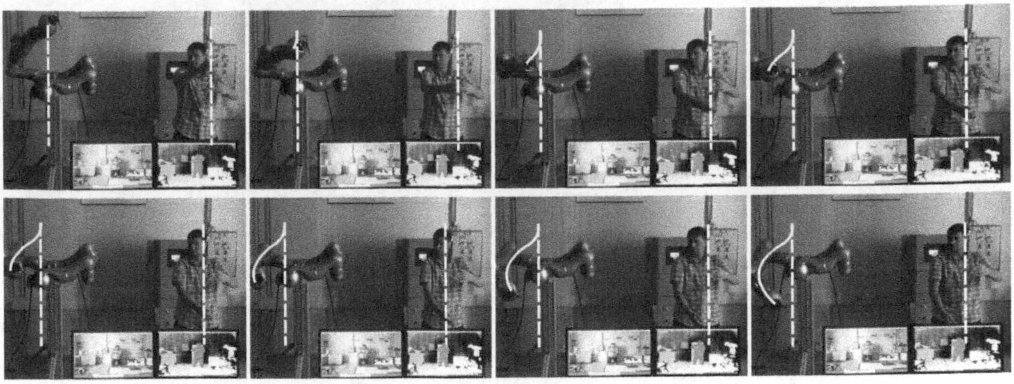

Fig. 16. A sequence of still photographs shows the movement of two Kuka LWR robots, while they successfully avoids each other. The desired movement for the robots is imitated in real time using the Microsoft Kinect sensor for the tracking.

human physiology. The posture of each arm may be described by four angles - three angles in the shoulder joint and one in the elbow. The shoulder joint enables the following motion (Hayes et al., 2001): arm flexion, arm abduction and external rotation. These angles are calculated from the data obtained with Microsoft Kinect.
A sequence for successful self collision-avoidance is shown in Fig. 16. Here, we can see that the robot angles are similar when the humans hands are away from the threshold, i.e., the robots are not close together. On the other hand, when close together, the robots properly adapt their motion to prevent a collision.

6.3 Contact forces

First we tested the behavior of the manipulator when no sensors were used to detect the obstacles. The desired task was to track the circular path and the motion of the manipulator was obstructed by an obstacle (the rod; see Fig. 13). To be able to monitor the contact forces the rod was mounted on a force sensor. Note that the force information measured by this sensor was not used in the close loop controller.
The controller was based on the algorithm (7) with the task controller (8) and null space controller (9) and we compared three sets of null space controller parameters. In the first case (a) the controller assured very stiff null space behavior, in the second (b) the controller assured medium stiffness in null space, and in the last case (c) the manipulator was compliant in the null space. To prevent high impact forces, the bumper was covered with soft material and the manipulator joint velocities were low.
The experimental results are shown in Fig. 17. First we can see that in case (a) the manipulator pushes the bumper more and more and finally, the task has to be aborted due to the very large contact forces. Next, comparing the responses one can see that although the motion is almost equal in both cases, the forces are lower in case (b) (compliant in null-space). Summarizing, the experimental results proved that the contact forces can be decreased by increasing the null space compliance.

6.4 Tactile sensors

Next we tested the efficiency of the tactile sensors. In our experiments we used simple bumpers on each link as tactile sensors. The sensor can detect an object when it touches the

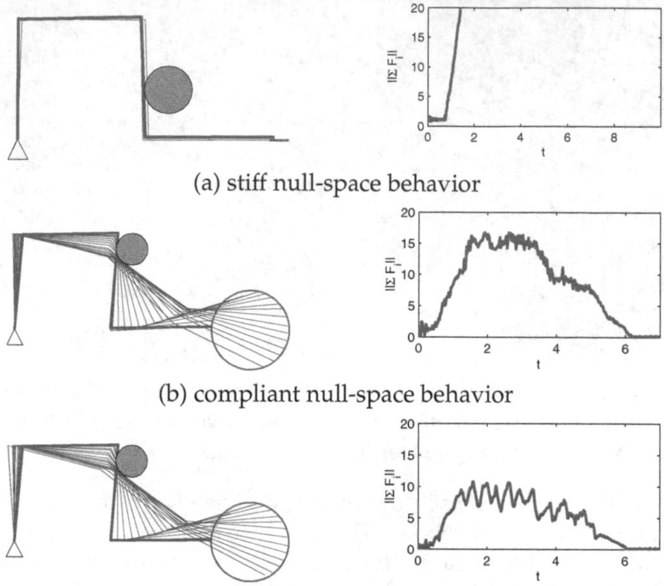

Fig. 17. Planar 4DOF manipulator tracking a circle while avoiding obstacles without any sensor for obstacle detection

object. In our case each switch needs a force of $2N$ to trigger it. This means that the sensor can detect an object if the contact force is greater then $\sim 2 - 4N$, depending on the particular contact position on the bar. The main drawback of this type of sensor is that it can only detect the link and the side of the link where the contact occurs, and not the exact position of the contact.

The desired task in these experiments was tracking the linear path and as before, the motion of the manipulator was obstructed by the rod. The virtual forces were calculated using Eqs. (49) – (50). As our sensor can detect only the side of the link where the contact occurs and not the exact position of the contact, we used the middle of the link as the approximation for the contact point and the avoiding direction was perpendicular to the link.

The experimental results are shown in Fig. 18. The figures show the contact forces norm $\|\boldsymbol{F}_{\text{ext}}\|$ and the configurations of the manipulator. The results clearly show that the manipulator avoids the obstacle after the collision. The behavior of the manipulator after the contact with the obstacle depends mainly on the particular nominal virtual force f_o and the time constants T_i and T_d. Tuning these parameters results in different behaviors of the system. With experiments we have found that it is possible to tune these parameters so that the manipulator slides along the obstacle with minimal impact forces and chattering.

6.5 Virtual forces

Finally, we tested the VF strategy. The desired task was tracking the linear path and the motion of the manipulator was obstructed by a different number of obstacles. To detect the obstacles a vision system was used. In the current implementation the vision system used a simple USB WebCam, which can recognize the scene and output the position of all obstacles in less than

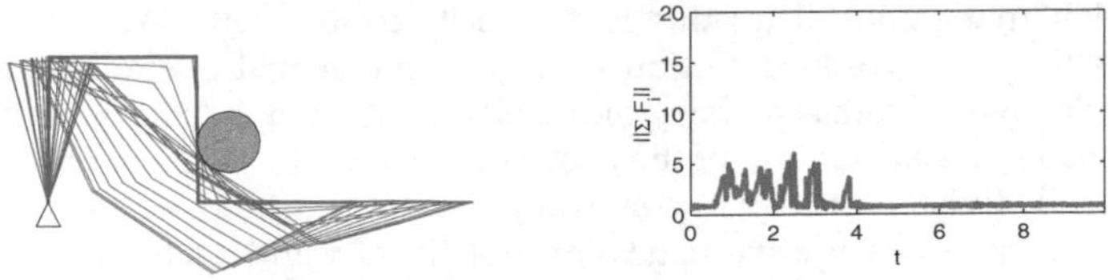

Fig. 18. Planar 4DOF manipulator tracking a circle while and avoiding obstacles using bumper and a virtual forces controller

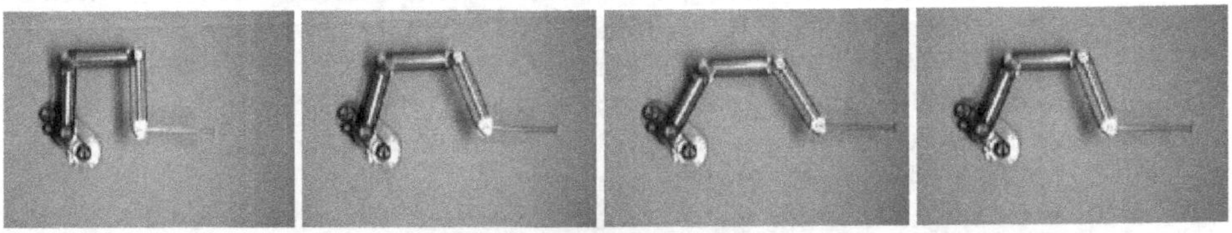

a) no obstacles

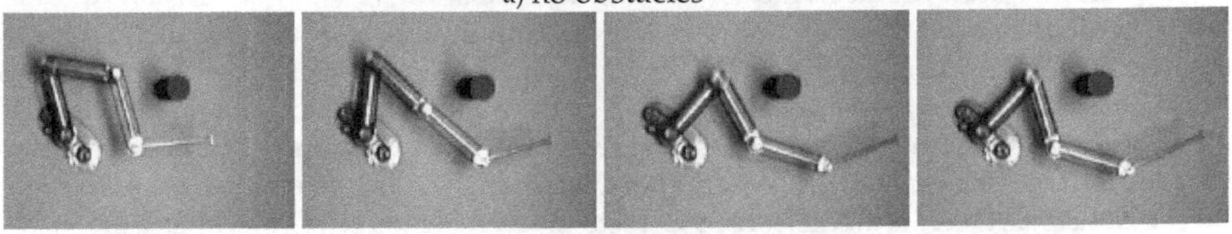

b) one obstacle

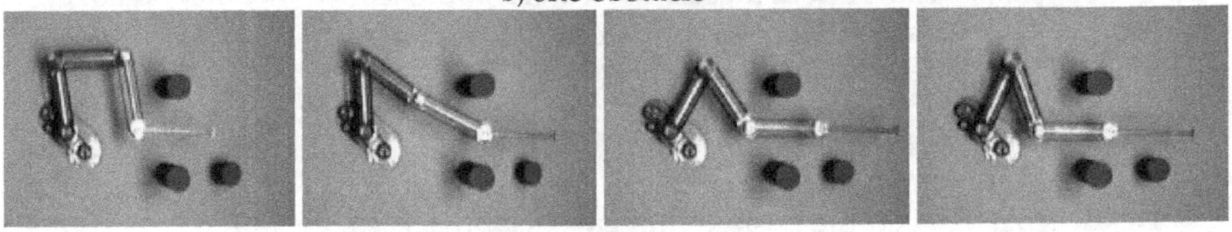

c) three obstacles

Fig. 19. Planar 4DOF manipulator tracking a path in an unstructured environment using a vision system to detect obstacles

0.04s. To avoid the obstacles the proximity sensor distance was selected as $d_m = 0.18m$ and the nominal virtual force as $f_o = 10N$. The rate of the torque controller was 400Hz (the necessary virtual forces for the avoiding motion were calculated in less than 0.5ms). The experimental results are given in Fig. 19. We can clearly see the difference in the motion when no obstacles are present and when one or more obstacles are present.

7. Conclusion

The presented approaches for on-line obstacle avoidance for redundant manipulators are a) based on redundancy resolution at the velocity level or b) considering also the dynamics of the manipulator. The primary task is determined by the end-effector trajectories and for the obstacle avoidance the internal motion of the manipulator is used. The goal is to assign each point on the body of the manipulator, which is close to the obstacle, a motion component in a direction that is away from the obstacle.

In the case of kinematic control (a) this is a velocity component. We have shown that it reasonable to define the avoiding motion in a one-dimensional operational space. In this way some singularity problems can be avoided when not enough "redundancy" is available locally. Additionally, the calculation of the pseudoinverse of the Jacobian matrix $\mathbf{J}_o$ is simpler as it includes scalar division instead of a matrix inversion. Using an approximate calculation of the avoiding velocities has its advantages computationally and it makes it easier to consider more obstacles simultaneously.

The second group of control algorithms (b) used in case b) is based on real or virtual forces. We compare three approaches regarding the sensors used to detect the obstacles: proximity sensors or vision, tactile sensors and no sensors. When proximity sensors are used we propose virtual forces strategy, where a virtual force component in a direction away from the obstacle is assigned to each point on the body of the manipulator, which is close to an obstacle. The algorithm based on the virtual forces avoids the problem of singular configurations and can be also easily applied when many obstacles are present. Additionally, the proposed control scheme enables us to use the null-space velocity controller for additional subtasks like the optimization of a performance criterion. Next, we have shown that under certain conditions obstacle avoidance can also be done without any information about the position and the size of the obstacles. This can be achieved by using a strategy that utilizes the self-motion caused by the contact forces to avoid an obstacle after the collision. Of course, an obstacle can be avoided only after a contact. The necessary prerequisite for this strategy to be effective is that the manipulator is backdrivable. As an alternative for the stiff systems we propose the use of tactile sensors. Here, a tactile sensor detects an obstacle and the controller generates the avoiding motion. The drawback of the last two control approaches is that they do not prevent the collision with the obstacle. Hence, they can only be applied if the collision occurs at a low speed so that the impact forces are not too high.

For the tasks where end-effector tracking is not essential for performing a given task we proposed a modified prioritized task control at the velocity level. The proposed approach enables a soft continuous transition between two different tasks. The obstacle-avoidance task only takes place when the desired movement approaches a given threshold, and then smoothly switches the priority of the tasks. The usefulness of this approach was shown on a two Kuka LWR robot to prevent a collision between them.

The computational efficiency of the proposed algorithms allows real-time application in an unstructured or time-varying environment. The simulations of highly redundant planar manipulators and the experiments on a four-link planar manipulator confirm that the proposed control algorithms assure an effective obstacle avoidance in an unstructured environment.

8. References

Brock, O., Khatib, O. & Viji, S. (2002). Task-Consistent Obstacle Avoidance of Motion Behavior for Mobile Manipulation, *Proc. IEEE Conf. Robotics and Automation*, Washington D.C, pp. 388 – 393.

Chiaverini, S. (1997). Singularity-robust task-priority redundancy resolution for real-time kinematic control of robot manipulators, *IEEE Trans. on Robotics and Automation* 13(3): 398 – 410.

Colbaugh, R., Seraji, H. & Glass, K. (1989). Obstacle Avoidance for Redundant Robots Using Configuration Control, *J. of Robotic Systems* 6(6): 721 – 744.

Egeland, O. (1987). Task-space tracking with redundant manipulators, *Robotics and Automation, IEEE Journal of* 3(5): 471 –475.

Featherstone, R. & Khatib, O. (1997). Load Independance of the Dynamically Consistent Inverse of the Jacobian Matrix, *Int. J. of Robotic Research* 16(2): 168 – 170.

Glass, K., Colbaugh, R., Lim, D. & Seraji, H. (1995). Real-Time Collision Avoidance for Redundant Manipulators, *IEEE Trans. on Robotics and Automation* 11(3): 448 – 457.

Guo, Z. & Hsia, T. (1993). Joint Trajectory Generation for Redundant Robots in an Environment with Obstacles, *J. of Robotic Systems* 10(2): 119 – 215.

Hayes, K., Walton, J. R., Szomor, Z. R. & Murrell, G. A. (2001). Reliability of five methods for assessing shoulder range of motion., *The Australian journal of physiotherapy* 47(4): 289–294.

Hogan, N. (1985). Impedance Control: An Approach to Manipulation: Part 1: Theory, Part 2: Implementation, Part 3: Applications, *Trans. of ASME J. of Dynamic Systems, Measurement, and Control* 107: 1 - 24.

Khatib, O. (1986). Real-Time Obstacle Avoidance for Manipulators and Mobile Robots, *Int. J. of Robotic Research* 5: 90 – 98.

Khatib, O. (1987). A Unified Approach for Motion and Force Control of Robot Manipulators: The Operational Space Formulation, *IEEE Trans. on Robotics and Automation* 3(1): 43 – 53.

Kim, J. & Khosla, P. (1992). Real-Time Obstacle Avoidance Using Harmonic Potential Functions, *IEEE Trans. on Robotics and Automation* 8(3): 338 – 349.

Žlajpah, L. & Nemec, B. (2003). Force strategies for on-line obstacle avoidance for redundant manipulators, *Robotica* 21(6): 633 - 644.

Lee, S., Yi, S.-Y., Park, J.-O. & Lee, C.-W. (1997). Reference adaptive impedance control and its application to obstacle avoidance trajectory planning, *Intelligent Robots and Systems, 1997. IROS '97., Proceedings of the 1997 IEEE/RSJ International Conference on*, Vol. 2, pp. 1158 – 1162 vol.2.

Lozano-Perez, T. (1983). Spatial Planning: A Configuration space approach, *IEEE Trans. on Computers* C-32(2): 102 – 120.

Maciejewski, A. & Klein, C. (1985). Obstacle Avoidance for Kinematically Redundant Manipulators in Dynamically Varying Environments, *Int. J. of Robotic Research* 4(3): 109 – 117.

Mansard, N., Khatib, O. & Kheddar, A. (2009). A unified approach to integrate unilateral constraints in the stack of tasks, *Robotics, IEEE Transactions on* 25(3): 670 –685.

McLean, A. & Cameron, S. (1996). The Virtual Springs Method: Path and Collision Avoidance for Redundant Manipulators, *Int. J. of Robotic Research* 15(4): 300 – 319.

Nakamura, Y., Hanafusa, H. & Yoshikawa, T. (1987). Task-Priority Based Redundancy Control of Robot Manipulators, *Int. J. of Robotic Research* 6(2): 3 – 15.

Nemec, B. (1997). Force Control of Redundant Robots, *in* M. Guglielmi (ed.), *Preprits of 5th IFAC Symp. on Robot Control, SYROCO'97*, Nantes, pp. 215 – 220.

Newman, W. S. (1989). Automatic Obstacle Avoidance at High Speeds via Reflex Control, *Proc. IEEE Conf. Robotics and Automation*, Scottsdale, pp. 1104 – 1109.

O'Neil, K. (2002). Divergence of linear acceleration-based redundancy resolution schemes, *Robotics and Automation, IEEE Transactions on* 18(4): 625 – 631.

Park, J., Chung, W. & Youm, Y. (1996). Design of Compliant Motion Controllers for Kinematically Redundant Manipulators, *Proc. IEEE Conf. Robotics and Automation*, pp. 3538 – 3544.

Raibert, M. H. & Craig, J. J. (1981). Hybrid Position/Force Control of Manipulators, *Trans. of ASME J. of Dynamic Systems, Measurement, and Control* 102: 126 – 133.

Sciavicco, L. & Siciliano, B. (2005). *Modelling and Control of Robot Manipulators (Advanced Textbooks in Control and Signal Processing)*, Advanced textbooks in control and signal processing, 2nd edn, Springer.

Sentis, L., Park, J. & Khatib, O. (2010). Compliant control of multicontact and center-of-mass behaviors in humanoid robots, *Robotics, IEEE Transactions on* 26(3): 483 –501.

Seraji, H. & Bon, B. (1999). Real-Time Collision Avoidance for Position-Controlled Manipulators, *IEEE Trans. on Robotics and Automation* 15(4): 670 – 677.

Sugiura, H., Gienger, M., Janssen, H. & Goerick, C. (2007). Real-time collision avoidance with whole body motion control for humanoid robots, *Intelligent Robots and Systems, 2007. IROS 2007. IEEE/RSJ International Conference on*, pp. 2053 –2058.

Volpe, R. & Khosla, P. (1990). Manipulator Control with Superquadratic Artificial Potential FUnctions: Theory and Experiments, *IEEE Trans. on Systems, Man, Cybernetics* 20(6).

Žlajpah, L. (2001). Integrated environment for modelling, simulation and control design for robotic manipulators, *Journal of Intelligent and Robotic Systems* 32(2): 219 – 234.

Woernle, C. (1993). Nonlinear Control of Constrained Redundant Manipulators, *in* J. A. et al. (ed.), *Computational Kinematics*, Kluwer Academic Publishers, pp. 119 – 128.

Xie, H., Patel, R., Kalaycioglu, S. & Asmer, H. (1998). Real-Time Collision Avoidance for a Redundant Manipulator in an Unstructured Environment, *Proc. Intl. Conf. On Intelligent Robots and Systems IROS'98*, Victoria, Canada, pp. 1925 – 1930.

Yoshikawa, T. (1987). Dynamic Hybrid Position / Force Control of Robot Manipulators Description of Hand Constraints and Calculation of Joint Driving, *IEEE Trans. on Robotics and Automation* 3(5): 386 – 392.

11

Singularity-Free Dynamics Modeling and Control of Parallel Manipulators with Actuation Redundancy

Andreas Müller and Timo Hufnagel
University Duisburg-Essen, Chair of Mechanics and Robotics
Heilbronn University
Germany

1. Introduction

The modeling, identification, and control of parallel kinematics machines (PKM) have advanced in the last two decades culminating in successful industrial implementations. Still the acceptance of PKM is far beyond that of the well-established serial manipulators, however. This is mainly due to the limited workspace, the drastically varying static and dynamic properties, leading eventually to singularities, and the seemingly more complex control. Traditionally the number of inputs equals the mechanical degree-of-freedom (DOF) of the manipulator, i.e. the PKM is non-redundantly actuated.

Actuation redundancy is a means to overcome the aforementioned mechanical limitations. It potentially increases the acceleration capability, homogenizes the stiffness and manipulability, and eliminates input singularities, and thus increases the usable workspace as addressed in several publications as for instance Garg et al. (2009); Gogu (2007); Krut et al. (2004); Kurtz & Hayward (1992); Lee et al. (1998); Nahon & Angeles (1989); O'Brien & Wen (1999); Wu et al. (2009).

These advantages are accompanied by several challenges for the dynamics modeling and for the PKM control. A peculiarity of redundantly actuated PKM is that control forces can be applied that have no effect on the PKM motion, leading to mechanical prestress that can be exploited for different second-level control tasks such as backlash avoidance and stiffness modulation Chakarov (2004); Cutkosky & Wright (1986); Kock & Schumacherm (1998); Lee et al. (2005); Müller (2006),Valasek et al. (2005). This also means that the inverse dynamics has no unique solution, which calls for appropriate strategies for redundancy resolution. The implementation of the corresponding model-based control schemes poses several challenges due to model uncertainties, the lack of globally valid parameterizations of the dynamics model, as well to the synchronization errors in decentralized control schemes calling for robust modeling and control concepts Müller (2011c); Müller & Hufnagel (2011).

The basis for model-based control are the motion equations governing the PKM dynamics. Aiming on an efficient formulation applicable in real-time, the motion equations are commonly derived in terms of a minimum number of generalized coordinates that constitute a (local) parameterization of the configuration space Abdellatif et al. (2005); Cheng et al. (2003); Müller (2005); Nakamura & Ghodoussi (1989); Yi et al. (1989). A well-known problem of this formulation is that these minimal coordinates are usually not valid on all of the configuration space, and configurations where the coordinates become invalid are called parameterization singularities. That is, it is not possible to uniquely determine any PKM configuration by one specific set of minimal coordinates. The most natural and practicable choice of minimal coordinates is to use the actuator coordinates. Then, the parameterization singularities are also input singularities. An ad hoc method to cope with this phenomenon is to switch between different minimal coordinates us discussed in Hufnagel & Müller (2011); Müller (2011a). This is a computationally complex approach since it requires monitoring the numerical conditioning of the constraint equations, and the entire set of motion equation must be changed accordingly. Instead of using a minimal number of generalized coordinates together with the switching method, a formulation in terms of the entire set of dependent coordinates of the PKM was proposed in Müller (2011b;c). This gives rise to a large system of redundant equations. Despite the large system of motion equations this formulation is advantageous since it does not exhibit singularities, i.e. it is valid in all feasible configurations of the PKM. Now the peculiarity of RA-PKM is that input singularities can be avoided by means of the actuation redundancy. Consequently, the actuator coordinates represent a valid set of redundant parameters that may give rise to another system of redundant motion equations, but of smaller size, as discussed in the following.

In this chapter the applicability of the standard minimal coordinates formulation is discussed and an alternative formulation in terms of actuator coordinates is presented. The latter formulation is globally valid as long as the PKM does not encounter input singularities, which is the aim of applying redundant actuation. Upon this formulation an amended augmented PD (APD) and computed torque control (CTC) scheme is proposed. These model-based control schemes are shown to achieve exponentially stable trajectory tracking. Experimental results are shown for a prototype implementation of a 2-DOF redundantly actuated PKM.

2. Kinematics of PKM with actuation redundancy

Denoting the joint variables of the PKM with $q^1,\ldots,q^n$, the PKM configuration is represented by the joint coordinate vector $\mathbf{q} \in \mathbb{V}^n$. The kinematic loops of the PKM give rise to a system of r geometric and kinematic constraints

$$\mathbf{0} = \mathbf{h}(\mathbf{q}),\ \mathbf{h}(\mathbf{q}) \in \mathbb{R}^r \tag{1}$$

$$\mathbf{0} = \mathbf{J}(\mathbf{q})\,\dot{\mathbf{q}},\ \mathbf{J}(\mathbf{q}) \in \mathbb{R}^{r,n}. \tag{2}$$

The configuration space (c-space) of the PKM model is the set of admissible configurations, defined by the geometric constraints (1)

$$V := \{\mathbf{q} \in \mathbb{V}^n | \mathbf{h}(\mathbf{q}) = \mathbf{0}\}. \tag{3}$$

Presumed $\mathbf{J}$ has full rank, the PKM has the DOF $\delta := n - r$. Consequently δ joint variables can be selected as independent coordinates representing a minimal set of generalized coordinates for the PKM and the (2) can be solved in terms of the corresponding independent velocities. Instead of using such independent minimal coordinates a solution can be expressed in terms of actuator coordinates.

To the actuated joints can be associated m joint coordinates summarized in the vector $\mathbf{q}_{\mathrm{a}}$ so that the vector of joint coordinates can be rearranged as $\mathbf{q} = (\mathbf{q}_{\mathrm{p}}, \mathbf{q}_{\mathrm{a}})$, where $\mathbf{q}_{\mathrm{p}}$ comprises the coordinates of passive coordinates. The constraints (2) can then be rearranged as

$$\mathbf{J}_{\mathrm{P}}\dot{\mathbf{q}}_{\mathrm{p}} + \mathbf{J}_{\mathrm{a}}\dot{\mathbf{q}}_{\mathrm{a}} = \mathbf{0} \tag{4}$$

with $r \times (n-m)$ matrix $\mathbf{J}_{\mathrm{P}}$.
The form (4) allows to identify two types of singularities. Assume that in regular configurations of the PKM (i.e. no c-space singularities) the constraint Jacobian $\mathbf{J}$ has full rank r. Further assume that the number m of actuators is at least $\delta = n - r$. Then if rank $\mathbf{J}_{\mathrm{P}} = n - m$ and rank $\mathbf{J}_{\mathrm{a}} = \delta$, the PKM motion is always determined by the m actuator coordinates and (4) can be resolved as $\dot{\mathbf{q}}_{\mathrm{P}} = -\mathbf{J}_{\mathrm{P}}^{+}\mathbf{J}_{\mathrm{a}}\dot{\mathbf{q}}_{\mathrm{a}}$ so that

$$\dot{\mathbf{q}} = \widetilde{\mathbf{F}}\dot{\mathbf{q}}_{\mathrm{a}} \tag{5}$$

with

$$\widetilde{\mathbf{F}} = \begin{pmatrix} -\mathbf{J}_{\mathrm{P}}^{+}\mathbf{J}_{\mathrm{a}} \\ \mathbf{I}_m \end{pmatrix} \tag{6}$$

that satisfied $\mathbf{J}\widetilde{\mathbf{F}} \equiv \mathbf{0}$, where $\mathbf{J}_{\mathrm{P}}^{+} = \left(\mathbf{J}_{\mathrm{P}}^{T}\mathbf{J}_{\mathrm{P}}\right)^{-1}\mathbf{J}_{\mathrm{P}}^{T}$ is the left-pseudoinverse. In (5) it is assumed that $\dot{\mathbf{q}}_{\mathrm{a}}$ satisfies the constraints. Moreover, δ of them can serve as minimal coordinates.
Since actuation redundancy can eliminate input-singularities it is assumed in the following that the PKM does not encounter input-singularities in the relevant part of the workspace. A configuration $\mathbf{q}$ is called a c-space singularity if rank $\mathbf{J}$ changes in $\mathbf{q}$.

3. Motion equations in actuator coordinates

A PKM is a mechanism with kinematic loops, and the corresponding constraint forces are incorporated via the generalized constraint forces. The Lagrangian motion equations of the PKM can be represented in the standard form

$$\mathbf{G}(\mathbf{q})\ddot{\mathbf{q}} + \mathbf{C}(\mathbf{q},\dot{\mathbf{q}})\dot{\mathbf{q}} + \mathbf{Q}(\mathbf{q},\dot{\mathbf{q}},t) + \mathbf{J}^{T}(\mathbf{q})\boldsymbol{\lambda} = \mathbf{u} \tag{7}$$

where $\boldsymbol{\lambda}$ is the vector of Lagrange multipliers, $\mathbf{G}$ is the generalized mass matrix of the unconstrained system, $\mathbf{C}\dot{\mathbf{q}}$ represents generalized Coriolis and centrifugal forces, $\mathbf{Q}$ represents all remaining forces (possibly including EE loads), and $\mathbf{u}(t)$ are the generalized control forces. In the following a formulation of the equations of motion governing the PKM dynamics are derived in terms of actuator coordinates. Since the columns of $\widetilde{\mathbf{F}}$ constitute a basis for the null-space of $\mathbf{J}$, the generalized constraint reaction forces in (7) can be eliminated by premultiplication with $\widetilde{\mathbf{F}}^{T}$ at the same time reducing the number of equations. On a kinematic level the expression (5) and $\ddot{\mathbf{q}} = \dot{\widetilde{\mathbf{F}}}\dot{\mathbf{q}}_{\mathrm{a}} + \widetilde{\mathbf{F}}\ddot{\mathbf{q}}_{\mathrm{a}}$, where

$$\dot{\widetilde{\mathbf{F}}} = \begin{pmatrix} -\mathbf{J}_{\mathrm{P}}^{+}\dot{\mathbf{J}}\widetilde{\mathbf{F}} \\ \mathbf{0}_m \end{pmatrix}, \tag{8}$$

allow substitution of the generalized velocity and acceleration vector $\dot{\mathbf{q}}$ and $\ddot{\mathbf{q}}$, by actuator velocities and accelerations, respectively. This gives rise to the reduced system of motion equations

$$\widetilde{\mathbf{G}}(\mathbf{q})\ddot{\mathbf{q}}_{\mathrm{a}} + \widetilde{\mathbf{C}}(\mathbf{q},\dot{\mathbf{q}})\dot{\mathbf{q}}_{\mathrm{a}} + \widetilde{\mathbf{Q}}(\mathbf{q},\dot{\mathbf{q}},t) = \mathbf{c} \tag{9}$$

with

$$\widetilde{\mathbf{G}} := \widetilde{\mathbf{F}}^{T}\mathbf{G}\widetilde{\mathbf{F}},\ \widetilde{\mathbf{C}} := \widetilde{\mathbf{F}}^{T}(\mathbf{C}\widetilde{\mathbf{F}} + \mathbf{G}\dot{\widetilde{\mathbf{F}}}),\ \widetilde{\mathbf{Q}} := \widetilde{\mathbf{F}}^{T}\mathbf{Q} \tag{10}$$

and $\mathbf{c}$ represents the actuator forces being part of the overall vector $\mathbf{u}$.
This is a reduced system of m equations. It is crucial to notice that only δ of these m equations are independent. That is, if the PKM is redundantly actuated ($m > \delta$), the projected mass

matrix is $\widetilde{\mathbf{G}}$ is singular. However, the advantage of the formulation (9) is that it already represents a solution for the inverse dynamics problem, in contrast to the minimal coordinate formulations Müller (2005). This will be beneficial for the model-based control.

4. Parameterization singularities of non-redundantly actuated PKM

The selection of actuator coordinates induces a parameterization of the PKM configuration. Commonly a minimal set of δ independent coordinates, denoted $\mathbf{q}_2$, are selected giving rise to δ motion equations. It is well-known, however, that such minimal coordinates are not globally valid in the sense that there are configuration where a particular set of of coordinates does not uniquely determine the PKM motion. Such configurations are called parameterization singularities. For non-redundantly actuated PKM $\delta = m$ actuator coordinates are usually taken as independent coordinates, so that $\mathbf{q}_2 = \mathbf{q}_\mathrm{a}$ and the parameterization singularities are exactly the input singularities. In these cases $\mathbf{J}_\mathrm{a}$ in (4) is rectangular and invertible as long as the non-redundantly actuated PKM does not encounter input singularities.
To explain this phenomenon consider the planar 2<u>R</u>RR/<u>R</u>R PKM shown in figure 1. This system has a DOF $\delta = 2$, but is actuated by the $m = 3$ actuators at the base joints. Hence it is redundantly actuated with a degree of redundancy of $\rho = m - \delta = 1$. This PKM is naturally parameterized in terms of two of these three actuator coordinates. Now, as outlined above, there are configurations where the PKM motion is cannot be prescribed by $\delta = 2$ coordinates. These parameterization-singularities can be observed if the joint angles of two of these actuators are used as independent coordinates. Figure 2 shows the EE locus corresponding to the input-singularities for three different choices of independent coordinates. The EE traces of the singularities in workspace are the coupler curves of the 4-bar mechanism formed by fixing the middle joint so that the middle links keep aligned.
To cope with the lack of a globally valid parameterization of the PKM configuration a switching method for RA-PKM was proposed in Hufnagel & Müller (2011). The basic idea of this method is to switch to a different set of δ (actuator) coordinates whenever the current set fails. The drawback of this method is that for each set of independent coordinates different set motion equations must be invoked, which increases the implementation effort. Ideally

Fig. 1. Prototype of a planar 2<u>R</u>RR/<u>R</u>R RA-PKM.

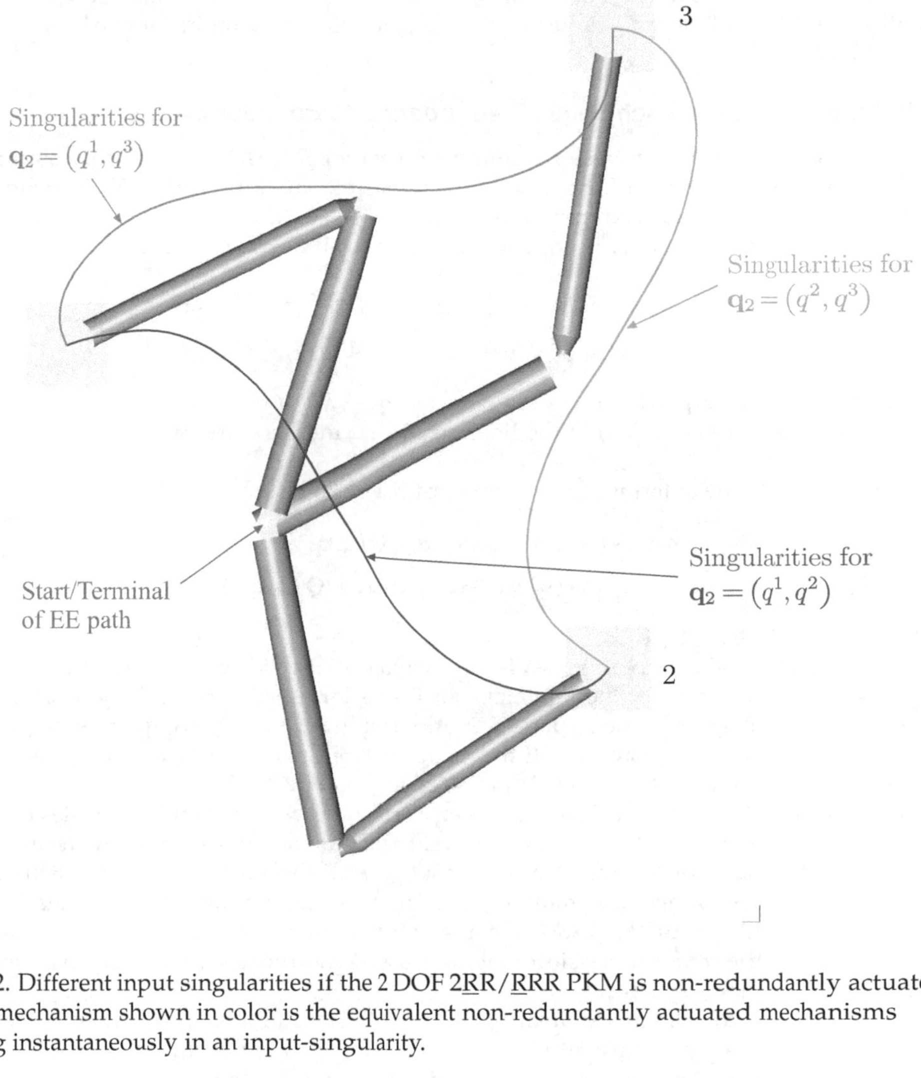

Fig. 2. Different input singularities if the 2 DOF 2RR/RRR PKM is non-redundantly actuated. The mechanism shown in color is the equivalent non-redundantly actuated mechanisms being instantaneously in an input-singularity.

a formulation should be used that is free from any parameterization singularities within the range of motion of the PKM. It often suggested to use EE coordinates as independent generalized coordinates, so that (9) would govern the PKM dynamics in workspace. It turns out, however, that even this choice suffers from parameter singularities.

Now the main motivation for introduction of redundant actuation is that most, or possibly all, of the input singularities of the non-redundant PKM can be eliminated. Assumption that the RA-PKM does not possess input-singularities, i.e. it can always be controlled by some δ out of the m actuator coordinates, the m input coordinates $\mathbf{q}_{\mathrm{a}}$ constitute feasible coordinates to parameterize the PKM motion. Hence the system (9) is globally valid for the entire motion range. In summary, the redundant formulation (9) is globally valid in the entire motion range of a RA-PKM where it does not exhibit input-singularities. If $m = \delta$, i.e. non-redundant actuation, the formulation (9) reduces to the classical formulation in minimal coordinates Müller (2005).

5. Model-based control schemes in redundant actuator coordinates

While the solution of the inverse dynamics problem of RA-PKM in minimal coordinates involves the pseudoinverse of the $m \times \delta$ control matrix, the formulation (9) in redundant actuator coordinates is already an inverse dynamics solution that can be immediately employed for the feedforward. Therewith an APD control scheme can be introduced as

$$\begin{aligned} \mathbf{c} &= \widetilde{\mathbf{G}}(\mathbf{q})\,\ddot{\mathbf{q}}_{\mathrm{a}}^{\mathrm{d}} + \widetilde{\mathbf{C}}(\dot{\mathbf{q}},\mathbf{q})\,\dot{\mathbf{q}}_{\mathrm{a}}^{\mathrm{d}} + \widetilde{\mathbf{Q}}(\mathbf{q},\dot{\mathbf{q}},t) - \mathbf{K}_{\mathrm{P}}\mathbf{e}_{\mathrm{a}} - \mathbf{K}_{\mathrm{D}}\dot{\mathbf{e}}_{\mathrm{a}} \\ &= \widetilde{\mathbf{F}}^{T}\left(\mathbf{G}(\mathbf{q})\,\ddot{\mathbf{q}}_{\mathrm{a}}^{\mathrm{d}} + \mathbf{C}(\mathbf{q},\dot{\mathbf{q}})\,\dot{\mathbf{q}}_{\mathrm{a}}^{\mathrm{d}} + \mathbf{Q}(\mathbf{q},\dot{\mathbf{q}},t)\right) - \mathbf{K}_{\mathrm{P}}\mathbf{e}_{\mathrm{a}} - \mathbf{K}_{\mathrm{D}}\dot{\mathbf{e}}_{\mathrm{a}} \end{aligned} \tag{11}$$

where $\mathbf{q}_{\mathrm{a}}^{\mathrm{d}}(t)$ is the target trajectory and $\mathbf{e}_{\mathrm{a}} := \mathbf{q}_{\mathrm{a}} - \mathbf{q}_{\mathrm{a}}^{\mathrm{d}}$ is the tracking error. The gain matrices $\mathbf{K} = \mathrm{diag}\,(K_1, \ldots, K_m)$ in the linear feed-back measure the errors in the m actuator coordinates.

Further a CTC scheme in terms of actuator coordinates can be introduced as

$$\begin{aligned} \mathbf{c} &= \widetilde{\mathbf{G}}(\mathbf{q})\,\mathbf{v}_{\mathrm{a}} + \widetilde{\mathbf{C}}(\dot{\mathbf{q}},\mathbf{q})\,\dot{\mathbf{q}}_{\mathrm{a}} + \widetilde{\mathbf{Q}}(\mathbf{q},\dot{\mathbf{q}},t) \\ &= \widetilde{\mathbf{F}}^{T}\left(\mathbf{G}(\mathbf{q})\,\mathbf{v}_{\mathrm{a}} + \mathbf{C}(\mathbf{q},\dot{\mathbf{q}})\,\dot{\mathbf{q}}_{\mathrm{a}} + \mathbf{Q}(\mathbf{q},\dot{\mathbf{q}},t)\right) \end{aligned} \tag{12}$$

with $\mathbf{v}_{\mathrm{a}} = \ddot{\mathbf{q}}_{\mathrm{a}} - \mathbf{K}_{\mathrm{P}}\mathbf{e}_{\mathrm{a}} - \mathbf{K}_{\mathrm{D}}\dot{\mathbf{e}}_{\mathrm{a}}$.

It is crucial that these control schemes lead to exponentially stable trajectory tracking. This property can be shown as for the minimal coordinates formulation by projection of the error dynamics to a δ subspace of the c-space V. Notice that in $(11)_2$ and $(12)_2$, $\mathbf{q}_{\mathrm{a}}^{\mathrm{d}}(t)$ and $\mathbf{q}_{\mathrm{a}}(t)$ are presumed to satisfy the constraints. If this is not ensured, when using measured values for instance, the formulation $(11)_1$ and $(12)_1$, respectively, is to be used.

At this point it should be remarked that the errors of all $m > \delta$ actuator coordinates must be used in the linear feedback. If only δ independent actuator coordinates are used, the feedback term does not account for the overall error in configurations where the motion is not uniquely determined by these δ actuator motions, i.e. in parameterization-/input-singularities of the non-redundantly actuated PKM. On the other hand the redundant feedback causes counteraction of the m actuators since only δ actuator coordinates are independent but the m feedback commands are not.

While the actuator coordinate formulation offers globally valid motion equations it does not involve the components of the control forces that correspond to the null-space of the control matrix as the inverse dynamics solution in minimal coordinates does.

6. Experimental results

The prototype of the planar 2 DOF 2<u>R</u>RR/<u>R</u>R PKM in figure 1 has been used as testbed for the proposed control schemes in redundant actuator coordinates and the classical formulation in minimal coordinate. The testbed is equipped with a dSPACE DS1103 real-time system operating with a sampling rate of 2.5 kHz. The controller gains were set to $K_{\mathrm{P}} = 500$ and $K_{\mathrm{D}} = 8$, in accordance to the actuator dynamics. The three base joints are actuated by Maxon Re30 DC motors. They are mounted at the vertices of an equilateral triangle with 400 mm

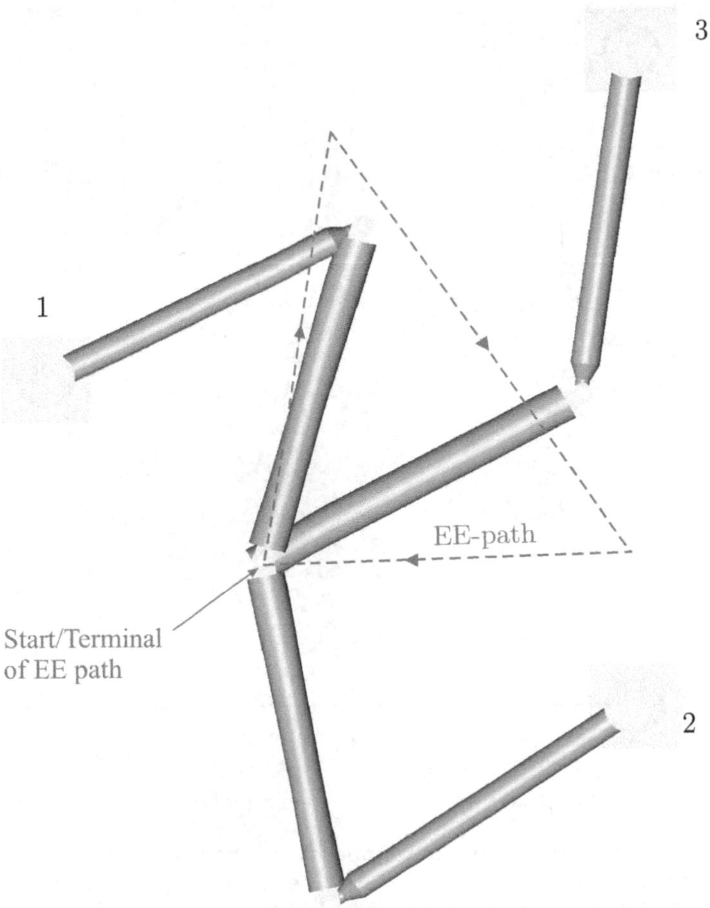

Fig. 3. EE-path and singular curves of the 2DOF RA-PKM.

lateral lengths. Each arm segment has a length of 200 mm, and the total weight of one arm is 134 g. The motion equations (7) are derived in terms of relative coordinates by opening the two kinematic loops. This gives rise to a PKM model (7) comprising $n = 6$ equations in terms of the $n = 6$ joint angles subject to $r = 4$ cut-joint constraints. Hence the DOF of the PKM is $\delta = 2$. the formulation (9) yields $m = 3$ equations in terms of the redundant actuator coordinates (of which $\delta = 2$ are independent). In the experiment the manipulator is controlled along the EE-path in figure 3. Denote with q^1, q^2, q^3 the joint angles of the actuated base joints so that $\mathbf{q}_a = (q^1, q^2, q^3)$. The joint trajectory is determined from the EE-path by solving the inverse kinematics where velocity and acceleration limits are taken into account. The EE path passes all singularity loci shown in 2. That is, a non-redundantly actuated PKM, of which only $\delta = 2$ joints are actuated, exhibits input-singularities for any combination of two actuator coordinates.

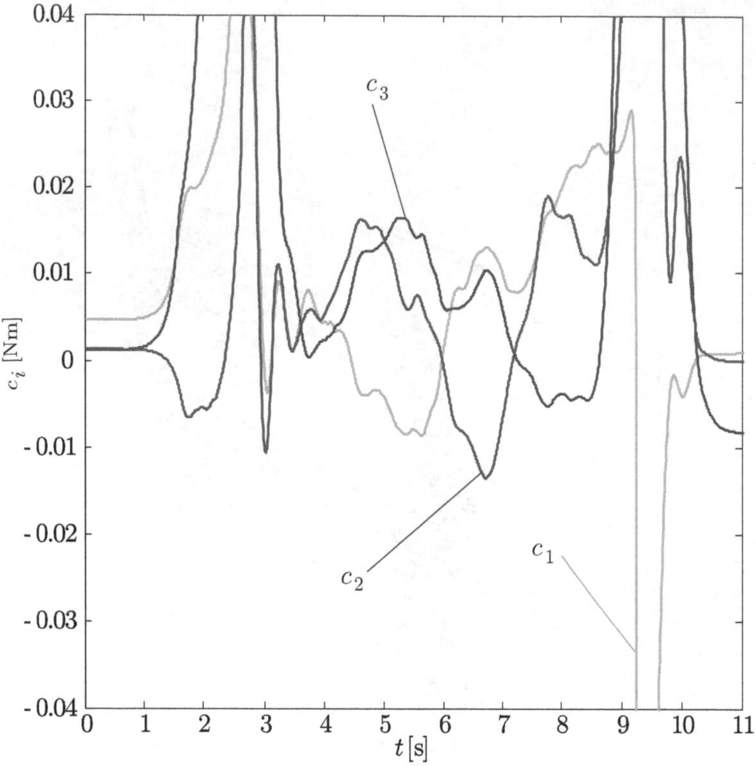

Fig. 4. Actuator torques computed from the CTC in minimal cooridnates if $\mathbf{q}_{\mathrm{a}} = (q^1, q^2)$.

In order to show the effect of parameter singularities of the minimal coordinate formulation figure 4 shows the computed control commands when only the two coordinates $\mathbf{q}_{\mathrm{a}} = (q^1, q^2)$ are used in (9), which corresponds to a non-redundantly actuated PKM. Since for this EE path the PKM has to cross the singularity curve for this combination (red curve in figure 2) once at the start and once just before the end of the EE-trajectory the control torques tend to infinity at these points. Consequently at these points the model does not admit to compute any sensible control torques and leads to instabilities.

In contrast, when controlling the RA-PKM using the CTC (12) in redundant actuator coordinates leads to the drive torques in figure 5. Figure 6 sows the corresponding joint tracking errors. Apparently the redundant coordinate formulation achieves a smooth motion and torque evolution unaffected by any singularities.

In this experiment it is clearly visible that there are non-zero drive torques even if the RA-PKM is not moving. This is a peculiar phenomenon that can only be observed in RA-PKM. It can partially be attributed to the interplay of measurement errors and finite encoder resolutions with actuation redundancy. Due to the actuation redundancy the control forces are not independent so that the RA-PKM attains a configuration that is not determined by the measurement error (as for non-redundant actuation) but by the static equilibrium of the control torques that do not affect the RA-PKM motion, but rather lead to internal prestress. This is a general problem for RA-PKM that was addressed in Müller & Hufnagel (2011).

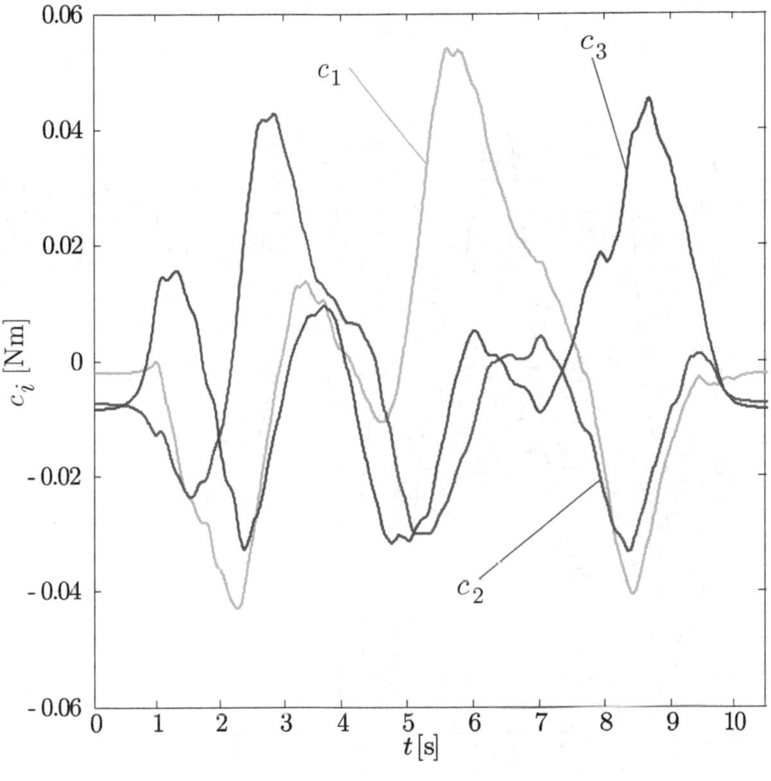

Fig. 5. Actuator torques when the RA-PKM is controlled along the EE-path of figure 3 by the CTC in redundant coordinates.

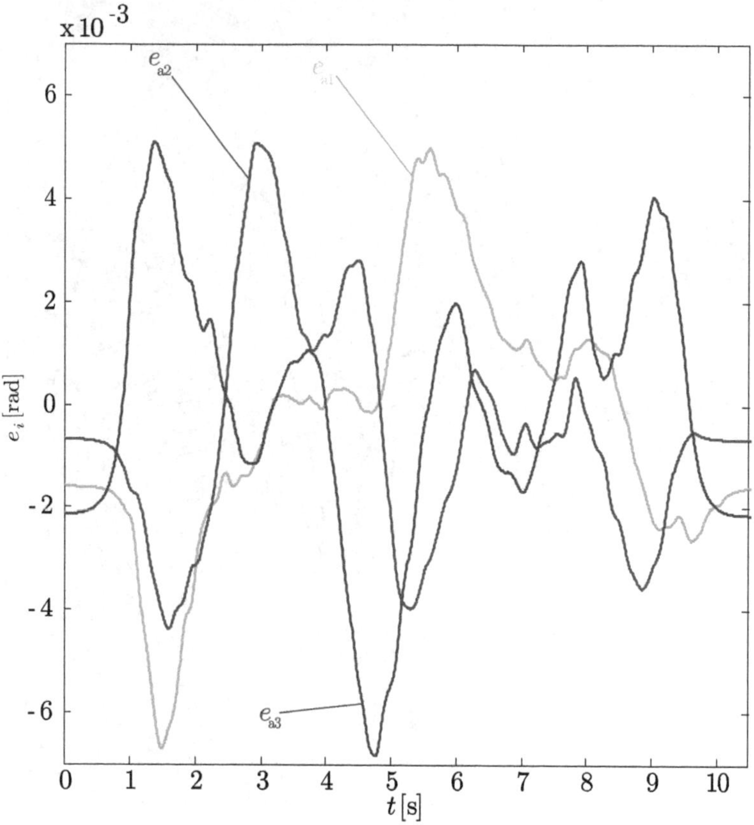

Fig. 6. Joint tracking errors when the RA-PKM is controlled by the CTC in redundant coordinates.

7. Summary

PKM are commonly characterized by inhomogeneous distribution of kinematic and dynamic properties within the work space, and eventually the existence of input-singularities. Actuation redundancy allows to eliminate these singularities, and thus to extend the usable workspace. The PKM motion equations are commonly formulated using a set of minimal coordinates. Moreover, actuator coordinates are usually used as minimal coordinates. Thus input-singularities are also parameterization-singularities of the PKM model, which is critical for any model-based control. In this chapter an alternative formulation of motion equations in terms of redundant actuator coordinates has been proposed, and the corresponding amended form of an augmented PD and computed torque control scheme were presented. The applicability of the proposed methods is confirmed by the experimental results for a planar redundantly actuated PKM.

8. References

H. Abdellatif et al. (2005) High Efficient Dynamics Calculation Approach for Computed-Force Control of Robots with Parallel Structures, IEEE Conference on Decision and Control and European Control Conference (CDC-ECC), 12-15 Dec. 2005, pp. 2024-2029

H. Asada & J.J. E. Slotine (1986) Robot Analysis and Control. New York, Wiley

D. Chakarov (2004) Study of the antagonistic stiffness of parallel manipulators with actuation redundancy, Mech. Mach. Theory, Vol. 39, 2004, pp. 583-601

H. Cheng et al. (2003) Dynamics and Control of Redundantly Actuated Parallel Manipulators, IEEE/ASME Trans. on Mechatronics, Vol. 8, no. 4, 2003, pp. 483-491

M.R. Cutkosky & P.K. Wright (1986) Active Control of a Compliant Wrist in Manufacturing Tasks, ASME J. Eng. for Industry, Vol. 108, No. 1, 1986, pp. 36-43

Garg, V.; Noklely, S.B. & Carretero, J.A. (2009). Wrench capability analysis of redundantly actuated spatial parallel manipulators, *Mech. Mach. Theory*, Vol. 44, No. 5, 2009, pp: 1070-1081

Gogu, G. (2007). Fully-isotropic redundantly-actuated parallel wrists with three degrees of freedom, *Proc. ASME Int. Design Eng. Tech. Conf. (IDETC)*, Las Vegas, NV, 2007, DETC2007-34237

T. Hufnagel & A. Müller (2011) A Realtime Coordinate Switching Method for Model-Based Control of Parallel Manipulators, ECCOMAS Thematic Conference on Multibody Dynamics, Juli 4-7, 2011, Brussels, Belgium

S. Kock & W. Schumacher (1998) A parallel x-y manipulator with actuation redundancy for high-speed and active-stiffness applications, Proc. IEEE Int. Conf. Robot. Autom. (ICRA), Leuven, pp. 2295-2300

Krut, S.; Company, O. & Pierrot, F. (2004) Velocity performance indicies for parallel mechanisms with actuation redundancy, *Robotica*, Vol. 22, 2004, pp. 129-139

Kurtz, R.; Hayward, V. (1992) Multiple-goal kinematic optimization of a parallel spherical mechanism with actuator redundancy, *IEEE Trans. on Robotics Automation*, Vol. 8. no. 5, 1992, pp. 644-651

Lee, J.H.; Li, B.J.; Suh, H. (1998) Optimal design of a five-bar finger with redundant actuation, *Proc. IEEE Int. Conf. Rob. Aut. (ICRA)*, Leuven, 1998, 2068-2074

S.H. Lee et al. (2005) Optimization and experimental verification for the antagonistic stiffness in redundantly actuated mechanisms: a five-bar example, Mechatronics Vol. 15, 2005, pp. 213-238

Müller, A. (2005). Internal Prestress Control of redundantly actuated Parallel Manipulators - Its Application to Backlash avoiding Control, *IEEE Trans. on Rob.*, pp. 668 - 677, Vol. 21, No. 4, 2005

A. Müller (2006). Stiffness Control of redundantly actuated Parallel Manipulators, Proc. IEEE Int. Conf. Rob. Automat. (ICRA), Orlando, May 15-19, 2006, pp. 1153 - 1158

A. Müller (2011). A Robust Inverse Dynamics Formulation for Redundantly Actuated PKM, *Proc. 13th World Congress in Mechanism and Machine Science*, Guanajuato, Mexico, 19-25 June 2011

A. Müller (2011). Motion Equations in Redundant Coordinates with Application to Inverse Dynamics of Constrained Mechanical Systems, *Nonlinear Dynamics*, DOI 10.1007/s11071-011-0165-5

A. Müller (2011) Robust Modeling and Control Issues of Parallel Manipulators with Actuation Redundancy, in A. Müller (editor): Recent Advances in Robust Control - Volume 1: Theory and Applications in Robotics and Electromechanics, InTech, Vienna, Austria, 2011

A. Müller, T. Hufnagel (2011) Adaptive and Singularity-Free Inverse Dynamics Models for Control of Parallel Manipulators with Actuation Redundancy, 8th International Conference on Multibody Systems, Nonlinear Dynamics, and Control, ASME 2011 International Design Engineering Technical Conferences, August 28-31, 2011, Washington, DC

A. Müller, T. Hufnagel (2011) A Projection Method for the Elimination of Contradicting Control Forces in Redundantly Actuated PKM, IEEE Int. Conf. Rob. Automat. (ICRA), May 9-13, 2011, Shanghai, China

Nahon, M.A.; Angeles, J. (1989) Force optimization in redundantly-actuated closed kinematic chains, *Proc. IEEE Int. Conf. Rob. Automat. (ICRA)*, Scottsdale, USA, May 14-19, 1989, pp. 951956

Y. Nakamura & M. Ghodoussi (1989) Dynamics Computation of Closed-Link Robot Mechanisms with Nonredundant and Redundant Actuators, IEEE Tran. Rob. and Aut., Vol. 5, No. 3, 1989, pp. 294-302

O'Brien, J.F.; Wen, J.T. (1999) Redundant actuation for improving kinematic manipulability, *Proc. IEEE Int. Conf. Robotics Automation*, 1999, pp. 1520-1525

Valasek, M. et al. (2005) Design-by-Optimization and Control of Redundantly Actuated Parallel Kinematics Sliding Star, *Multibody System Dynamics*, Vol. 14, no. 3-4, 2005, 251-267

Wu, J. et al. (2009) Dynamics and control of a planar 3-DOF parallel manipulator with actuation redundancy, *Mech. Mach. Theory*, Vol. 44, 2009, pp. 835-849.

B.Y. Yi et al. (1989) Open-loop stiffness control of overconstrained mechanisms/robot linkage systems, Proc. IEEE Int. Conf. Robotics Automation, Scottsdale, 1989, pp. 1340-1345

Permissions

The contributors of this book come from diverse backgrounds, making this book a truly international effort. This book will bring forth new frontiers with its revolutionizing research information and detailed analysis of the nascent developments around the world.

We would like to thank Assoc. Prof. PhD. Serdar Küçük, for lending his expertise to make the book truly unique. He has played a crucial role in the development of this book. Without his invaluable contribution this book wouldn't have been possible. He has made vital efforts to compile up to date information on the varied aspects of this subject to make this book a valuable addition to the collection of many professionals and students.

This book was conceptualized with the vision of imparting up-to-date information and advanced data in this field. To ensure the same, a matchless editorial board was set up. Every individual on the board went through rigorous rounds of assessment to prove their worth. After which they invested a large part of their time researching and compiling the most relevant data for our readers. Conferences and sessions were held from time to time between the editorial board and the contributing authors to present the data in the most comprehensible form. The editorial team has worked tirelessly to provide valuable and valid information to help people across the globe.

Every chapter published in this book has been scrutinized by our experts. Their significance has been extensively debated. The topics covered herein carry significant findings which will fuel the growth of the discipline. They may even be implemented as practical applications or may be referred to as a beginning point for another development. Chapters in this book were first published by InTech; hereby published with permission under the Creative Commons Attribution License or equivalent.

The editorial board has been involved in producing this book since its inception. They have spent rigorous hours researching and exploring the diverse topics which have resulted in the successful publishing of this book. They have passed on their knowledge of decades through this book. To expedite this challenging task, the publisher supported the team at every step. A small team of assistant editors was also appointed to further simplify the editing procedure and attain best results for the readers.

Our editorial team has been hand-picked from every corner of the world. Their multi-ethnicity adds dynamic inputs to the discussions which result in innovative outcomes. These outcomes are then further discussed with the researchers and contributors who give their valuable feedback and opinion regarding the same. The feedback is then collaborated with the researches and they are edited in a comprehensive manner to aid the understanding of the subject.

Apart from the editorial board, the designing team has also invested a significant amount of their time in understanding the subject and creating the most relevant covers. They scrutinized every image to scout for the most suitable representation of the subject and create an appropriate cover for the book.

The publishing team has been involved in this book since its early stages. They were actively engaged in every process, be it collecting the data, connecting with the contributors or procuring relevant information. The team has been an ardent support to the editorial, designing and production team. Their endless efforts to recruit the best for this project, has resulted in the accomplishment of this book. They are a veteran in the field of academics and their pool of knowledge is as vast as their experience in printing. Their expertise and guidance has proved useful at every step. Their uncompromising quality standards have made this book an exceptional effort. Their encouragement from time to time has been an inspiration for everyone.

The publisher and the editorial board hope that this book will prove to be a valuable piece of knowledge for researchers, students, practitioners and scholars across the globe.

List of Contributors

Mohammad H. Abedinnasab
Door To Door Company, Sharif University of Technology, Iran

Yong-Jin Yoon
Nanyang Technological University, Singapore

Hassan Zohoor
Sharif University of Technology, The Academy of Sciences of IR Iran, Iran

Serdar Küçük
Kocaeli University, Turkey

Zafer Bingul and Oguzhan Karahan
Mechatronics Engineering, Kocaeli University, Turkey

Khalifa H. Harib and Kamal A.F. Moustafa
UAE University, United Arab Emirates

A.M.M. Sharif Ullah
Kitami Institute of Technology, Japan

Salah Zenieh
Tawazun Holding, United Arab Emirates

R. Tapia Herrera, Samuel M. Alcántara, J.A. Meda-Campaña and Alejandro S. Velázquez
Instituto Politécnico Nacional, México

Alessandro Cammarata
University of Catania, Department of Mechanical and Industrial Engineering, Italy

David Úbeda, José María Marín, Arturo Gil and Óscar Reinoso
Universidad Miguel Hernández de Elche, Spain

Emna Ayari, Sameh El Hadouaj and Khaled Ghedira
High Institute of Management, Tunis, Tunisia

Selçuk Kizir and Zafer Bingul
Mechatronics Engineering, Kocaeli University, Turkey

Leon Žlajpah and Tadej Petri
Jožef Stefan Institute, Ljubljana, Slovenia

Andreas Müller and Timo Hufnagel
University Duisburg-Essen, Chair of Mechanics and Robotics, Heilbronn University, Germany